The Consumer Society and the Postmodern City

The fact that we inhabit a consumer society has incredibly far-reaching implications. It affects the way we live our lives, the things we value, the direction society is taking. The city has always been a prime site of social change. The relationship between the city and the consumer society is the focus of this book.

Drawing on the work of the consumer society's most influential theorists, Jean Baudrillard and Zygmunt Bauman, the book examines the nature of consumption and its increasing centrality to the postmodern city by:

- considering the development of consumerism as a central facet of urban social life;
- demonstrating that social inequalities are increasingly structured around consumption;
- uncovering the hidden consequences of consumerism;
- pondering the meaning of lifestyle;
- revealing how the nature of reality is changing in an age of globalization.

Working through the often controversial ideas of Baudrillard and Bauman, the book assesses the ways in which consumerism is reshaping the nature and meaning of the city. Employing a sustained and engaging theoretical analysis, the book ranges across a variety of sometimes unexpected topics. It represents an impassioned plea for everyone interested in the social life of cities to take the notion of the consumer society, and the arguments of its major theorists, seriously.

David B. Clarke is a Senior Lecturer in Human Geography at the University of Leeds.

The Consumer Society and the Postmodern City

David B. Clarke

Routledge
Taylor & Francis Group

LONDON AND NEW YORK

First published 2003
by Routledge
11 New Fetter Lane, London EC4P 4EE

Simultaneously published in the USA and Canada
by Routledge
29 West 35th Street, New York, NY 10001

Routledge is an imprint of the Taylor & Francis Group

Typeset in Times by Taylor & Francis Books Ltd
Printed and bound in Great Britain by TJ International Ltd, Padstow,
Cornwall

British Library Cataloguing in Publication Data
A catalogue record for this book is available from the British Library

Library of Congress Cataloging in Publication Data
A catalog record for this book has been requested

ISBN 0–415–20514–X (hbk)
ISBN 0–415–20515–8 (pbk)

Consumption becomes a larger element in the standard of living in the city than in the country.

(Veblen 1994, 55)

Just as industrial concentration results in an ever increased production of goods, so urban concentration results in a limitless promotion of needs.

(Baudrillard 1998a, 65)

Now will the city have to fill and swell with a multitude of callings which are not required by any natural want.

(Plato, *The Republic* 2, 373b [Jowett 1871])

Contents

Acknowledgements

I have accrued an enormous number of debts in bringing this work to completion. The most direct I owe to two individuals with whom I have had the privilege of working most closely: Mike Bradford, who first sparked (and for a while nurtured) my interest in consumption; and Marcus Doel, a long-term collaborator whose unique brand of poststructuralist geography has proved enormously beneficial to my own thought processes. I should also like to acknowledge a debt to the late Antony Easthope and to Rob Lapsley, whose extra-mural course at the University of Manchester, 'Under French eyes', provided an extraordinarily pleasurable initiation into the rigours of poststructuralist thought. The present text will undoubtedly reveal the enormity of my debt to Zygmunt Bauman, whom, on occasion, has offered invaluable words of encouragement. In thanking him, I can only hope that the use to which I have put his ideas might do justice, in some small measure, to his own remarkable body of work.

In a somewhat different context, I would like to thank two Heads of Department at Leeds, John Stillwell and Graham Clarke (no relation), both of whom have been supportive of my work; even though it is, I suspect, relatively inscrutable from their own ardently positivist perspective. My sincere thanks also to everyone at Routledge, past and present, for their assistance, support, and, above all, patience. Sarah Carty, Andrew Mould and Ann Michael deserve my gratitude. I should also like to acknowledge the support of an ESRC Fellowship (award no. H52427002294), without which the time and means to explore the topic of consumption would not have been forthcoming. More formally, I acknowledge permission to include the following work as chapters of the present book: a version of Chapter 2 previously appeared in the *International Journal of Urban and Regional Research*, 21(1), 218–237. Thanks to Blackwell Publishers Ltd for permission to include this material, in amended form, here. Likewise, thanks to Carfax Publishers Ltd for permission to reprint the material comprising Chapter 3 from the review issue on 'Urban consumption' of *Urban Studies*, 35(5 and 6), 865–888, again in slightly amended form. I am once again grateful to Mike Bradford for his permission to include our jointly authored work as part of this book. Much of the material included here has been presented as seminars at a variety of institutions. Thanks to all those who offered valuable suggestions and posed challenging questions on such occasions. Thanks also to several generations of students taking my master's modules at Leeds, especially for turning up new references every time I thought I was on top of the literature. The staff of the Graphics Unit of the School of Geography, University of Leeds, and their counterparts at the University of Manchester, skilfully prepared Figures 1 and 4, and Figure 3, respectively. Thanks to all concerned for their efforts.

Last, but by no means least, I would like to thank a number of individuals who have read and offered comments on various parts of the manuscript, particularly Martin Purvis and Marcus Power. Thanks also to Jon Goss for providing perceptive comments on an earlier draft.

Introduction

Just over a fortnight after the 11 September 2001 terrorist attacks on the twin towers of New York's World Trade Center, the Pentagon in Washington DC, and the failed attack on the White House, the *Daily Telegraph* ran the front-page headline: 'Britain "needs you to shop" '. 'Tony Blair asked people to go shopping and take holidays to prevent the economy going into recession … after the terrorist attacks in the United States', the article continued (Jones and Smith 2001, 1). The Prime Minister's exhortation might have appeared somewhat surreal – if not, to some, downright distasteful – but there could be no doubting its seriousness. The so-called War on Terror, declared in the wake of the al-Qaeda strikes, would have been fatally compromised had consumers failed to keep on consuming, thereby plunging the world economy into recession. James Arnold of the BBC went to great pains to explain the importance of 'consumer confidence', recording similar sentiments to the *Telegraph* with subheadings such as: 'Shopping: your patriotic duty' (Arnold 2001). What is intriguing in all of this, however, is that there was no mention of the fact that precisely the opposite sentiment was promulgated throughout the Second World War. Dig for Victory, Savings Bonds, and a general ethos of 'tightening one's belt' to 'get us through' figured amongst the patriotic duties of the 'citizen of the free world' at that time. Production, not consumption, was then the order of the day. What clearer indication could there be than the edict to 'Shop for Victory' that *consumption* has become the most significant factor in the reproduction of the capitalist West?

Alas, things are rarely as straightforward as they appear. Thus, for example, the patriotic duties of the loyal citizen in the US and in Britain during the Second World War differed markedly with respect to consumption – perhaps sufficiently so as to undermine the contrast between 'then' and 'now' suggested by the *Telegraph* headline. Britain's Second World War effort was better served, for instance, by obtaining revenue from the export of Scotch whisky to the United States than it was by boosting morale on the home front with a ready supply of spirits. But if it appears churlish to undercut so striking an example as the call to shop in the face of adversity with caveat and qualification – not to mention perverse for a book intending to make the argument that ours is a society in which consumption is increasingly central – it should equally be apparent that there is little to be gained from oversimplification. Consumption has always been an important aspect of human society, in different ways at different times and in different places. And yet, despite appearances, there is no contradiction in accepting this point and arguing that Western societies are increasingly *consumer societies* – in a sense far more fundamental than the trivial observation that 'people have always consumed' either grasps or is capable of countering. And, of course, this point can hardly be fully expounded by recourse to a simple one-off anecdote about our 'duty to shop'.

That ours is a *consumer society* (with equal emphasis laid on both those words) is the central contention of this book. It will take a good deal of effort to pin down precisely what this means, but it should be noted at once that it puts it in opposition to a significant body of recent work on consumption. Miller (2000, 210), for example, resolutely refuses the (allegedly) 'homogenizing term "consumer culture" or "consumer society"' in favour of undertaking 'the original task of [reconstructing (?)] an empathetic experience of moments of consumption'. For me, however, this kind of thinking is in grave danger of missing precisely what is most important about consumption today. We will broach this issue more fully in Chapter 1, but to state baldly what is involved at the outset, it might be proposed that it is the way in which consumption has increasingly assumed a central *systemic* role in the reproduction of capitalist society that is crucial. The implications of this deceptively simple contention are immensely wide reaching. What is certain, however, is that they can hardly be grasped by limiting our concerns to 'empathetic experiences' of consumption alone. A far more rigorous conceptualization of the consumer society is vital to understanding both the nature and consequences of consumption today. The kind of conceptualization I have in mind has been developed to its fullest extent in the early work of Jean Baudrillard and the recent work of Zygmunt Bauman. For this reason, the present book leans especially heavily on their writings. Such an approach may not, of course, be to everybody's taste. Nonetheless, the level of insight their work generates is, for me, sufficient justification for a close examination and sustained deployment of their arguments. Part I of the book provides a detailed assessment of precisely what is at stake in the conception of consumption developed by these authors. Let me state at once, however, that I am not slavishly following their ideas out of some naive conviction that 'they know best'. Whilst some might perceive an undue reliance on the authority of others in the strategy of close textual reading adopted here, such a perception would be mistaken. Such a strategy is intended to do nothing more than to bring out the strength of the arguments themselves. I should add that, whilst the book hopefully manages to provide a coherent exposition of many of their arguments, I am not necessarily an entirely constant disciple of either Baudrillard or Bauman. There is a considerable degree of selectivity in the portions of their arguments I choose to deploy, and a certain degree of spin put on particular lines of thought that it would be inappropriate to attribute to the authors themselves.[1] Furthermore, whilst the conception of the consumer society derived from Baudrillard and Bauman forms a vital part of the book, it is nonetheless intended primarily as a means to an end. It is subsequently put into service, in Parts II and III, to provide a better understanding of the so-called 'postmodern city'.

The postmodern city is, perhaps, already in danger of becoming an outmoded concept. Yet this notion – however much longer the particular terminology may last – signals something crucial for urban theory. An aphorism gleaned from Calvino (1974, 56) captures its significance: '*The city ... has a simple secret: it knows only departures, not returns*'. The impress of consumerism on the city, I would argue, renders all previous conceptual frameworks increasingly problematic. Though all theoretical developments ultimately face the same fate, it is vital to ensure that we do not linger – with undue nostalgia – over conceptions of the city that are no longer particularly apposite. Equally, of course, one must remain sensitive to the dangers of throwing out the baby with the bathwater. In balancing these opposing tensions, the underlying contention of the present book is that *the consumer society has remade the city in its own image*. Let me therefore state, unequivocally, that the consumer society is still very much a capitalist society; whilst at the same time insisting that it involves a form of capitalism that departs significantly from its earlier incarnations. The implications this holds for the city are legion. In one guise or another, the notion of the 'postmodern city' has effectively sought to capture this situation. Many of the implications it holds have already been spelled out with considerable force and clarity. Even so, I would argue that a good deal of previous work on the topic has failed to theorize its nature and significance satisfactorily. In order to do so, I propose, the relations between 'postmodern society' and 'consumer society' need to be assessed more adequately. Admittedly, the importance of consumption to the city has been addressed with increasing frequency over recent years.[2] This is evident across a diverse range of literature, from work in social theory (Paolucci 2001), to the work of urban economists whom, in the past, have maintained something of a fixation on production (Glaeser *et al.* 2001). This notwithstanding, the present book seeks to develop this connection along lines that are not especially well trodden. In drawing on the work of Baudrillard and Bauman, it takes the notion of postmodernity entirely seriously (though certainly not uncritically), and proposes a particularly strong association with consumerism. But rather than spell out the nuances of this approach at this stage, it is preferable to let its particularity emerge during the course of the book as a whole. It is worth pausing, however, to mark out some of the basic contrasts between the conception of postmodernity developed here and certain other high-profile accounts. Although the present book retains the hard-fought recognition that the urban process under capitalism produces the capitalist city as an historically specific form, it does not follow Harvey (1989a) in effectively dismissing postmodernity as an unnecessary (and ultimately conservative) distraction. Nor does it resort to the kind of dubious eclecticism that has led Soja (1989) to identify Los Angeles as the paradigmatic postmodern city. To repeat, therefore, the postmodern city, and the postmodern condition itself, is regarded here as being intimately related to the refiguration of capitalist society as consumer society.

Simmel once remarked that the city 'is not a spatial entity with sociological consequences, but a sociological entity that is formed spatially' (cited in Frisby 1984, 131; cf. Simmel 1997; Lechner 1991). This resonates strongly with the well established truth that the city is never simply a neutral container of social processes. The city is constantly made and remade in relation to all manner of social processes, from which its particular character emerges – whilst, at the same time, this emergent character serves to mediate or inflect particular social processes in distinctive ways. If this conception has become something of a truism, it is vital to reinvigorate the truth it contains. For me, this is precisely where the conjunction of consumption and the city becomes important. Dealing with this conjunction inevitably raises all manner of difficulties, both conceptual and presentational. As with all social processes, consumption cannot effectively be treated as a discrete category and then considered, unproblematically, 'in relation to' the city: neither 'consumption' nor 'the city' is amenable to full and final examination in isolation, prior to their 'interaction effects' being considered. Rather, they need to be considered in a relational manner from the very start. Whilst the difficulties this raises might appear insurmountable, it nonetheless determines that the initial conception of consumption to which we hold must be broad enough for its urban dimensions to be considered in due course. The tack taken here, therefore, is to begin with a thoroughgoing account of consumption itself (Part I), before moving on to the city. If this comes perilously close to suggesting an 'unrelational' treatment, it at least provides an expedient means of exposition (the constraints of linear presentation being what they are). Accordingly, Chapter 1 offers a general way in to the topic, whilst Chapter 2 provides a far fuller engagement with the deceptively simple-sounding notion of consumption, following closely in the footsteps of Baudrillard.

Part II begins to consider the relations between consumption and the city in earnest. Chapter 3 provides what amounts to a kind of minimalist statement of these relations. It employs a number of urban archetypes – principally the stranger and the *flâneur* – as a device for considering the most direct relations between the city and consumption. Many of the themes we encounter here will reassert themselves as the overall course of the book progresses. Chapter 4 further assesses the centrality of consumption to the city, from a slightly different angle – attempting to repudiate the disappointingly common belief that consumption is not a particularly serious topic for urban studies. It is sometimes supposed that a sort of tinseltown characterization of the city proceeds from the kind of approach that affords consumption significance. By reconsidering the notion of collective consumption – once an undisputed centrepiece of urban theory – it places the social inequalities structured around consumption to the forefront of attention. The consumer society has particularly devastating consequences in terms of social justice. The specific mode of gestation of these inequalities is, however, all too often either overlooked or mischaracterized. More often than not, such inequalities are accounted for in terms that, strictly speaking, no longer apply. Chapter 4, then, is an attempt to elucidate the specific mechanisms involved, to demonstrate their association with consumption, and to reveal their departure from earlier forms of social inequality.

Part III, which again comprises two chapters, considers the full-blown implications of consumerism for the city. Chapter 5 aims to develop a clearer appreciation of the origins and the consequences of consumerism, highlighting the currency of the term 'lifestyle'. Whilst that term is often dismissed as essentially frothy and meaningless, the aim here is to show that it is crucial to the nature of consumerism. The chapter contrasts the self-proclaimed ease of living in a consumer society with its manifest dis-ease, concentrating particularly on the way in which the consumer is expected to turn consumerism into a way of life. It suggests that the city has become the prime site for the kinds of activity this involves. Building on this discussion, Chapter 6 brings the nature of the postmodern city, postmodern space, and postmodernity *per se* centre stage, to consider the fate of place in the consumer's world (cf. Augé 1995; Meyrowitz 1985; Relph 1976; Sack 1988; 1993). The notions of 'hyperreality' and 'glocalization' frequently feature amongst the defining features of the postmodern world. Here, their inherent connections with consumerism are emphasized. Whilst this chapter is, in many respects, the culmination of what has gone before, it nonetheless marks a departure as much as an arrival. This is evident, for example, in the way in which it draws on Baudrillard's later work. All too often there is a lamentable failure to see the connection between Baudrillard's work on 'simulation' and 'hyperreality' and his early work on the consumer society. Yet when Baudrillard speaks of hyperreality, the situation he describes is a direct consequence of the logic of consumerism. If the book has a final, overall message, therefore, it is to be found here. That message is that the processes unleashed by consumerism are ultimately responsible for reconfiguring society, space, and time in ways that we have not witnessed before. The recognition that 'we have not been here before' is vital for understanding the nature of the contemporary city. This kind of conclusion is, without doubt, far removed from the kind of analysis that persists in regarding consumption as a narrow process, which affects only certain places, in relatively superficial ways. There is no denying, of course, that consumption creates certain kinds of spaces for its own particular purposes. But this should not be considered the end of the matter. The final chapter of the book, therefore, aims to demonstrate the broadest effects of the consumer society on the city, space, and time.

It remains to make a few remarks of a more pragmatic nature. First, as far as possible, I have endeavoured to ensure that each chapter can be read more or less independently of the rest. I have attempted to adopt a style that avoids simple repetition, yet develops themes introduced in earlier chapters as the work proceeds. Reading the book from cover to cover, in other words, should present the reader with both a consistent thread of continuity and a reasonable semblance of sequential development. Those merely dipping into the book, however, should not face the kind of insurmountable difficulties that are sometimes met when trying to read chapters in isolation or out of sequence. Second, I have taken the opportunity of using the endnotes to add supplementary detail of a more technical nature; detail that either assumes a more specialist knowledge or which anticipates arguments to be developed subsequently. These (occasionally detailed) notes amount to the knotted, tangled threads that allow the embroidering, chapter by chapter, of a recognizable picture of the relations between the consumer society and the city. They may be passed over without fear of losing the overall course of the argument, though they are, nonetheless, constitutive of it.

Finally, I should add that I have explicitly not attempted to be comprehensive in terms of the book's coverage. It might be regarded as idiosyncratic, for instance, that a book on the present topic should fail to give proper consideration to the process of gentrification and the wider contours of what Ley (1996, 1) calls 'the embourgeoisement of inner-city and downtown landscapes' (cf. Butler 1997; Smith 1996; Zukin 1988; 1995; 1998). Gentrification is, undeniably, a significant area of inquiry. By the same token, it has already generated an immense scholarly literature – to the extent, one suspects, that diminishing returns may have set in some time ago. In my estimation, moreover, the definitive work on gentrification – assuming there could ever be such a thing – has already been produced. I refer to the highly original and convincing explanation put forward by Redfern (1997a; 1997b; cf. Lyons 1998; Redfern 1998), which places changes in domestic consumption technologies at the heart of the process. There are, undoubtedly, many other interesting and important topics that the reader may be disappointed not to find included within these pages. I can only hope that this selectivity on my part is forgivable.

Part I
Consumption

1 Consumption controversies

> postmodern society engages its members primarily in their capacity as consumers rather than producers. That difference is seminal.
>
> (Bauman 2000, 76)

> no more 'productive' or 'unproductive' consumption, only a *reproductive* consumption.
>
> (Baudrillard 1993a, 28)

Introduction

According to Wyrwa (1998, 432), the Middle Ages had no single term corresponding to what we now matter-of-factly refer to as 'consumption': 'Only with the advent of mercantilism, devoted to the fiscal interests of the sovereign, does the Latin derivative "consumption" begin to take on real meaning'. Consumption only came to be recognized as such when the opportunities it afforded for taxation were recognized. From this point on, consumption has commanded ever-greater attention, though never before has consumption been seen as meriting the level of attention it is currently afforded. Over the past decade or so, there has been a veritable explosion of interest in the topic, putting it firmly on the map of concepts of fundamental importance to the social scientist, of whatever stripe or disciplinary background (Miller 1995). Some may, of course, regard this turn towards consumption as little more than academic fad and fashion. There is, no doubt, an element of truth in such an assessment. Nonetheless, the current wave of interest in the topic arguably marks a far more profound recognition that consumption is increasingly vital to the survival of capitalism. Within the social sciences, consumption has shifted from a position where it could, without too many difficulties, be effectively sidelined, to one where an acknowledgement, appreciation and understanding of its role has become increasingly important. Whilst it is true that consumption has been subject to cyclical attention-swings in the past, the current level of attention seems to be of an entirely different order. To claim that contemporary capitalist societies need to be understood, first and foremost, as 'consumer societies' is not, of course, uncontroversial. Even more contentious is to reserve the term 'consumer society' for contemporary capitalist societies alone. This is, nonetheless, I will argue, a valid proposition.

Slater (1997) makes the case that it is possible to take the time-horizon of the 'consumer society' back more or less indefinitely. Thus, we find Sherratt (1998, 15) referring to changes in social organization occurring in Europe around 1300 BC as marking 'a consumer revolution', and pointing to the importance of so-called 'secondary products' in even earlier times (Sherratt 1981). It is demonstrably possible, in other words, to highlight the incredibly long-standing importance of consumption to human social arrangements. However, such observations do not necessarily pronounce on the appropriateness of reserving the term 'consumer society' to present-day societies (or, at least, remarks such as Sherratt's are not intended to make such an adjudication). An ever-growing body of work has highlighted the particular centrality of consumption to early modern social arrangements (Agnew 1993; Appleby 1993; Bermingham and Brewer 1995; Breen 1988; Dyer 1989; Lemire 1990; Shammas 1990; Thirsk 1978; Weatherhill 1988). In the light of such diligent research, which reveals the extent and variety of past consumption practices to be far more complex than was hitherto appreciated, is it not misleading and inappropriate to reserve the term 'consumer society' for today's 'consumerist' kind of consumer society? Again, one must answer in the negative. In fact, were we to pin down the emergence of 'consumption' as a meaningful concept in painstaking detail – particularly in relation to the evolution of political-economic thought – or to document the immense variety of consumption practices characterizing past and present societies in different parts of the world, none of this would invalidate the specific sense in which the 'consumer society' label is appealed to when applied to today's capitalist societies (Baudrillard 1996; 1998a; Bauman 1987; 1998a). The appropriateness of the label, as intimated in the Introduction, centres on the *systemic* role increasingly being played by consumption.[1] It is this systemic role – which coincides with consumption assuming a 'mythic' status – that defines our society as a fully fledged consumer society. As Baudrillard (1998a, 194) remarks: 'The historic emergence of the *myth* of consumption in the twentieth century is radically different from the emergence of the technical concept in economic thinking or science, where it was employed much earlier'. This historic emergence broadly coincides with what many have attempted to grasp under the rubric of 'postmodern' society. Clearly, it will take some effort to establish this connection, and to demonstrate the significance of the sense in which the 'consumer society' label is appropriate to postmodern society in a way that it is not, for instance, to early modern or prehistoric society. This is the primary purpose of this chapter.

Consumer society, casino culture

To pursue the suggestion that the defining feature of a consumer society is the systemic role taken on by consumption, let me begin by recalling a discussion at an open-air bar in the central square of the Estonian capital, Tallinn – the city where the European Sociological Association's 'Sociology of Consumption Working Group' met in 1996. The discussion centred on different ways of purchasing drinks: specifically, whether one pays for drinks as and when they are obtained (the common practice in British pubs); or whether a tally is kept by the bartender and the bill settled at the end of the night (as in the café-bars of many continental European cities). These different methods of payment are, in and of themselves, interesting. They reveal a distinctive geography of consumption, a set of situated cultural differences that persist despite the oft-repeated claims that the world is becoming ever more homogeneous as a result of the standardization of consumption. Yet matters are considerably more complex than this. In Britain, for instance, drinks obtained along with a meal in a restaurant are usually paid for at the end of the evening, whilst food purchased in a pub is usually paid for as and when it is ordered. Perhaps the remnants of a class difference are discernible in the need to demonstrate one's ability to pay (with 'ready cash') in certain contexts, and the element of trust (for the moment unsullied by vulgar monetary considerations) implicit in others. Both systems, however, are functional: that is, they are capable of sustaining or reproducing themselves over time (and, within culturally defined limits, space).

A very different system operates in the casino complexes of Las Vegas. There, drinks are supplied free of charge; providing, of course, one indulges in a spot of gambling, and exempting the fact that a gratuity is generally expected. In this system, the income the casino receives from gambling allows the drink – which in any case fuels the gambling – to be treated as a loss leader. Evidently, the customers perceive benefits from the nominally free drinks their gambling subsidizes: they get both 'free' drinks *and* the chance to win fabulous amounts of money. The casino operators also derive significant benefits from the set-up: supplying drinks has become very much secondary to the gambling side of the operation; the supplementary activity has long since supplanted its initial basis. This expanded system is also capable of reproducing itself over time, since the casino is in a position to ensure that, in the long run, the odds are massively stacked against the casino itself ever going broke. What is most significant about this system, however, is the way in which it seems capable of conjuring up an increased 'total amount of benefit' – for both consumers and casino operators – simply by diverting the system of supplying drinks through the supplementary detour of games of chance.

Day after day, of course, the casinos generate losers as well as winners; deal out misery as well as happiness. Counterintuitively, perhaps, winning and happiness may not even go hand-in-hand (Frank 1999). If the dream of hitting the jackpot comes true, it can all too easily turn into a nightmare – as the occasional press coverage of the winner who goes 'off the rails' serves to confirm. This provides little consolation, of course, for those who have never found themselves in the winning seat. Nor is it likely to deter others from wanting to see how they would cope with such a situation. Invariably, however, the downside of gambling is carefully hidden from view – literally so, in Las Vegas, by a glitzy, fantasy facade. This is not, though, the only dissimulative aspect of the gambling industry. The unpalatable truth lying behind the dream of instant riches is also carefully concealed by the impregnable logic of games of chance: one always *might* win; one has as good a chance as anyone else. In this regard, the casino is ruthlessly egalitarian, shamelessly democratic.

It might reasonably be objected that it is patently untrue that everyone has an equal chance of hitting it rich at the casino. Are not some, those already wealthy, more equal than others? Indeed they are, insofar as the marginal utility of money – the value of another pound or dollar – differs starkly between rich and poor. But the logic of the casino is impregnable precisely because it is tautologous. Fortunes *can* be made with minimal stakes. There is no prior requirement a winner must first fulfil. Sheer good luck is all it takes. The chance is open to all, and fate will run its course with absolute indifference. The implication, of course, is that there is no point in being a sore loser. 'Losers have no less reason than winners to wish that the game goes on, and that its rules stay in force; and no more reason to want the game to be proscribed or its rules to be overhauled' (Bauman 1993a, 44). With a fiercely commercial benevolence, the casino automatically offers a second bite of the cherry, ensuring that the misfortunes of the past are not brought to bear on the opportunities of the present. Today's loser may yet become tomorrow's winner; yesterday's loser is as welcome again today as anyone else – with the sole exception that those who incur debts they are unable to honour can expect no mercy. The sinister threat of Las Vegas' gangster past still hangs in the air; an ambient warning to those who would leave town without first clearing their debts. Those who prefer to disappear when payback time comes around, the suspicion remains, might be given a helping hand in precisely that direction. It is surely no coincidence that Las Vegas is surrounded by desert (Baudrillard 1988; Bauman 1988a).

Learning from Las Vegas has, of course, rapidly become an outmoded activity. Even gambling looks set to be supplanted by other (equally profitable) leisure activities, from theme park rides to art galleries. Nonetheless, I want to suggest – without too much hesitation or qualification – that the casino system provides a near-perfect analogy for the consumer society. The way in which it is capable of conjuring up, seemingly from thin air, an increased amount of 'total happiness' is very much like the way in which the consumer society itself operates. And, whilst there can be no doubting its capitalist character, the consumer society is as unlike earlier forms of capitalist society as Las Vegas is from other systems of buying and selling drinks. Indeed, one would probably be hard pressed to characterize the Las Vegas casinos as simply another system for supplying drinks, on a par with continental European café-bars and British pubs. As we have already had cause to remark, the supplementary dimension of gambling renders the supply of drink a minor (if nonetheless significant) subsystem of a far larger whole. As with Las Vegas, so with the consumer society. Consumption is no longer just one aspect of society amongst others. In a fully fledged consumer society, consumption performs a role that keeps the entire social system ticking over. It is this sense of 'consumer society' that is both appealed to and explicated in this book, and from which the understanding of 'consumption' it both develops and applies derives. All societies, of course, have seen consumption feature amongst the activities that are necessary to sustain and reproduce them. A consumer society, however, sees this common, everyday activity elevated to new heights. And lest it be thought that this is merely a minor modification to the existing structure of capitalism, it is worth recalling Derrida's (1973, 89) definition of the 'strange structure of the supplement' – which identifies the supplement as 'a possibility [that] produces that to which it is said to be added on'. The consumer society represents a development the potential for which was there from the very start.

Consumption, theory, postmodernity

'Many diagnoses of our "postmodern condition" hinge upon debates about consumption', note Miller and Rose (1997, 1). This connection is, in fact, a source of considerable contention. Thus *The Dictionary of Human Geography* (Johnston *et al.*, 2000, s.v. 'consumption, geography of') berates accounts of consumption that appeal to the postmodern as little more than overblown theoreticism:

> In the most apocalyptic of postmodern pronouncements, the chief reason for existence has become consumption; signs of the commodity have become more important than the commodity itself and people have begun to lose their identity in the mêlée of consumption (Baudrillard 1998[a]; Bauman 1993 [*sic.* – 1996a]; Clarke 1997[a]). But the careful empirical research carried out by geographers over the course of the 1990s (e.g. Miller, Jackson, Thrift, Holbrook and Rowlands 1998) must give considerable pause to this kind of depiction.

This is, sadly, very much a caricature of understandings of consumption that posit a strong connection between the consumer society and postmodernity. It needs to be set against a more accurate depiction.[2] The very notion of 'postmodernity' is, of course, an essentially contested concept. Many faulty conceptions of postmodernity remain in circulation. Little wonder, then, that the notion should so frequently evoke confusion and incredulity. A more adequate account of what the notion entails is therefore of the utmost importance. The way in which postmodernity relates to the consumer society must also be specified, particularly given that an historically naive formulation frequently emerges from the linking of the two concepts. These two issues – the nature of postmodernity itself, and its relation to the consumer society – are closely entwined, but they may be tentatively separated, if only for ease of exposition.

Although the 'postmodern thesis' contains what might be regarded as an historical dimension, the issue is rendered infinitely complex by the fact that the postmodern programmatically disrupts the sense of history bequeathed to us by modernity. All too often, however, the term 'postmodern' is mistakenly formulated as referring to an historical periodization, to a discrete, diachronic sequence: we were once modern, we are now post-modern. Such an unproblematized notion of historical succession is evident, for instance, in accounts of postmodern society that posit a relatively recent consumerist 're-enchantment' of a 'disenchanted' modern world (Firat and Venkatesh 1995; Ritzer 1999). This sense of re-enchantment was, however, intrinsic to modernity (Heelas 1996), and cannot, in and of itself, be seen as marking the irruption of a 'postmodern era'. In a similar vein, Dear (2000, 140) suggests that one of 'the most innovative aspects of recent debates on the postmodern condition is the notion that there has been a radical break from past trends in political, economic, and socio-cultural life'. The postmodern is, however, more complex than accounts that appeal to such a 'radical break' allow. The postmodern marks, in a very particular sense (which will only finally be pinned down in the final chapter), the *end of history*, where 'history' itself is *part and parcel of the modern way of viewing the world*. Thus, as Lyotard (1984, 81) insists, to make sense of the notion, the '*Post modern* would have to be understood according to the paradox of the future (*post*) anterior (*modo*)' – it *will have been*.[3] Lyotard's suggestion involves a complex argument about the processes set in sway by modernity and their unintended consequences. Specifically, it involves a sense in which the modern progression of history works towards its own liquidation, revealing the shape and nature of modernity only in retrospect (the Owl of Minerva's crepuscular habits being well known). It also amounts, however, to a recognition that the onset of a 'postmodern condition' does not entail that modernity is simply done away with or left behind us (as accounts of the 'radical break' variety tend to maintain). As Bauman (1993a, 38) puts it, in the midst of postmodernity, modernity 'remains very much with us and around us – perhaps never more than now, in its posthumous life'.[4] Or in Kuspit's (1990, 60) words: 'the term "postmodern" implies contradiction of the modern without transcendence of it'. Accordingly, perceptions of an 'apocalyptic' element to the postmodern – which are invariably invoked to dismiss the notion as far-fetched (Wynne and O'Connor with Phillips 1998) – tend to misspecify entirely the relationship between the postmodern and the modern, not least in assuming a thoroughly conventional understanding of 'apocalypse' (Doel and Clarke 1998). The full implications of the sense of 'postmodernity' and its connection to the 'consumer society' will emerge as the book proceeds. It would be premature to say much more before a fuller account of the consumer society has been provided. Suffice it to say, for the moment, that the conception of the consumer society appealed to by way of the casino analogy has a good deal to do with the notion of postmodern society; but only insofar as this is understood as a kind of 'decomposition' of modern society – as its posthumous form.[5]

Even if many commonly peddled versions of postmodernity are indeed misleading, the concern – again voiced in the definition contained in *The Dictionary of Human Geography* – that 'postmodernism' legitimates the latest brand of armchair theorizing, is certainly not disingenuous. Thrift (2000a, 1), for instance, fears that the floodgates to the practice of 'letting theory outrun the data' have been opened. The consequences are reputed to be especially dangerous in promoting the unsubstantiated pronouncements of particular theorists. Thus, for Thrift (*ibid.*, 5), too many 'geographers' use of theory … remains resolutely eucomistic and representational; and there-fore in danger of simply retracing steps already made by others'. The present book, insofar as it follows closely in the footsteps of Baudrillard and Bauman, may well fall foul of Thrift's objection. There is, however, a ques-tionable logic in Thrift's assessment, which comes close to suggesting that theory is essentially *opposed* to data – that theory simply gets in the way. Theory, however, always marks out a path. Attempting to retrace the trails blazed by others, examining the routes followed and the vistas offered, is an established practice not because of some fearful and unheroic need to stick to the beaten track, but because routes are only ever established by repeated use. The work of theory can never simply be abandoned in the naive hope of getting straight to the facts of the matter. 'Whoever believes that one tracks down some *thing*?' (Derrida 1973, 150). On the contrary: 'One tracks down tracks' (*ibid.*; cf. Calle/Baudrillard 1988).

The approach favoured by Thrift seems very close to that recommended by Miller, who, as noted in the Introduction, wishes to discard the notion of the 'consumer society' entirely. Miller also apparently regards the notion as little more than an obstacle to engaging with the real 'experience' of consumption. Miller's work, however, not only frequently mischaracterizes other theorists; his own (largely unacknowledged) theoretical precepts also risk missing aspects of consumption that may well turn out to be vital. Consider, for instance, Miller's (1988) study of the modification and decora-tion carried out by council tenants to their kitchens. Miller has subsequently drawn upon this research to challenge what he regards as the mistaken approach of theorists such as Baudrillard:

> According to those who condemn consumption as an expression of a materialist obsession with goods, it is assumed that there is an opposi-tion between a concern with social relations and with material goods. In the kitchen study, exactly the opposite conclusion arose. Doing up the kitchen was not the work of isolated housewives who had lost their social contacts. Such housewives were much more likely to be living in

kitchens they disliked and yet felt unable to change. It was those who were socially most involved and supported by friends and families who had the confidence and support to transform their environment and to create their own aesthetic.

(Miller 1997, 23–24)

Miller (1997, 22) assumes that Baudrillard's critique of consumerism holds that

> in this modern materialistic and capitalist world, an orientation towards objects has so replaced an orientation towards persons that, instead of objects symbolizing people, we have now become merely 'lifestyles' – that is, the passive carriers of meanings that are created for us in capitalist business.

This is far from an accurate characterization of Baudrillard's work (and there is certainly an odd conception at work in the suggestion that people 'become' lifestyles). Whilst the kitchen study is interesting, it is spurious to use it to dismiss a 'straw version' of Baudrillard. Miller's kitchen study found that consumer goods tend to mediate social relations. Surprisingly, perhaps, this is not a million miles from Baudrillard. Yet the implications of this situation are far more complex than Miller allows – which is precisely what Baudrillard's work demonstrates. This returns us to the role of theory *per se*.

Thrift (2000a, 1, citing Strathern 1999, 199) suggests the desirability of 'cleav[ing] to a "certain brand of empiricism, making the data so presented apparently outrun the theoretical effort to comprehend it" [*sic.*]'. If we consider this suggestion with respect to Miller's (1988) kitchen study, which would seem to adhere to this particular brand of empiricism, the question that arises is, on what basis are the kind of conclusions Miller draws the only ones consonant with the evidence? Douglas (1996) questions whether Miller's interpretation of those council tenants who had not done up their kitchens as 'isolated' and 'alienated' stands up to closer scrutiny. The type of housewife that Miller identifies as 'isolated' and 'alienated', Douglas (1996, 82) suggests, might be thought of in an entirely different way: 'it might be a positive value for her to have a colour on the wall that she considers ghastly, a reminder of the war against the others and their despised way of life, on behalf of herself and hers'. Douglas writes from a perspective which holds that consumer choices are made more in terms of what they are *not* than what they *are*: people choose the items they feel would not be chosen by the kind of people they themselves would not want to be, the people they would want to distance themselves from (Douglas and Isherwood 1996). 'What colours did she [the unhappy council tenant] wear?', asks Douglas (1996, 82): 'No shopper himself, the anthropologist does not report on her other antipathies'. Empirical evidence is always theory-laden, in the sense emphasized by Sayer (1984). The framework Douglas advocates dictates the necessity of gathering such evidence, to assess whether or not the theoretical conviction that choices tend to exhibit similar characteristics across very different categories of goods – food, clothing, home-decor, health-care, and so forth – holds true (Laumann and House 1970). Although Miller claims that the superiority of his own work lies in its basis in empirical evidence, the evidence on which he bases his interpretation is ineluctably partial. Whilst claiming that his approach is superior to theoretically orientated work such as Baudrillard's, his attempt to eschew given theoretical frameworks in favour of 'empathic experience' has a curious blind-spot with respect to its own inadequacies.[6]

As a further example, consider the conclusions reached by Miller *et al.*'s (1998) study of shopping practices in North London, referred to in the *Dictionary of Human Geography* entry cited above. On the basis of this work, Miller (1997, 46) suggests that 'shopping is usually experienced as a moral project based on expressing one's concerns for others rather than oneself'. The finding is aimed against the theoretical proposition that the morality of consumption is 'a purely *formal* morality: not *how* to consume; just *to consume*' (Clarke and Doel 2000; 223). For me, however, the interesting empirical question would be, *how* is the tension between the 'formal morality' of consumerism and the moralities of everyday life negotiated? This formal morality is real enough: it cannot be dismissed as simply the invention of social theorists with overactive imaginations. Indeed, although he sets it up as a purely epistemological issue, this basic ontological tension is very much in evidence in Miller's own work. In particular, Miller is concerned to show that shopping is rarely about unadulterated hedonism, and more often a form of 'saving'. Miller is surely right to point out that the everyday realities of consumption hardly correspond to the paroxysms of pleasure portrayed in advertising – but he is on much shakier ground if he believes that social theorists such as Baudrillard and Bauman are guilty of taking such representations at face value. Moreover, there are all manner of unacknowledged problems with Miller's alternative formulation in terms of 'saving' – despite his belief that it represents a significant advance on work such as Baudrillard's. Miller employs the term 'saving' in a rather idiosyncratic (and somewhat slippery) sense, not in the sense used by the economist – neither semantically (where savings and consumption are inversely proportional: what is not spent on consumption is available for saving, and vice-versa) nor ontologically (the 'desire to save is not based on the rationalizing of budgets but on an essentially moral project of self-sacrifice', says Miller [1997, 46]). Thus, for Miller, a moral project of consumption as 'saving' takes shape:

> Amongst the most avid savers are pensioners, who may live at a level of consumption which causes suffering and shortage for themselves in order to give a very large present to a descendant whom they hardly ever

see, and may little appreciate the gift, but who manages to represent the concept of family continuity for that pensioner. The middle class is often offended by the way working-class shoppers save money on what they would regard as proper priorities, such as healthy food, in order to spend large sums on what they regard as improper purchases, such as Sega and Nintendo games for their children.

(*ibid.*)

This is, evidently, the kind of situation where those of a Marxist persuasion would wish to bring ideas of 'false consciousness' into play (rather than remaining strictly neutral over the issue of whether it is in anyone's but the manufacturer's interest that a Nintendo or Sega console has priority over a healthy diet). It would, of course, be nonsensical to deny that computer games can be a source of pleasure. Yet, in Miller's eyes, acknowledging this point seems to put an end to the matter. He is evidently conscious of the contradictions this kind of situation brings to light, but rather than explore them further, Miller seems content with simply remaining more liberal-minded about such issues than those who would wish to appeal to notions of 'false consciousness' or 'false needs'. Whilst there are undoubtedly untenable aspects to such conceptions, Miller's alternative can hardly be judged as superior – and his premature dismissal of those who have gone on to develop a clearer insight into this situation is, to say the very least, unhelpful.

The ultimate conclusion Miller (1997, 46) draws from his account of the self-sacrifice made by pensioners and working-class parents is that 'there is no relation between income and a concern for savings – such a concern is expressed as fully by households that might be regarded as wealthy as by those who would be deemed poor'. The particular sense in which Miller employs the word 'saving' perhaps overcomes the obvious objection that saving is far easier for the rich than the poor. The particular phraseology employed is presumably intended to cover this objection (such that the stated commonality of *expression* need not be read as denying or precluding *actual* variability; the statement refers only to 'a *concern* for saving'). But there is an invidious slippage here. It is, in the final instance, dangerous to make assertions of this kind. Indeed, one would have to conclude that Miller's statement is simply untrue – by any stretch of the imagination. Is this not a prime case of letting the data outrun the theory – to highly unfortunate effect?[7] Bauman's (1998a) discussion of the way in which the consumer society recasts the poor in a new structural position, considered further in Chapter 4, is avowedly theoretical in approach, but reaches far more sustainable conclusions than Miller is able to muster.

The real issue that reveals itself in the examples discussed here is the way in which the consumer society holds in place a formidable range of contradictions, tensions and paradoxes, within a dynamic but reproducible social formation. *Contra* Miller, it is the need to develop the critical faculties capable of disclosing the way in which this is achieved – and the consequences to which it gives rise – that necessitates a thoroughgoing theorization of the consumer society. The politics of consumption are undoubtedly complex.[8] It is a serious abrogation of the responsibilities of thought to fail to engage fully with the kind of theoretical work that strives to bring out these complexities.

Work and play

A substantial amount of effort has already been channelled into the task of theorizing the consumer society. Taking this work seriously pays enormous dividends. Let us begin to do so by considering the political-economic conception of consumption. It should be evident, in the light of the previous discussion, that being attentive to the subtleties of different theoretical positions is vital. Consider, in this regard, Marx's (1973, 92) remarks on the relations between production and consumption:

> Production ... not only creates an object for the subject, but also a subject for the object. Thus production produces consumption (1) by creating a material need for it; (2) by determining the manner of consumption; and (3) by creating the products, initially posited by it as objects, in the form of a need felt by the consumer.

Various interpretations have been put on these words. If they seem to imply a one-way relationship, with production unilaterally determining consumption, this is largely a consequence of wrenching them out of context.[9] Even accepting this point, interpretations vary. One notable interpretation, which seems especially close to what Miller attributes to Baudrillard, is provided by Marcuse (1964, 5), who proposes the following:

> We may distinguish both true and false needs. 'False' are those which are superimposed upon the individual by particular social interests in his repression: the needs which perpetuate toil, aggressiveness, misery and injustice. ... Most of the prevailing needs to relax, to have fun, to behave and consume in accordance with advertisements, to love and hate what others love and hate, belong to this category of false needs.

There are, undoubtedly, serious problems with such a formulation and, although Miller rides roughshod over the fact, Baudrillard (1981, 136) is fully aware of its inadequacy: 'Marx says: "Production not only produces goods; it produces people to consume them, and the corresponding needs". This proposition is most often twisted in such a way as to yield simplistic ideas like "the manipulation of needs" and denunciations of "artificial needs" '. Rejecting such a reading, Baudrillard (*ibid.*) continues to say that it is necessary to grasp that what produces the commodity system in its general form is the *concept* of need itself, as constitutive of the very structure of the individual – that is, the historical concept of a social being who, in the rupture of symbolic exchange, autonomizes himself [*sic.*] and rationalizes his desire, his relation to others and to objects, in terms of needs, utility, satisfaction and use value.

(ibid.)

Even if, at first sight, this seems unnecessarily obfuscatory, it is clearly light-years away from the conception that Miller attributes to Baudrillard. A better understanding of Baudrillard's meaning emerges if one considers the general tack of his work. Unfortunately, this is difficult without a full grasp of the terminology involved, which is introduced in Chapter 2. At the risk of preempting the definitional clarifications contained therein, the sense of Baudrillard's theoretical endeavour can be given in broad-brush outline here.

Both classical political economy (Ricardo 1951; cf. Walsh and Gram 1980; Sawyer 1989), and Marx's (1967) critique (cf. Howard and King 1975), concerned themselves with the analysis of the production and reproduction of an economic surplus, defined as 'a flow of production in excess of *necessary* consumption' (Robinson 1980, xii). Necessary consumption, in this context, is afforded a purely analytical definition: what is 'necessary' is whatever is deemed necessary within the society in question (in the sense that historically or geographically *contingent* factors assume, without contradiction, the status of socially *necessary* ones). The status of 'necessary consumption' is, nonetheless, affected by its relation to the production of an economic surplus. Hence Marx's analytical decomposition of consumption into

> *productive consumption*, which includes both consumption of consumer goods by producers, and consumption of the means of production in the productive process; and *unproductive consumption*, which includes all consumption of goods which do not enter the reproduction process, do not contribute to the next cycle of production.
>
> (Bottomore et al., (1987) A Dictionary of Marxist Thought,
s.v. 'consumption')[10]

This distinction is, in and of itself, unproblematic. It is, nonetheless, irredeemably partial. Production and consumption are dialectically related insofar as they amount to different moments in the circulation of capital. Consumption, under capitalist conditions, amounts specifically to *commodity* consumption; that is, to the conversion of exchange-values into use-values. But raw materials are also consumed in the labour process; that is, in the conversion of use-values into exchange-values.[11] Indeed, the perpetual transfer of use-values into exchange-values, and of exchange-values into use-values, underpins the entire expanded reproduction of the capitalist system. However, the sense in which this analytical framework retains the trace of the historical conditions in which it first suggested itself is far more problematic than is usually allowed. Whatever degree of analytical spin is put on the sense of 'necessary' consumption, it still carries a distant echo of the distinction between 'necessary' and 'luxury' consumption, thus maintaining a rigid distinction that capitalism itself long since set about erasing. Luxury consumption – conspicuously *unproductive* consumption – plays equally as crucial a part in the reproduction of capitalism as does 'productive consumption'. So-called 'unproductive' consumption involves, in Baudrillard's terminology, the conversion of 'exchange-values' into 'sign-values'. And this process of conversion has come to play an increasingly fundamental role in the reproduction of capitalism. The particular sense in which this is so will not be pinned down in any rigorous way until the end of Chapter 2. Suffice it to say, for now, that Baudrillard's (1981) clarion call amounts to a plea for the recognition that the reproduction of capitalism involves something more than an economic process that exhausts itself in terms of exchange-value and use-value alone. Accordingly, for Baudrillard (1981, 133n), 'there is no fundamental difference between 'productive' consumption (direct destruction of utility during the process of production) and consumption by persons in general' – even where this seemingly involves 'unproductive consumption' – since the 'individual and his [*sic.*] needs are produced by the economic system like unit cells of its reproduction'. The imbrication of sign-value within the capitalist system renders *all* consumption vital to that system's reproduction. Such a conclusion is not, in fact, entirely original to Baudrillard's analysis. It is already evident, in embryonic form, in the work of a number of earlier commentators, including Veblen, Mauss, Bataille, and, arguably, Benjamin. Since these first three are considered in some detail in the next chapter, let us briefly consider Benjamin in a little more detail – not least because Benjamin returns us to the casino.

'Where would one find a more evident contrast than the one between work and gambling?' asks Benjamin (1975a, 177), in rhetorical vein. The contrast is by no means as clear-cut as it might first appear. Unskilled factory work, Benjamin (*ibid.*) notes, 'lacks any touch of adventure, of the mirage that lures the gambler. But it certainly does not lack the futility, the emptiness, the inability to complete something which is inherent in the activity of a wage-slave in a factory'. The seemingly automatic, mechanical hand-movement of the factory worker at the machine is near identical to that of the gambler at the card-table. What is more, Benjamin (*ibid.*, 178) concludes, 'the mechanism to which the participants in a game of chance entrust themselves seizes them body and soul, so that even in their private sphere, and no matter how agitated they may be, they are capable only of a reflex action'. Gambling is, perhaps, the best example of a situation outside the world of work where individuals are seemingly *forced* to behave in a certain way. Yet, despite the similarities Benjamin points up, there would appear to be at least one significant difference between worker and gambler.

That difference lies in the nature of the force that elicits the behaviour in question. Simplifying to the extreme, one might say that the worker is pushed into work, whereas the gambler is pulled into gambling. These different forces are, no doubt, each present, to some degree, in both situations. Indeed, one might argue that people are more and more *lured* into working for a living – on the basis of the increased purchasing power and opportunities to consume this gives them – rather than being forced into working in quite the same way as their forebears. In other words, the world of work has moved closer to the strategy of the casino. It is a moot point whether gamblers gamble against their will. Since gambling has a truly addictive quality, it is fair to say that some do. More generally, however, we might say that people most frequently gamble not against their will, but against their own better judgement (not to mention against the wishes of their families and loved ones, without the approval of friends or institutions like the Church, and so on, in many cases). Arguably, therefore, the force forcing them to gamble is seductive rather than coercive in nature: the gambler is lured, tempted, enticed, seduced – and, having tasted the forbidden fruit, the more likely it is that they can be lured into playing again. Benjamin's remarks were indeed prescient, since the irresistible force of seduction increasingly appears to operate, more efficiently and effectively than the brute force of coercion, in areas of society much broader than gambling. In particular, members of the consumer society are seduced into adopting the role of the consumer. As a member of the consumer society, one is set the non-too-arduous task of indulging in pleasurable pursuits. For the silent majority, the cost of this membership – earning enough money to consume by selling one's labour power – is no greater than if one were simply forced to work by coercive means. In effect, the consumer society encourages people to work by means of the 'carrot' rather than the 'stick', diverting an initial state of affairs through a supplementary detour that expands the system as a whole. It operates by offering its 'consumer citizens' (Christopherson 1994) a seemingly unobjectionable freedom – yet in doing so, it is inherently duplicitous, since it admits of no alternative. It represents, one might say, the kind of offer 'one cannot refuse'.

Again, this is not the whole story.[12] The casino analogy retains its validity insofar as the consumer society also automatically generates losers as well as winners. And it, too, dissimulates the fact behind an impregnable, tautologous logic. Failed consumers, however, like failed gamblers, can expect no mercy: they have been given their chance, even if the odds were stacked against them. It is notable that the pervasive slogan of the consumer society – 'Enjoy!' – maintains the foreboding threat of coercion: 'Imagine, putting pleasure in the imperative like that' (Calle 1999, 246). But if the threat is 'once removed' for some, those who fail to live up to the pleasurable role they have been assigned can expect to experience that threat in the most direct of senses. It is no accident that the demonization of certain excluded groups, together with an increasingly uncharitable attitude to the 'undeserving poor', proceeds apace with the development of a society that engages its members primarily in their capacity as consumers (Bauman 1998a). 'Beggars', as the old adage has it, 'can't be choosers'. Today, the conventional wisdom is reversed: those who can't be choosers can expect no more than to be treated as beggars. Those who fail to find themselves on the winning side can expect no more than to be castigated for their failure – despite the fact that they are dealt a consistently poor hand, typically from a marked deck. As Offe (1996, 33) has noted,

> the 'game' of accumulation and 'exploitation' that pits capital against labour – a game which shows a positive sum in the form of a 'growing pie' – has been replaced by a negative-sum game of 'collective self-injury' ... everyone inflicts injury on himself or herself as well as on everyone else without any net gain.

Such is the nature of a 'winner-takes-all' society (Frank and Cook 1996).

These prefatory remarks leave us with the basic proposition that the consumer society is the continuation by other means of the earlier system of productive capitalism. The essential features of the consumer society were already inherent to the former system. The consumer society amounts, nonetheless, to an entirely different figuration – one that reproduces itself by subtly yet seminally different means.[13] The notion of consumer capitalism as a 'casino culture' reveals the extent to which the ideology of consumption hides the full consequences of the triumph of the market. The consumer society both cultivates and thrives upon a sense that one can, ultimately, always buy oneself out of trouble. It proposes a market solution to each and every one of life's problems. It declares that all one needs to ensure, finally, is that the private resources for coping with such eventualities are in place. This is a particularly insipid, anaemic, and regressive vision of society.

Consumption defined

It would not be too gross an exaggeration to suggest that the term 'consumption' has generally been deployed in a thoroughly familiar sense in the social sciences, in the English-speaking world, at least. In this respect, the term differs significantly from comparable concepts, such as 'labour', 'power', 'gender' or 'production', all of which have received a constant barrage of analytical effort designed to unpack their meaning. Such concepts are, as a consequence, widely used in a manner that crystallizes the results of such a concerted effort. The fact that 'consumption' has generally retained its narrow, commonsensical definition in the social sciences represents both a serious deficiency and a missed opportunity. It signals a widespread failure to recognize the extent to which the apparent simplicity of the term belies its actual complexity. This is most evident, perhaps, in the stark contrast thrown up by different traditions of work on the topic. Thus, for instance, what might broadly be termed the Anglo-Saxon tradition has, by and large, routinely misunderstood the insights of what could, equally crudely, be labelled the Continental tradition. There has always been a tendency to translate the insights of the latter into more familiar terms. In the process, much of the significance of these insights has been lost. A substantial effort is required to recoup what has been occluded by the inadequate translation of concepts from one context to another. The fact that the French *consommer* carries a dual implication (of both 'fulfilment' and 'annulment'), whereas the English language commonly employs two words – 'consumption' (from the Latin *consumere; con sumere*: 'to use up entirely, which involves destruction of matter' [Falk 1994, 93]); and 'consummation' (Latin *consumare; con summa*: 'to sum up, to carry to completion' [*ibid.*]) – is sometimes held to explain or at least underlie these different traditions. There is more than a grain of truth in this observation. The implication is that the English term is more limited than the French, carrying one-sided implications, related to the destructive sense of consumption as the 'using up' of material resources. The fact that the French term embodies two, contradictory meanings – lending the term a thoroughly ambivalent sense – comes far closer to grasping the true nature of consumption. Consumption is always destructive and creative; entropic and negentropic; a process of 'creative destruction' and 'destructive creation'. This duality is itself internalized within the modern system of consumption. Yet the potency of this ambivalence is all too frequently hidden behind the view that consumption is merely a concrete operation, concerned with using up goods to satisfy needs.

As Baudrillard (1981, 134–135n) stresses, in the consumer society, 'what is consumed isn't the product itself, but its utility … consumption is not the destruction of products but the destruction of utility'. This destruction of utility is, at one and the same time, its consummation. Where 'it appears to consume (destroy) products, production only consummates their utility. Consumption destroys objects as substance the better to perpetuate this substance as a universal, abstract form' (*ibid.*). The 'consummation' involved in modern consumption amounts, furthermore, to 'a labor of expanded reproduction of use value as an abstraction' (*ibid.*). To this extent, capitalist consumption mirrors capitalist production. Marx showed that, in a capitalist mode of production, 'production is no longer in its present finality the production of "concrete" goods, but the expanded reproduction of the exchange value system' (*ibid.*). Baudrillard stresses the parallel process at work in consumption – which remained hidden, both in Marx's time and Marx's analysis, behind the supposedly 'concrete' process consumption 'naturally' represents (defined in terms of use-value). Hence Baudrillard contrasts *consumption* as a process geared, despite appearances, to the expanded reproduction of the system of use-values (the capitalist system of objects), which is a necessary corollary of capitalist production, with *consummation* of a radically different kind. This latter – exemplified in the pre-modern form of gift exchange – marks out a line of flight from the capitalist system as a whole. 'Only consumation (*consummation*) escapes recycling in the expanded system of the value system – not because it is the destruction of substance, but because it is a transgression of the law and finality of objects' (*ibid.*). The aim of the next chapter is to render the significance of this distinction more intelligible, since it is vital to understanding both the nature of modern consumer capitalism and what it is that this system disavows.

2 Everything you ever wanted to know about consumption

(but were afraid to ask Baudrillard)

> Collective representations are exterior to individual minds...they do not derive from them as such but from the association of minds, which is a very different thing...private sentiments do not become social except by combination under the action of the *sui generis* forces developed in association.
>
> (Durkheim 1953, 25)

> That wants are, in fact, the fruits of production will now be denied by few serious scholars.
>
> (Galbraith 1958, 154)

> The truth is not that 'needs are the fruits of production', but that *the system of needs* is *the product of the system of production*. This is quite different. ... [N]eeds are not produced one by one, in relation to the respective objects, but are produced as a *consumption power*, as an overall propensity within the more general framework of productive forces.
>
> (Baudrillard 1998a, 74)

A preliminary defamiliarization

> Today the promises of the old society of production are raining down on our heads in an avalanche of consumer goods that nobody is likely to call manna from heaven.
>
> (Vaneigem 2001, 84)

An enormous amount of energy has been expended over recent years in showing that consumption increasingly saturates every corner of modern life; that consumption is about far more than one-off purchasing decisions or even the aggregate economic patterns these create. Indeed, a *culture* – or plurality of *cultures* – of consumption seems to be the order of the day (Featherstone 1991; Hearn and Roseneil 1999; Jackson 1993; 1999; Lury 1996; MacClancy 1992; McCracken 1990; Miller 1987; Mort 1996; Nava 1992; Slater 1997; Storey 1999; Wernick 1991a). Yet this still begs the question of what, precisely, 'consumption' is. Accordingly, Campbell (1995, 102) offers a 'working definition', which 'identifies consumption as involving the selection, purchase, use, maintenance, repair and disposal of any product or service'. Such a list of the elements normally subsumed under a single term seeks to overcome the limited attitude, synonymous with neo-classical economics, that equates consumption with the act of purchase (or, from the perspective of the marketeer, with sales).[1] This economistic definition is, in fact, so inadequate that it is remarkable that it could have ever commanded much credence. The strength of Campbell's working definition lies in its departure from the abstraction of the economist to inject something of the concrete fullness of real processes at work in the world. It does so, however, in a resolutely commonsensical manner – and to this extent remains rather stunningly superficial.

Consider, by way of contrast, Baudrillard's (1981, 30–31) insistence that 'the fundamental conceptual hypothesis for a sociological analysis of "consumption" is *not* use value, the relation to needs, but *symbolic exchange* value, the value of social prestation, of rivalry and, at the limit, of class discrimination'. This would immediately seem to fly in the face of common sense. Consumption, we are told, has nothing whatsoever to do with satisfying individual needs; nothing to do with an individual's relation to an object or objects capable of providing physiological and/or psychological sustenance.[2] Instead, we are asked to accept that consumption relates to something else entirely; to 'symbolic exchange', 'social prestation', 'rivalry' and 'class discrimination' – unexpected and (to Anglo-Saxon eyes) mysterious categories that will command a good deal of our attention in due course. The gulf separating these two conceptions, Campbell's and Baudrillard's, is enormous. Consequently, understanding the different positions they mark out is a much more onerous task than is often assumed. As an illustration of the difficulties involved, let us look briefly at the way in which Campbell dismisses work such as Baudrillard's as unnecessarily obfuscatory; as an affront to clear thinking.

According to Campbell (1995, 103), Baudrillard

> modifies Marx's original usage [of the 'commodity' concept] by drawing upon semiotics in order to stress the significance of the 'commodity-sign' rather than the commodity itself. Thus, he argues that in capitalist societies, consumption should be understood as a process in which only the signs attached to goods are actually consumed, and hence that commodities are not valued for their use but understood as possessing a meaning that is determined by their position in a self-referential system of signifiers.

For those unfamiliar with the terminology, 'signs' are, loosely speaking, the 'units of meaning' studied by semiotics (the 'science of signs' established by Saussure [1959]). 'Commodities' are, equally loosely, material objects produced for sale on the open market.[3] In this light, Campbell's thumbnail sketch does not seem unreasonable. Baudrillard, like Debord (1977), points to a 'union' between sign and commodity (Wernick 1991b). More problematically, however, Campbell (1995, 115) suggests that Baudrillard's theorization adheres to a 'communicative act or expressive paradigm', which carries the consequence – and for Campbell this is a dire consequence –

> that consumer actions are not viewed as real events involving the allocation or use of material resources (or even as transactions in which money is exchanged for goods or services) so much as symbolic acts or signs: acts which do not so much 'do something' as 'say something', or more properly, perhaps, 'do something through saying something'.

Campbell acknowledges that consumption of this kind is a (limited) possibility. He notes that *gifts*, for instance, might accord to such a model: 'One can indeed "say it with flowers"' (*ibid.*). Thus, as Sturrock (1979, 7) notes, flowers 'can be and are used as signs: when they are made into a wreath and sent to a funeral, for example. In this instance, the wreath is the signifier whose signified is, let us say, "condolence"'.[4] In acknowledging such possibilities, however, Campbell effectively delimits the area where 'communicative acts' assume validity only to show that, such special cases aside, it is little short of nonsensical to assert that consumption has much (or anything at all) to do with conveying messages or meanings. Specifically, Campbell (1997a) argues, meanings need not amount to messages, and generally do not do so when it comes to consumption. If one is prepared to overlook the fact that Campbell loses sight of McLuhan's point about the medium *being* the message, it seems as if he is making an incisive argument. He goes on to imply, moreover, that a whole host of other theorists are caught up in the same unfortunate 'communicative paradigm'. In the face of this supposed threat, Warde (1996, 304) adds his backing, suggesting that 'an implausible account of consumer behaviour arises from giving excessive weight to a notion of consumption as the communicative action of the individual'. If Campbell and Warde are right, the only question remaining is, why should Baudrillard and his ilk ever have been so muddle-headed? How could they have overlooked such basic problems with their approach as those Campbell brings to light? Unless one is content to accept that Baudrillard's intellectual reputation is, along with that of many other Continental theorists, a case of the Emperor's new clothes, there is surely something amiss here. We can begin to show that something is indeed amiss, and that such a response to Baudrillard's work is far from satisfactory: the exasperation implicit in Campbell's reading suggests a basic failure to see what Baudrillard is getting at, where he is coming from, or both. It is important to get beyond such frequently peddled misunderstandings.[5]

In order to do so, one could raise various objections to Campbell's reading of Baudrillard. The idea of a 'communicative act', for instance, is one that Baudrillard would not particularly recognize, and certainly not as a characterization of his own position. The implications of Baudrillard's 'drawing upon' semiotics must also be treated with caution: there is little sympathy towards the 'semiotic' in Baudrillard's (1981, 163) vehement declaration that 'signs must burn'.[6] Indeed, Campbell's elision of 'symbol' and 'sign' fails to grasp that Baudrillard (1981, 149n) deploys the notion of 'symbolic exchange' *against* the semiotic notion of the sign: he explicitly uses 'the term symbol (the symbolic, symbolic exchange) in opposition to and as a radical alternative to the concept of the sign and of signification'. But we have already run ahead of ourselves, and rather than impute such misunderstandings onto Campbell, we would be better occupied in working towards a more satisfactory understanding of the position Baudrillard does in fact outline. Suffice it to say, for the moment, that where such criticisms are levelled against Baudrillard, they fundamentally misconstrue his argument. This is not, of course, to claim that Baudrillard is immune from critique. Many perfectly legitimate objections may be levelled against his work. They can only be effectively targeted, however, on the basis of a more adequate grasp of what his account of consumption actually entails. To achieve this will require an engagement with a formidable range of social theorists: Mauss, Bataille, Veblen, and Saussure. The remainder of this chapter traces some key elements of their thought – with an eye towards the inadequacies of their respective formulations, which Baudrillard attempts to overcome in his account of consumption.

Exchange: gift and commodity

> Exchange is not a complex edifice built on the obligations of giving, receiving and returning, with the help of some emotional-mystical cement. It is a synthesis immediately given to, and given by, symbolic thought, which, in exchange as in any other form of communication, surmounts the contradiction inherent in it; that is the contradiction of perceiving things as elements of dialogue, in respect of self and others simultaneously, and destined by nature to pass from the one to the other. The fact that those things may be *the one's* or *the other's* represents a situation which is derivative from the initial relational aspect.
>
> (Lévi-Strauss 1987, 58–59)

Given that Campbell (1995) locates Baudrillard's alleged misunderstanding of consumption as lying within the realm of the 'symbolic' (which, for Campbell, means: *concerned with 'communicative acts'*), it is useful to begin our reconsideration from here. As we have already noted, Baudrillard's use of the term 'symbolic' is, *pace* Campbell, distinct from the way in which consumption might operate as a 'signifying practice' (that is, where consumer goods serve as signs, and hence convey meanings). Baudrillard's symbolic resolutely *does not* equate to the signifying qualities of objects consumed (or even, in any simple sense, to 'ways' or 'styles' of consuming).

Yet 'signs' and 'symbols' are frequently elided by authors grappling with the fact that consumption involves both material objects and immaterial meanings – which, of course, it does (Douglas and Isherwood 1996). Baudrillard's argument, however, involves something other than this particular duality. Most authors do employ the term 'symbolic' in a thoroughly colloquial and unproblematized sense, where its distinction from 'signifying' is minimized, if not entirely effaced (Hirschman and Holbrook 1981; McCracken 1990). For Baudrillard, however, the distinction between the two is paramount. A sign always points away from itself. But as Luhmann (1988, 96) notes: 'Symbols are not signs, pointing to something else'.[7] It is at least in part from Lacan's psychoanalytical conception that Baudrillard's sense of the 'symbolic' derives.[8] In order to trace his understanding of the term more fully, it is necessary to return to the case of the gift. Here, though, we must proceed with caution, for by making it a special case, as in 'saying it with flowers',[9] Campbell more or less inverts the significance of the gift in the tradition of thought to which Baudrillard belongs; a tradition that can be traced back to Mauss' seminal essay on *The Gift*.

Synthesizing a range of early ethnographic observations, Mauss derived a very particular understanding of gift exchange in 'archaic' (pre-modern) societies. Put another way, Mauss developed an understanding of the *kind of society* such systems of gift exchange necessarily entail. This is important because, as Carrier (1991) shows, the specificity of Mauss' argument has frequently been lost in Anglo-Saxon renditions of his work.[10] Before unravelling the significance of his argument, two prefatory remarks are in order. First, if 'symbolic exchange' is best broached by first considering Mauss' analysis of 'gift exchange', we shall need to defer any further consideration of the sign until such time as it forces its way back into the picture. Second, despite what has already been said about the lineage between Mauss and Baudrillard, there are important differences in their respective formulations. For ease of exposition, these will remain largely unexplicated until the point where their divergence becomes crucial to the argument. Without further ado, therefore, let us turn to Mauss' work in earnest.

Mauss was, above all, concerned to demonstrate the existence of a system of exchange fundamentally different to that operative in societies dominated by *economic* exchange (that is, exchange involving an abstract measure of value and means of exchange, in the form of money). We need to be clear about what such a demonstration entails; it entails a fundamental problematization of the cultural baggage we usually bring to the notion of exchange. Even the sense of the 'economic' as a separate sphere, differentiated from other areas of social life, has to be recognized as a peculiarly *modern* distinction (however difficult it may be to 'unthink' what we normally take for granted). It is a decidedly modern practice to divide social life into purportedly separate 'economic', 'cultural', 'political' and 'moral' spheres. The importance of Mauss' analysis is that it undercuts the familiarity that permits modern assumptions to masquerade as unchanging, universal attributes of human society. It brings to the fore, for instance, the tacit assumption that there actually *is* such a thing as an 'economy'. Campbell, of course, also draws our attention to gift exchange in order to demonstrate its divergence from purely economic exchange. However, where Mauss' focus on gift exchange serves to defamiliarize economic exchange *as a whole*, Campbell merely points to its *exceptional* status: as a 'special case', gift exchange may be of interest, but it should not, Campbell maintains, be afforded undue emphasis in conceptualizing consumption more generally.

To make headway, here, it is necessary to scrutinize the nature of the gift in more detail. Let us begin by considering the old adage that, where gifts are concerned, 'it's the thought that counts'. The implication is that gift-giving involves a calculus of *qualities*, irreducible to the calculus of *quantities* characterizing economic exchange. Even if the words 'it's just what I wanted' are sometimes mouthed with hollow conviction, we do not evaluate gifts primarily in monetary terms (or, at least, convention dictates that we shouldn't). It is more than likely, in this day and age, that any gift we receive will be a mass-produced item, the price of which can be readily ascertained on our next shopping trip. But precisely because it has been given as a gift, the object is *something more* than its basic, material status (and monetary value) implies.[11] It embodies an excess, above and beyond its objective qualities. In the case of the gift-object, therefore, 'once it has been given – and *because* of this – it is *this* object and not another' (Baudrillard 1981, 64). Accordingly, one can never buy a gift from someone (just as one cannot buy friendship).[12] A gift can only be given – and, of course, received. This is highlighted by the way in which gifts, even when they are mass-produced items, are always personalized: a gift is not simply 'given'; it is given *by* and given *to*. The various rituals surrounding the selection, wrapping, labelling and presenting of gifts make this all the more evident. A gift, then, expresses a social bond: it *symbolizes* (and crystallizes) a social relationship with another. And the particular object constituting the gift is, in line with the old adage, less important than the bond it symbolizes. Accordingly, the gift-object itself 'is arbitrary, and yet absolutely singular' (*ibid.*). The symbolic meaning it holds cannot be found in any other object, however similar its objective nature may be.

One last, vital point to note is the way in which receiving a gift seemingly sets up an *obligation* to reciprocate – to give back a gift in return. One does not, of course, return the same gift that one was given; one 'returns' a gift the same sense as one 'returns' a favour: one 'gives back, in the place...of the thing itself, a symbolic equivalent' (Derrida 1992, 13). The gift, in other words, solicits a *counter-gift*, which serves to sustain social relations over time. Indeed, the temporality of the gift is of particular significance (Bourdieu 1977; 1987). For instance, gift-giving is invariably accompanied by a sense of occasion, and frequently follows a cyclical rhythm (in many cases, one only offers a gift as and when the appropriate time comes around). More importantly still, this cycle of gift exchange has an open-ended, aporetic nature. Once a gift has been given, there is no foreseeable end to the cycle programmed in advance. Any sense of a 'final settlement' is infinitely deferred (Derrida 1992). This brings us to perhaps the most disconcerting aspect of the gift. We are accustomed to thinking of gift-giving as a selfless, altruistic act, yet the obligation to reciprocate set up by the gift instils a sense of indebtedness (Simmel 1950a). Receiving a gift places one in debt to another, whilst giving a gift puts one in a position of power over another. Gift-giving is a form of power play and the relation it establishes has a decidedly agonistic character; it initiates, in Schivelbusch's (1993, 173) apposite phrase, 'a tourney of generosity'.

Although gifts carry such manifold implications, they are frequently portrayed as simply more complex, 'emotionally invested' versions of the same basic situation implied by economic exchange. According to Gouldner (1960, 164), for example, understanding the gift relationship 'requires investigation of the mutually contingent benefits rendered and of the manner in which the mutual contingency is maintained'. Just as in barter or money transactions, gift exchange, as far as Gouldner is concerned, entails mutual individual benefit. Likewise, Blau (1964, 91) considers that gift transactions involve the 'voluntary actions of individuals who are motivated by the returns they are expected to bring and typically do in fact bring from others'. Whilst Gouldner and Blau are right to see gifts as involving something other than mere altruism, they are wrong to attribute a fundamentally economic calculus to gift exchange.[13] The same problem reveals itself in Malinowski's (1978) classic characterization of the Kula Ring. Malinowski famously portrayed this cycle of gift exchange amongst the Trobriand Islanders as involving 'essentially *dydactic* transactions between *self-interested individuals*, and as premised on some kind of *balance*' (in Parry's [1986, 454] terse summation). The way in which gift exchange is finally revealed as encapsulating a strictly economic logic – a rationality of 'individualized equivalence' – is precisely what Mauss sought to subvert. Of course, Malinowski's intent was laudable enough. He intended to overcome invidious Western suppositions about the 'irrationality' of 'primitive' societies by pointing to their essential *rationality* (thus placing them both on a par, so to speak, rather than permitting the view that one society is advanced, the other inferior and backward). Mauss, however, sought to achieve much the same thing from the opposite direction: it is certainly not the case that modern societies are rational whilst 'primitive' societies are irrational, but it is equally mistaken to assume that the same universal rationality is at work in different kinds of society. The error becomes glaringly obvious when it is recognized that the kind of 'universal rationality' being appealed to bears more than a passing resemblance to the Enlightenment vision of 'free enterprise'.

Grasping the logic of gift exchange still requires a leap of thought, since our contemporary understanding is invariably refracted through a *naturalized* view of economic exchange. There is an entrenched tendency, in modern

Western thought, to view social facts as natural ones. We need to be on our guard in order to ensure that our familiarity with one state of affairs does not lead us to mistake it for an unchanging, universal one – one that could never be, and never have been, otherwise. Nowhere is this tendency more evident than in the notion of 'economic rationality'. Anything departing from this seemingly insurmountable rationality, or failing to adhere to its strict logic, seems to register automatically as irrational. Appearances, however, can be deceptive: the 'rationality' of economic exchange between individuals, both acting in accordance with their own self-interest *and* their mutual advantage, is in no way a natural, neutral, or self-evident truth – however much it proposes and promotes itself as such. It is the result not of a scientifically disinterested account of social relations but of an historically specific ideology, handed down from the seventeenth and eighteenth centuries and associated, in particular, with the rise of 'civil society' (Althusser 1969; 1972; Appleby 1978; 1993; Dumont 1977; 1986; Hirschman 1977; Marx 1973; Xenos 1989). Mauss was able to subvert this kind of thinking by locating its historical origins.

In 1796, on his archetypally modern voyage of discovery in the Pacific, Captain Cook observed the practice of gift exchange amongst the native Polynesians (Carter 1987; Cobbe 1979). These were taken to be primitive instances of barter, early stirrings of the recognition of the possibility of the mutual benefit to be had in exchange. The error Cook committed was to project back onto 'primitive' societies the (self-legitimating) 'myth of origin' dreamt up by early modern economic discourse. That discourse had established – to its own satisfaction, at least – that simple barter represented the origins of the complex modern Western economic system: individuals, living in a pre-social 'state of nature', initially came together in acts of barter that would, given time, evolve into the kind of fully developed, 'civilized' social system found in the West. In viewing the Polynesians through this historically and geographically specific framework, Cook failed to recognize that elaborate systems of gift exchange were *already* well established in such societies. What he witnessed was not an embryonic precursor to the modern market economy, but an entirely different social system. Indeed, the very idea that the individual precedes society, with society evolving out of the interaction between self-interested individuals at some later stage, is, as Mauss insisted, a thoroughly modern myth. There were never any 'pre-social' individuals living in a Hobbsian 'state of nature'. Humans are always already social *before* they are individual: 'the one-for-the-other exists before the one *and* the other exist for themselves' (Baudrillard 1981, 75; cf. Levinas 1981).[14] Gift exchange serves to forge social groupings and to allow society to sustain and reproduce itself in a sense detached from any individual member of society.

It is against the hidden assumptions of 'economic rationality' that Mauss' work assumes its full significance. The modern economic system is predicated, above all else, on the *accumulation* of wealth. Viewed against this familiar backcloth, gift giving would certainly not seem to be a sufficiently robust basis on which to found an entire social system. Indeed, a social system based on 'giving away' rather than 'accumulating' one's wealth – on the expectation that others will do likewise – might seem a highly unlikely principle for any society seeking to reproduce itself over time. Yet this is precisely the principle Mauss claimed to have discovered. The principle of gift giving, Mauss proposed, set up a network of obligations that, in turn, called forth a series of counter-gifts. In one way or another, this principle lay at the basis of all pre-modern societies. From our own perspective, as Harland (1987, 44) notes, such a system would seem to depend 'on something quite insubstantial'. We 'cannot see what there is to stop anyone from just shrugging the obligation off. By contrast, we see our present-day system as sound and natural, based on the real substance of the solid property that an individual possesses' (*ibid.*). There is, though, an answer to this seemingly obvious objection, for 'solid property does not count for much if it can be taken away by mere superior fighting force' (*ibid.*). Under archetypal pre-modern circumstances, possession depended upon the power to *enforce* possession:

> A tribal chief or feudal lord was powerful in proportion to the number of men he could call upon to fight for him; which is to say, the number of men he had placed under an obligation to him; which is also to say, the number of men who had received favours or grants or gifts from him. In such a period, there were very sound – and even selfish – reasons for giving away goods and wealth, rather than accumulating them. Our present-day system is possible only because the threat of mere superior fighting force has been removed.
>
> (*ibid.*)

The preconditions for the modern system of economic exchange are somewhat more complex than Harland implies. Prerequisite to such a system is what effectively amounts to an 'alternative' to the sense of obligation (and principle of reciprocity) underpinning gift exchange. To cut to the quick, this resides in the notion of private property rights – and the concomitant notion of the free individual, able to possess property rights over particular resources (a situation MacPherson [1962] aptly describes in terms of 'possessive individualism'; cf. Abercrombie *et al.* 1986; Macfarlane 1978).[15] Under such conditions, Harland (1987, 45) notes, there is

> no longer the need to prevent fighting by direct social bonding. As Mauss observes, the gift-and-obligation system makes sense when the state of society is the simple sole alternative to the state of conflict. But our present day system actually incorporates conflict within society. This involves a changed form of conflict, which is now limited to economic competition between individuals. But economic competition between

individuals is still incompatible with direct social bonding between individuals.

Again, Harland is simplifying for the sake of exposition. Conflict in modern society is far from having been *solely* displaced into economic competition.[16] Neither is economic competition strictly incompatible with social bonding: a good deal of 'economic' activity – from cronyism and old boy's networks to lavish business lunches – is directed at establishing and nurturing such relationships. The broad change Harland finds Mauss delineating is, nonetheless, fundamental to modern society. No longer tied to the form of sociality established by gift exchange, we moderns 'bond ourselves to Society as a principle, Society as a whole, Society as founded upon the [mythical] long-ago transaction of an original social contract' (*ibid.*). In its specifically modern incarnation, 'Society' is recast in terms of the Law: 'It is the idea of the Law that keeps conflict within the bounds of economic competition; it is the legal right-of-possession that underlies the substance of solid property that an individual possesses' (*ibid.*). This legal right, it should be noted, is an equally insubstantial, abstract principle as the obligation to reciprocate. Nothing more than familiarity with the one system makes the other seem strange. Nonetheless, there is one further, all-important aspect of Mauss' theory to consider – which will ultimately allow us to understand why Baudrillard insists on speaking of *symbolic* exchange. As we will finally pin down in the next section, it relates to the metaphysical, rather than purely social, implications of the gift.

We can begin to broach this issue by formalizing Mauss' conception. According to Carrier (1991, 122), it has three distinct elements: 'These are that gift exchange is (1) the obligatory transfer of (2) inalienable objects or services between (3) related and mutually obliged transactors'. The implications this carries can be teased out by contrasting each component, in turn, with the defining characteristics of economic exchange. First, unlike market exchange, where individual transactors are formally *free* to buy or sell, gift exchange works on the basis that transactions are *obligatory*. This is not necessarily to say that they are somehow *enforced*. It is, however, to note that any 'refusal' of such an obligation amounts to the denial of, or curtails the possibility of, a social relation. Here, there is an evident resonance with our contemporary understanding of gifts as establishing and maintaining social relations. In fact, as we shall see, although gift and commodity crystallize out from fundamentally different forms of social arrangement, they are not, strictly speaking, alternatives, as Harland (1987) implies. Their relation is, rather, one of *displacement*. For in spite of the predominance of economic exchange, the vestiges of gift exchange continue to haunt modern society.

The second aspect of Mauss' formulation, which reinforces the first, is that gift exchange concerns *inalienable* objects or services. Again, there are resonances, here, with our everyday understanding of gifts. The singular, if arbitrary, nature of the gift-object has already been noted: once given, a gift-object is *that* object, inalienably so. This means that a gift is irreplaceable, and not simply one object amongst others. Paradoxically, as Baudrillard (1981, 64) states, 'the object is not an object: it is inseparable from the concrete relation in which it is exchanged, the transferential pact that it seals between two persons: it is thus not independent as such'. The gift-object thus differs fundamentally from the eminently alienable objects – mere *things* – transacted in economic exchange.[17] Objects of economic exchange, or commodities, are, says Carrier (1991, 126), 'abstract bundles of utilities and values that are precisely not unique'. They are, in a word, *fungible* – 'substitutable' or 'interchangeable' – because of the wholly abstract qualities that define them. The gift-object, in contrast, steadfastly retains its singularity – to the extent that the relationship it expresses is inalienable from the given object (the object itself assumes meaning 'in a symbolic relation with the subject' [Baudrillard 1981, 64]). There is, in other words, an enduring link between giver, gift, and recipient, which is the precise opposite of the transitory connections constituting the nebulous network of market exchange. In the world Mauss renders intelligible, therefore, 'objects' and 'subjects' are far less distinct than we moderns typically hold to be the case. 'We live in a society where there is a sharp distinction ... between persons and things', Mauss (1990, 47) wrote – but we should not therefore assume that the distinction travels well. The subject/object dualism is a direct reflection of our type of society, not a 'natural' perception or universal truth. 'Such a separation ... constitutes the essential condition for a part of our system of property, transfer, and exchange' (*ibid.*), which is why it appears so thoroughly natural to us today.[18] The logic of the gift allows us to glimpse the cultural relativity of different societies.

This brings us to the final element of Mauss' conception: the sense of a *mutual obligation* between transactors. Earlier, it was noted that the obligation to reciprocate plays the same role in the system of gift exchange as the principle of private property rights in the system of economic exchange. It is one thing to point out that these are broadly analogous principles. It is quite another, however, to confront head-on the nature of the obligation held by the parties involved in gift exchange. What causes a gift to solicit a counter-gift? What ensures the perpetuation of the cycle of exchange? '*What power resides in the object given that causes its recipient to pay it back?*' (Mauss 1990, 3). Mauss' reasoning points back to the inalienability of gift and giver. The object of *economic* exchange, once transacted, appears to have no further connection to the salesperson who sold it, still less to the labourer who produced it – the situation Marx (1967) characterized as the 'fetishism of commodities'. In the case of *gift* exchange, though, Mauss notes that it is as if the gift-object were *a part of the person who gives it*, who thereby continues to lay a claim over it. The counter-gift is not, then, as in economic exchange, a means of repayment ('settling up' the account once and for all, and terminating a legalistic relation between two nominally free individuals). It involves affirmation rather than cancellation. It serves to re-establish or reaffirm the bond between donor and donee, departing fundamentally from the situation of economic exchange, and giving gift exchange its indeterminate, aporetic character. Hence Parry's (1986, 456) insistence that it 'is not individuals ... but *moral persons* who carry on exchanges' of the kind Mauss analysed.

Parry's distinction is a pertinent one. A ready, if approximate, illustration of its importance is provided by the different conditions that apply to borrowing money from a bank and doing likewise from a relative (Bauman 1990a, 89–106). To the bank manager, it matters less *who* I am than *what* I am: 'Am I credit-worthy?', 'Have I a steady income?', 'Is it sufficient to cover the repayment of the loan?'. To, say, my sister, *who* I am is far more important than *what* I am. My economic circumstances do not assume the same priority: at least they are not the sole determinant of the decision on the loan. In contradistinction to the situation underpinned by a bond of kinship, economic exchange defines parties and objects in wholly abstract ways: fungible commodities, free individuals, a rational measure of value and (marginal) utility. Under such circumstances, the personalities of the seller (or moneylender) and the buyer (or would-be borrower) are more or less dissoluble from the commodity being transacted and the roles being played in conducting the transaction (Gregory 1982; Simmel 1978). This modern, abstract conception of the individual stems from the way in which modern society itself amounts to the abstract, disembodied *principle* of Society. Such a principle is always already predicated on the existence of such abstract agents as 'individuals', solely responsible for themselves and their actions (though judged and sanctioned in social terms). Simultaneously, our acceptance of the 'social contract' enshrined in the notion of Society grants us our status as individuals (an entirely tautologous logic, to be sure). Society and individual are, in Althusser's (1972) terminology, 'mirror categories': each entails the other, the notion of the 'individual' arising in tandem with the notion of 'Society'. The implication is that people are constituted as individuals by Society, rather than Society arising as a (rational) arrangement between pre-existing individuals.[19] As previously noted, the idea of a social contract between primordial individuals is very much an ideological construct – and one of relatively recent vintage, bearing all the hallmarks of modernity (Althusser 1969; 1972; Dumont 1977; 1986).

To sum up, therefore, where Maussian gift exchange implies reciprocity, inalienable objects, and moral subjects, economic exchange involves abstract economic rationalizations and rational economic abstractions (free individuals included). Hence Baudrillard's (1981, 133) insistence, against all received understandings of consumption, that far 'from the individual expressing his [*sic.*] needs in the economic system, it is the economic system that induces the individual function and the parallel functionality of objects and needs'. Here, we have reached the point where the limitations of Mauss' analysis cannot forego further scrutiny. They become apparent if we consider the reasoning behind Baudrillard's deployment of the term '*symbolic* exchange'. In essence, Baudrillard (1981, 64, 65n) insists that the structure of symbolic exchange, 'of which the gift is our most proximate illustration', 'is never that of simple reciprocity'. Mauss recognized the extent to which a *relation* is constitutive of the *relata* it brings into being. He successfully demonstrated that the reciprocity set up by gift exchange involves an *aporetic* form of power play between *relata*. Gift exchange does not, Mauss revealed, imply the kind of *symmetrical* relation of mutual advantage between free individuals assumed by economic exchange. What remains underdeveloped in Mauss' account of gift exchange, however, is the recognition that it 'is not two simple terms, but two *ambivalent* terms that exchange, and [that] the exchange establishes their relationship as ambivalent' (*ibid.*, 65n). To grasp this, Baudrillard (1993a, 1) maintains, we need to 'turn Mauss against Mauss' – to apply the insights Mauss uncovered to Mauss' own work, pushing it further in the direction he himself established. Baudrillard appears to find a predecessor in his desire to read Mauss 'against himself' in the work of Bataille (1988; cf. Pawlett 1997). Although he will ultimately part company with Bataille, opting to retain a lesson from Mauss that Bataille chose to disregard, Bataille, in Baudrillard's eyes, offers something vital.

Expenditure, or the sweet smell of excess

> *Homo oeconomicus* is not the human being who represents his own needs to himself, and the objects capable of satisfying them; he is the human being who spends, wears out and wastes his life in evading the imminence of death.
>
> (Foucault 1974, 257)

Bataille's (1988) work represents a consumption-oriented approach *par excellence*. It affords consumption a far more fundamental role than any previous social theory. Its relation to Mauss' thought is not, however, particularly straightforward. Bataille (1988, 193n) freely acknowledges that 'the studies whose results I am publishing here came out of my reading of the *Essai sur le don*.' According to Richardson (1994, 77), though, 'Bataille seems oblivious of what Mauss himself considered the most crucial feature [of the gift], which was the obligation created by exchange'. In fact, far from being oblivious to Mauss' argument, Bataille is intent on pushing it to its farthest extreme. Derrida (1992, 24) has observed that *The Gift* 'speaks of everything but the gift: It deals with economy, exchange, contract [*do ut des*], it speaks of raising the stakes, sacrifice, gift *and* counter-gift'. Bataille's work embodies a comparable conclusion. Where Mauss emphasized *reciprocity* – 'gift *and* counter-gift' – Bataille discerns a metaphysical principle of *expenditure* (cf. Bataille 1985). This is not, of course, understood in anything like the conventional economic sense of the term. It is more or less synonymous with 'squandering' or 'waste' – with giving *without* receiving. As Richardson (1994) implies, this does appear to reverse Mauss' argument. Nonetheless, Bataille claimed to have isolated the principle of expenditure from Mauss' discussion of the potlatch ceremony, in which a form of rivalry is played out in the excessive giving – and even the conspicuous destruction – of worldly goods. And the principle revealed by the potlatch, Bataille (1988, 25) maintains, lays the basis for a 'Copernican transformation: a reversal of thinking – and of ethics'.

Bataille's starting point, like Mauss', is that the principle of mutual advantage enshrined in economic exchange diverges fundamentally from pre-modern forms of social relations: 'Potlatch is, like commerce, a means of circulating wealth, but it excludes bargaining' (*ibid.*, 67–68). Though it results in a circulation of wealth comparable to that achieved by modern commerce, the calculated mutual advantage of barter is precisely what is *not* at stake in the potlatch:

> More often than not it [the potlatch] is the solemn giving of considerable riches, offered by a chief to his rival for the purpose of humiliating, challenging and obligating him. The recipient has to ease the humiliation and take up the challenge; he must satisfy the *obligation* that was contracted by accepting [the gift]. He can only reply, a short time later, by means of a new potlatch, more generous than the first: He must pay back with interest.

The potlatch thus gives rise to a form of *escalating* rivalry, since 'receiving prompts one – and obliges one – to give more' (*ibid.*, 71). Rather than producing some kind of steady-state equilibrium, the principle of obligation is to be understood primarily in terms of antagonism and challenge – a perpetual raising of stakes – which can only ever accomplish an 'unstable social equilibrium' (Schivelbusch 1993, 173). To this extent, there is no direct contradiction with Mauss (1990, 74), who explained the potlatch in precisely such terms:

> the reason for these gifts and frenetic acts of wealth consumption is in no way disinterested, particularly in societies that practice the potlatch. Between chiefs and their vassals, between vassals and their tenants, through such gifts a hierarchy is established. To give is to show one's superiority, to be more, to be higher in rank, *magister*. To accept without giving in return, or without giving more back, is to become client and servant, to become small, to fall lower (*minister*).

For Bataille, this process, correctly interpreted, reveals the fundamental necessity of expenditure. We can begin by offering a simple, almost tautological, interpretation of what he means: whoever manages to give most, to expend most, or to destroy most, comes out on top in the potlatch; what their superiority in the contest of rivalry reveals, therefore, is the basic power of expenditure. As Werbner (1996) suggests, occasions like Christmas still seem to preserve this logic – maintaining a strict hierarchy between adults and children, for instance. If one is prepared to follow Bataille's reasoning closely, overlooking its apparent circularity, a whole new conception of 'economy' is brought to light.

'Classical economy imagined the first exchanges in the form of barter', Bataille (1988, 67) reminds us. But why therefore assume – for it is but an assumption – that these exchanges were predicated on acquisition? Might they not have 'answered...the contrary need to lose or squander?' (*ibid.*). If the possibility appears, at first blush, unlikely, it is nonetheless highly suggestive. As Maffesoli (1997) points out, insofar as it possesses no finality or purpose, the unproductivity of expenditure parallels the polymorphous perversity of the infant. In other words, the origins and development of the modern economy would seem to parallel the origins and development of the modern subject described by Freud. Despite this suggestiveness, however, we will be ultimately forced to conclude, with Baudrillard (1987a, 61), that Bataille's manoeuvre remains, in the last instance, 'still too economic, too much the flip side of accumulation'. It merely replaces the foundational assumption of modern economic reasoning (scarcity) with its precise opposite (excess). Even so, the insights it achieves disclose something vital.

Bataille's (1988, 38) overriding concern is to develop a form of economic thought capable of taking into account a 'basic movement that tends to restore wealth to its function, to gift-giving, to squandering without reciprocation'. In modern society, as we have previously noted, wealth seems securely bound to a logic of accumulation. For Bataille, this reveals a fundamental modern malaise, which manifests itself in a profoundly *miserly* attitude. Not that Bataille is concerned with presenting a narrow normative judgement on the matter, however. In his eyes, the calculating attitude lying behind the accumulation of wealth not only sells short the sovereign potential of humanity, it also *disavows* a vital aspect of the nature of wealth: the necessity that it be squandered without reciprocation. Disavowal, of course, is always a dangerous attitude. If the attribution of such a function to wealth seems to err against common sense, Bataille insists that such a reaction belongs to a *restricted* modern perspective, which is guilty of adopting precisely such an attitude, predicated on *scarcity* and marked by a utilitarian calculus (Thompson 1997). Whether we accept the fact or not, Bataille insists that we live in a universe that is characterized by *excess*. And this excess, finally, 'must be spent, willingly or not, gloriously or catastrophically' (Bataille 1988, 21). Our failure to recognize this 'basic fact' merely 'deprives us of the choice of an exudation that might suit us' (*ibid.*, 21, 23–24). If we continue to deny the necessity of expenditure, it will occur anyway – the only difference being that our relation to it will be marked by absolute ignorance (modernity could, in fact, be defined as the triumph of this blinkered attitude). His analysis, Bataille (1988, 9) contends, develops 'the notion of a "general economy" in which the "expenditure" (the "consumption") of wealth, rather than production' is primary.

Insofar as it is posited in terms diametrically opposed to 'restricted economy', Bataille's (1985; 1988) 'general economy' is animated by abundance and excess. The premise of this state lies, for Bataille (1988, 21), in what might be termed a 'natural necessity' (though the term is somewhat paradoxical, insofar as 'necessity' is precisely what the perspective of general economy purports to free us from):

> The living organism, in a situation determined by the play of energy on the surface of the globe, ordinarily receives more energy than is necessary for maintaining life; the excess energy (wealth) can be used for the growth of a system (e.g., an organism); if the system can no longer grow, or if the excess energy cannot be completely absorbed in its growth, it must necessarily be lost without profit.[20]

Basing his analysis on the additional premise that growth faces an absolute limit – that 'the possibilities of life cannot be realized indefinitely' (*ibid.*, 31) – Bataille arrives at the proposition that a squandering of energy will *necessarily* take place (that only its *form* will be contingent). The laws of general economy following from these axioms contrast vividly with the perspective of restricted economy. 'From the *particular* point of view, ... problems are posed *in the first instance* by a deficiency of resources. They are posed *in the first instance* by an excess of resources if one starts from the *general* point of view' (*ibid.*, 39).

The modern, utilitarian attitude is problematical precisely because of its restricted viewpoint. 'Humanity exploits given material resources, but by restricting them as it does to a resolution of the immediate difficulties it encounters ... it assigns to the forces it employs an end which they cannot have' (Bataille 1988, 21). Whatever particular ends we might assign to things, this in no way amounts to something with which they are complicit. Whatever purpose we enlist them in achieving, this hardly exhausts – let alone defines – their possibilities. Marx (1967, 87), in fact, recognized as much when he wrote: 'Could commodities themselves speak, they would say: Our use-value may be a thing that interests men. It is no part of us as objects'. Only too readily, though, we regard the anthropocentric form of use-value as an attribute of the *natural* world.[21] Though we rarely reflect on the fact, our own particular ends are invariably 'vitiated by a determination of which [we are] ignorant', by a 'movement that surpasses them' (*ibid.*). For the most part unwittingly, and certainly beyond our 'immediate ends', our 'activity in fact pursues the useless and infinite fulfilment of the universe' (*ibid.*). The 'uselessness' (one might equally say the 'purposelessness' or 'endlessness') of the universe is precisely what our utilitarian, means/ends rationality has occluded and *repressed* (that is, relegated to the unconscious).

In the process, it has transformed it into an *accursed* share – an unruly, uncontrollable element that is destined to humiliate modernity's ambition. Faced with 'a play of energy that no particular end limits', our only real 'choice is limited to how the wealth is to be squandered'; to the form its 'useless consumption' should take (*ibid.*, 23). 'In no way can this inevitable loss be accounted useful', Bataille (1988, 31) insists: 'It is only a matter of acceptable loss, preferable to another that is regarded as unacceptable: a question of *acceptability*, not utility'. The implication of Bataille's vertiginous perspective can now be plainly stated: '*it is not necessity but its contrary, "luxury", that presents living matter and mankind with their fundamental problems*' (*ibid.*, 12).

A sense of the explosive, effusive, exuberance of a world teeming with life emerges from Bataille's conception – whilst 'luxury' becomes, almost self-evidently, a 'natural principle' once the perspective of general economy is adopted. Ultimately, Bataille (*ibid.*, 28) notes, all energy reaching the Earth has a solar origin: the sun 'dispenses energy – wealth – without any return', and thus 'gives without receiving'. The biochemical energy to which it gives rise utilizes only a tiny proportion of this solar energy for functional activity. Beyond this, energy is diverted into *growth* (Bataille provides the analogy of a fattened calf: its functional actions are restricted by confinement in order to promote the animal's growth). As previously noted, growth is itself restricted: by inter-species competition, for example; though, ultimately, 'space is its finite limit' (*ibid.*, 29). Up against this limit, the result is a certain pressure. If 'the space is completely occupied ... nothing bursts', but nonetheless, 'the pressure is there' (*ibid.*, 30):

> In a sense, life suffocates within limits that are too close; it aspires in manifold ways to an impossible growth; it releases a steady flow of excess resources, possibly involving large squanderings of energy. The limit of growth being reached, life, without being in a closed container, at least enters into ebullition: Without exploding, its extreme exuberance pours out in a movement always bordering on explosion.
>
> (*ibid.*)

The internal dynamics of this situation prompt a number of 'great luxurious detours that ensure the intense consumption of energy' (*ibid.*, 35). The food chain – or more accurately the food pyramid, since its hierarchical structure demonstrates the principle in question – is one such 'luxurious detour': the amount of excess energy expended increases up the hierarchy, as a result of the more 'burdensome character of the indirect development of living matter' (*ibid.*, 33). Eating meat is a luxury, for example, given the greater efficiency in terms of food productivity yielded by a comparable unit of land in arable production (growing grain on an acre of land could sustain more people than feeding the grain to a cow and drinking its milk or eating its flesh). In comparative terms, therefore, 'the herbivore relative to the plant, and the herbivore relative to the carnivore, is a luxury' (*ibid.*, 37), since each successive link of the food chain represents a more 'burdensome' form of life in terms of the energy it is able to consume.

Death is also a luxury. Indeed: '*Of all conceivable luxuries, death … is undoubtedly the most costly*' (*ibid.*, 34) in terms of the excess energy it serves to use up. The hierarchy of life forms just described already implies as much, insofar as eating invariably involves the death of the thing consumed. Apart from certain scissiparous life forms, which achieve immortality insofar as they reproduce by cloning, death is a strict inevitability. Restrictions of space render death necessary in order to make 'room for the coming of the newborn' (*ibid.*). As has often been pointed out, death in modern society differs from almost all other societies in that it is afforded an accursed status (Baudrillard 1993a; Bauman 1992b; 1992c; 1993a). Yet as Bataille (1988, 34) insists, we are 'wrong to curse *the one without whom we would not exist.*' Modernity's anguish over death is misplaced, since we ourselves would not be here were it not for the claim staked out in advance by our offspring: a claim that is always already implied by the successive position we occupy in the interminable cycle of life and death. Fatality only means something – only spells the end – from an *individual* point of view (hence the particular anguish to which it gives rise in the 'society of individuals' to which modernity amounts (Elias 1991; Bauman 2001a)). This individual perspective contrasts fundamentally with '*general* existence whose resources are in excess and for which death has no meaning' (Bataille 1988, 39).

Unsurprisingly, perhaps, Bataille (*ibid.*, 35) also regards sexual reproduction as 'the occasion of a sudden and frantic squandering of energy'. Sexual reproduction, he writes, 'accentuates that which scissiparity announced: the division by which the single being foregoes growth for himself [*sic.*] and, through the multiplication of individuals, transfers it to the impersonality of life'. Thus, albeit 'independently of our consciousness' (*ibid.*), sexuality is also a luxurious detour for the expenditure of excess energy. Sex and death are, of course, co-implicated: sexual reproduction is rendered necessary by death, just as death is rendered necessary by sexual reproduction. Bataille's strategy of drawing back and forth between particular and general perspectives reveals a kind of *reversibility* to which the latter perspective points, but which the former perspective is incapable of perceiving. This insight, as we will shortly be in a position to see, points Baudrillard towards his conception of 'symbolic exchange'. First, however, let us pursue Bataille's point that the restricted economy of modern societies disavows the principle of expenditure in a way that contrasts markedly with all previous societies. The most dramatic exemplification of this principle, which points up the contrast most vividly, is contained in the practice of sacrifice.

Sacrifice, Bataille contends, is an anti-utilitarian practice *par excellence.* A society that makes sacrificial offerings relates to the world in a fundamentally different way to modern society. We have already noted Mauss' observation that modern society is characterized by a sharp distinction between *persons* and *things*. 'A *thing,*' says Bataille (*ibid.*, 32), 'is what we know from without, what is given to us as a physical reality (verging on a utility, available without reserve)'. (He means, by this last phrase, that a thing is taken to be fully present to itself, without any excess over and above what it objectively is.) Thus we 'cannot penetrate a thing, and it has no meaning other than its material qualities, adapted or not to some useful purpose, in the productive sense of the word' (*ibid.*, 132). The modern world is, as Gregory (1982, 641) notes, a world geared towards ensuring 'the social conditions of the reproduction of *things*'. It is a disenchanted world, 'which commits mankind only to *do* – without any further aim – that which can be done *in the order of things*' (Bataille 1988, 130). It is, perhaps, unsurprising that, in such a world, '[man] estranges himself from his own intimate being', to assume 'himself the limits of a *thing*' (*ibid.*, 56). Under conditions of slavery, for example, *persons* are formally afforded the status of *thing*s. More generally, the reduction of the world to the '*order of things*' – for order is the condition by which *things* are so defined – severs the intimacy of our connection to the universe; it places 'a screen between the universe and me' (*ibid.*, 57). Sacrifice, however, maintains (and reveals) a fundamentally different relation to the universe.[22] Bataille (*ibid.*, 132) spells out the way in which sacrifice articulates this different relation: 'Intimacy is not expressed by a *thing* except on one condition: that this *thing* be essentially the opposite of a *thing*, the opposite of a product, of a commodity – a consumption and a sacrifice'. In other words, sacrifice ritually affirms a non-utilitarian relation to the universe, by *reversing the order of things*. The 'victim of the sacrifice cannot be consumed in the same way a motor uses fuel. What the ritual has the virtue of rediscovering is the intimate participation of the sacrificer and the victim, to which a servile use had put an end' (*ibid.*, 56). The meaning of the sacrificial offering, its *sacred* meaning, reveals an attempt to reach beyond the limited world of *things*. In the sacrificial rites of pre-modern societies, the 'victim was separated definitively from the profane world; it was *consecrated*, it was *sacrificed*, in the etymological sense of the word, and various languages have given the name *sanctification* to the act that brought that condition about' (Hubert and Mauss 1964, 35). The violence entailed by sacrifice, moreover, is its strictly necessary or defining feature. 'Destruction', as Bataille (1988, 56) suggests, 'is the best means of negating a utilitarian relation between man and the animal or plant'. It is through 'this act of destruction the essential action of the sacrifice was accomplished' (Hubert and Mauss 1964, 35).

Whilst our relation to the universe defined in terms of use (productive consumption) is a necessary condition of existence, it cannot be said to be a sufficient condition. For Bataille, this is precisely what the logic of the sacrifice reveals – and, by the same token, what modern society has sought to disavow. 'In former times', Bataille (1988, 29) writes, 'value was given to unproductive glory, whereas in our day it is measured in terms of production'. Modernity's programmatic commitment to accumulation has ensured that precedence 'is given to energy acquisition over energy expenditure' (*ibid.*). But where modernity steadfastly maintains a utilitarian relation to the universe, sacrifice, in reversing this relation, affirms that the offering is *something more* than its utilitarian status alone implies. This reversal is, for Bataille (*ibid.*, 193n), crucial: the 'squandering of energy is always the opposite of a *thing*, but it enters into consideration only once it has entered into the *order of things*, once it has been changed into a *thing*.' Hence Baudrillard's (1998a, 44) suggestion that it is 'this "something in it more than itself" by which a thing becomes what it is "in itself"' that is 'the law of symbolic value, which ensures that the essential is always beyond the indispensable' (translation amended after Butler 1999, 51). This law, Baudrillard (1998a, 44) states, 'is best illustrated in expenditure, in loss, but it may also be verified in appropriation, provided that the latter has the differential function of being "something more"'. We will return to the way in which the law of symbolic value might be verified in appropriation in due course. First, however, we need to examine more closely the kind of 'reversibility' that is affirmed by sacrifice. For here, we have finally reached a position from which to appreciate Baudrillard's notion of symbolic exchange.

Forewarning us that the term is 'rather deceptive', Baudrillard (1987c, 84) paradoxically pronounces that '[s]ymbolic exchange is the opposite of exchange'. He adds that 'symbolic exchange [is] fatal' (meaning that it is a matter of 'fate'), and emphasizes that fate 'is in the dividing line separating chance and necessity' (Baudrillard 1987c, 85). The fatal, therefore, and hence symbolic exchange itself, implies something that is *neither chance nor necessity*: Baudrillard (*ibid.*) refers to it as 'the order of events'. Although the distinction between chance and necessity is generally regarded as a particularly profound distinction, it is actually secondary to, and derivative of, a more basic contrast between fate (symbolic exchange) and exchange *per se*: 'There is an order of exchange and an order of fate. The only means of exorcizing fate is through exchange; in other words, through a contractual agreement. Where exchange is not possible, fate takes over' (*ibid.*, 84). Unlike contractual agreements, the defining characteristic of fate is that it is marked by a 'dual reversibility' (*ibid.*). With fate, or symbolic exchange, matters can go either way. Fate is marked by a fundamentally agonistic character. It is always a challenge to be taken up. But it therefore permits the possibility of things coming into existence. There are no foregone conclusions, no irreversible outcomes, no predetermined finalities. Fate is not a simple matter of determinacy. But nor is it simply a matter of chance. One can pursue one's fate even against the most overwhelming odds. Thus, although fate implies no finality, the outcome is never simply random or accidental. On the contrary, it 'is of the highest necessity' (*ibid.*, 85). Fate, therefore, implies something beyond determinacy and chance, which permits something new to appear on the scene. However: 'Symbolic exchange is no longer the organizing principle of modern society' (Baudrillard 1993a, 1). From its very inception, modernity declared war on fate, attempting to radically exclude the principle of symbolic exchange. Needless to say, modernity's attempt to instate determinacy – to make the world over in the image of order – inevitably produces contingency (disorder) as an uncontrollable remainder. Such a one-sided strategy cannot but produce such a residue. With symbolic exchange, however, 'everything is exhausted in the reciprocity of terms, in the reversibility of terms' and there is 'no residue' (Baudrillard 1995a, 87).

Invariably, the strategy pursued by modernity rests not simply on an opposition of terms but on the attempt to install one side of the opposition at the expense of the other. For example, order is opposed to chaos, good is opposed to evil, life is opposed to death, presence is opposed to absence, and man is opposed to woman. With all such oppositions, 'we are not dealing with the peaceful coexistence of a *vis-à-vis* but rather with a violent hierarchy. One of the two terms governs the other … or has the upper hand' (Derrida 1981, 41). In every case, the latter term is figured negatively, as the *other* of a purportedly fully self-present term. And the latter term is invariably defined as *lacking* what the former term possesses – as its *flawed* double.[23] If modernity has invariably attempted to annul the second term, this term nonetheless retains its strength. For it is not only the case that all determination is negation (that 'what a thing is' depends on 'what it is not'); it is also the case that all negation is determination (that 'what a thing is not' draws the limits defining 'what it is'). In other words, whereas the first term opposes itself to the other term, the second term implies the non-opposition of the two terms. Whereas the former term disavows its dependency on the latter, the latter term embodies an acknowledgement of their agonistic relation.

Life, for instance, seems to cling to a metaphysics of presence, positing a wholeness and unity all of its own. It would be fully present to itself, unthreatened by the foreboding absence marked by death. This is, however, a strict impossibility. For life is given, and must be given back – such is the symbolic exchange between life and death. Death cannot, finally, be cheated. Only its delay or deferral might, for a time, be accomplished. Life is never a self-defining principle; it is one moment in a reversible cycle. Where the attempt is made to deny this reversibility, death acquires the foreboding status of a residue – which demands constant vigilance, since the repressed is always poised to return. Under such conditions, which modernity has made its own, life's overstretched ambition faces the constant threat of the reappearance of mortality. Life becomes preoccupied with the vain attempt of warding off the inevitable. 'Fighting death is meaningless', says Bauman (1992c, 7). 'But fighting the causes of dying turns into the meaning of life' (*ibid.*). Little wonder, then, that this war is waged most fiercely in a society fixated on the cult of personal immortality, where the general, impersonal perspective is subjugated to the restricted and particular perspective of the individual. Time and again, modernity has striven in vain to instate one term as positive and irreversible, only to find itself facing 'the portentous power of the negative'.

Bataille's conception of 'general economy' recognizes the impossibility of modernity's attempt to outwit fate, uncovering the way in which modernity constantly acts to repress – and at the same time to curse – what other social formations have affirmed by ritual means. Even so, Baudrillard ultimately finds Bataille guilty of naturalizing the principle he recasts in terms of symbolic exchange. Baudrillard (1987a, 61) contends that it is the principle located in Mauss' account of the potlatch, not in Bataille's general economy, that successfully isolates the principle of symbolic exchange:

> The 'excess of energy' does not come from the sun (from nature) but from a continual higher bidding in exchange – the symbolic process that can be found in the work of Mauss, not that of the gift (that is the naturalist mystique into which Bataille falls), but that of the counter-gift. This is the single truly symbolic process.

The problem lying at the heart of Bataille's conception of general economy relates back to the planetary limits to growth from which the supposed 'natural necessity' of squandering energy derives. As Noys (2000, 114) points out, 'general economy is only possible because of the restrictions on the earth's surface, and thereby ... general economy is really a restricted economy'. Thus, as Bennington (1995, 48) remarks, ' "general economy" is not the other of "restricted economy" but is *no other than* restricted economy'. 'Another way of putting this', which itself accords to the reversibility of symbolic exchange, 'is that general economy is the economy of general *and* restricted economy' (*ibid.*). General economy and restricted economy are not, therefore, opposed terms, but reversible and agonistic concepts. They form, as it were, a single, continuous Möbius strip. The implication is that the restricted economic principles of modern exchange have never escaped the principles of general economy, of expenditure, potlatch, sacrifice, the gift and the counter-gift. Modernity remains haunted by what it has attempted to annihilate but succeeded only in repressing. Hence Baudrillard's insistence that the

> act of consumption is never simply a purchase (reconversion of exchange value into use value); it is also an expenditure (an aspect as radically neglected by political economy as by Marx); that is to say, it is wealth manifested, and a manifest destruction of wealth.
>
> (Baudrillard 1981, 112)

To appreciate the significance of this proposition, let us briefly return to Campbell's (1995) claim, cited in the opening pages of this chapter, that Baudrillard understands modern consumption in terms of 'communicative acts'. This is a basic misinterpretation, arising from the failure to appreciate the particular sense of the 'symbolic' in Baudrillard's terminology. Yet as we shall see by the end of the chapter, Campbell is not entirely wrong to suggest that Baudrillard's concerns relate to the role of the sign in modern consumption – even if the caricature he offers gives us little clue as to Baudrillard's real purpose. We can glean an initial sense of Baudrillard's (1981, 30) intention from his insistence that, 'behind all the superstructures of purchase, market, and private property, there is always the mechanism of social prestation which must be recognized in our choice, our accumulation, our manipulation and our consumption of objects'. To begin with, this suggests that consumption 'does not derive primarily from vital necessity or from "natural law", but rather from a cultural constraint' (*ibid.*). It also implies that consumption remains connected to the agonistic structuring of society, and that the 'mechanism of discrimination and prestige is at the very basis of the system of values and of integration into the hierarchical order of society' (*ibid.*). The fact that modern society hides this fatal principle behind the supposed rationality of economic exchange does not imply that it has been done away with once and for all. Consumption still accords to a reversible logic that lies between the mirror categories of 'Society' and 'individual' (which modernity duplicitously presents as exhaustive of the entire social body). It is towards this 'excluded middle' that we are drawn by the work of Veblen.

Pecuniary emulation and social differentiation

> while the regulating norm of consumption is in large part the requirement of conspicuous waste, it must not be understood that the motive on which the consumer acts in any given case is this principle in its bald, unsophisticated form. Ordinarily his motive is a wish to conform to established usage, to avoid unfavourable notice and comment, to live up to the canons of expected decency in the kind, amount, and grade of goods consumed, as well as in the decorous employment of his time and effort.
>
> (Veblen 1994, 71)

It is highly ironic that Veblen is usually approached in relation to an idle debate over consumer demand that has precious little to do with his own concerns, and which he had, in fact, already surpassed in his own writings. Veblen is frequently regarded as simply an unorthodox economist who argued that consumption is far from a wholly individualistic affair and invariably involves a sizeable dose of 'keeping up with the Joneses'. Unfortunately, the emphasis Veblen lay on 'pecuniary emulation' has allowed a crude, caricatured view of his work to enter into circulation. Typically, Veblen is unceremoniously ushered in to mediate between the polarized terms of the eternal debate over 'consumer manipulation' *vs.* 'consumer freedom'. Yet this hardly provides for an adequate understanding of Veblen, whose thought resonates particularly clearly with that of Mauss and Bataille. Veblen's underlying concern was to disclose the *honorific*, rather than utilitarian, logic of consumption, and to reveal the way in which modernity has transformed a 'tourney of generosity' into a rigged game. He specifically sought to highlight the way in which the conferral of 'merit' on members of society has, since the onset of modernity, been translated into one-dimensional, pecuniary terms. Like Simmel (1978), Veblen held that the complexity and anonymity of modern society rebounds in a reductive, single standard of value, with the consequence that people are no longer judged with respect to their character and deeds, but in terms of what they *appear* to be worth. Hence the emphasis he gave to 'conspicuous consumption': to the ostentatious display of wealth in the consumption of fine food, fashionable clothing, and, above all, the pursuit of leisure. Our eventual aim, here, is to recover a better understanding of the sense of Veblen's work, in order to see why Baudrillard should afford it such significance. Let us begin, however, by approaching Veblen within the narrow confines of the conventional debate.

Faithfully reflecting the interminable debate over free will and determinism, an extraordinary amount of attention has been devoted to considering the degree to which consumers are free to determine their own actions – as against the opposing view that the consumer is an essentially gullible creature, capable of being manipulated at will by the agents of capitalism. The former view, couched in terms of 'consumer sovereignty', is fully ingrained within neo-classical economics (Mohun 1977).[24] The latter is common in neo-Marxian work, which purports to show how capitalism, particularly through advertising, manipulates consumer demand (Baran and Sweezy 1966; Galbraith 1958; 1967; Packard 1957; cf. Ewen 1975).[25] This hotly contested issue, which also manifests itself in debates over the 'authority' of the consumer (Keat *et al.* 1994) and the 'authenticity' of needs (Marcuse 1964), has generated far more heat than light – for the simple reason that it is a false problem.[26] The positions defended in this debate are invariably frozen and polarized: the consumer is *either* sovereign and all-powerful, *or* a passive dupe, devoid of all volition. Campbell (1987), for instance, frames this debate in terms of an 'instinctivist' account of consumer demand (inherent to neo-classical economics) and a diametrically opposed 'manipulationist' tradition (reliant on notions of 'demand creation', which he associates primarily with Galbraith [1958]). Whilst briefly entertaining the possibility that Veblen successfully mediates between the usual polarized positions, Campbell is finally dismissive of the possibility. If we follow his reasoning more closely, we will see that this dismissal is somewhat peremptory.

Galbraith, Campbell notes, focused principally on 'active' rather than 'passive' demand creation, meaning that he placed a good deal of emphasis on the intentional 'conditioning' of demand supposedly achieved by advertising. (A more generous reading might stress Galbraith's attempt to refute, once and for all, the absurd notion that consumer preferences are exogenous to the economic system.) Campbell also notes the possibility of a less conspiratorial, 'passive' type of demand creation, which he associates primarily with Veblen. The Veblenesque consumer, Campbell suggests, actively generates his own demands, in relation to a socially engendered desire to display his 'pecuniary strength', and hence his social standing.[27] Veblen's consumer is not an abstract economic agent – an 'homogeneous globule of desire of happiness under the impulse of stimuli that shift him about ... but leave him intact', as Veblen (1961a, 73) wonderfully caricatured such a view. He is a thoroughly social creature, whose demands arise in relation to the social world. Since human nature is, for Veblen (1961a, 74), always a 'coherent structure of propensities and habits which seeks realization and expression in an unfolding', Veblen's consumer cannot be regarded as a mere dupe, hoodwinked into demanding certain things. Consumer demands are never simply involuntary, knee-jerk responses to business wishes. But nor does Veblen characterize consumers as simply free agents, who stand apart from society, rather than being thoroughly immersed in, and part of, society from the very start. On the face of it, therefore, Veblen does appear to overcome the limitations of the conventional, polarized positions. For Campbell, however, he does so only superficially, and ultimately fails to live up to his promise. As it stands, this assessment is not particularly fair to Veblen. It rests on an appropriation of his work, which seems almost primed in advance to discredit him. As we shall shortly see, Campbell's adjudication rests on firmer grounds than this alone, but before scrutinizing Campbell's judgement any further, we first need to obtain a better appreciation of Veblen on his own terms.

Veblen's entire theoretical enterprise needs to be understood as an attempt to swim against the current of thought established by Adam Smith. As far as Veblen was concerned, Smith's work was hardly the breakthrough in economic science it has always been taken to be. In fact, it amounted to little more than a form of intellectual escapism, abstracting from the harsh realities of the market, and putting a consoling gloss on 'the messy truth of the money circuit of capital' (Dyer 1997, 47). To Veblen's mind, the modern economy gestates an absurd reality, upheld in no small measure by its ideological justification in economic theory. A true economic science, he maintained, would not shrink from such judgements.[28] Once reached, they permit one to perceive the fundamental perversity of the way in which the modern economy induces a wholesale reliance on a single, pecuniary standard of worth, diverting the impulse that once manifested in the potlatch towards the dead-end of private accumulation. Marx (1967, 153) once remarked that the 'boundless greed after riches, [the] passionate chase after exchange value, is common to the capitalist and the miser; but whilst the miser is merely a capitalist gone mad, the capitalist is a rational miser'. The power of money over the human personality is ramified to the nth degree where money circulates as capital (money *plus* rationality). The fundamental irrationality of 'economic rationality' is also at the heart of Veblen's (1975) view of the modern economy, where all areas of life are forced to adjust to the imperative of pecuniary advancement. Under such conditions, the search for knowledge becomes imbricated with the search for profit; learning becomes an exercise in credentialism (Hirsch 1976); and publishing and the arts come to march to the beat of the advertisers' drum. Whatever the niceties of value theory might claim, therefore, it is the individual desire for pecuniary self-aggrandizement that determines the nature of economic exchange.

By far the most absurd consequence of this situation, which underlies all the others, is the way in which '[c]onspicuous abstention from labour ... becomes the conventional mark of superior pecuniary achievement and the conventional index of reputability' (Veblen 1994, 25), whilst productive activity – from which all wealth ultimately derives – is cast as a sign of social inferiority.[29] Conspicuous consumption is a direct manifestation of this. As exemption from productive activity becomes a mark of prestige, the means of demonstrating one's standing in such terms exerts its influence across the whole of society. The same situation is evident in 'vicarious consumption'. 'Veblen shows that even if the primary function of the subservient classes is working and producing, they simultaneously have the function (and when they are kept unemployed, it is their only function) of displaying the *standing* of the master' (Baudrillard 1981, 31). The very possibility of vicarious consumption reveals, moreover, the paucity of conceptions that reduce consumption to the satisfaction of individual needs by means of particular use-values.

> The end of acquisition is conventionally held to be the consumption of the goods accumulated. ... This is at least felt to be the economically legitimate end of acquisition, which alone it is incumbent on economic theory to take account of. ... But it is only in a sense far removed from its naive meaning that consumption of goods can be said to afford the incentive from which accumulation invariably proceeds. The motive that lies at the root of ownership is emulation; and the same motive of emulation continues active in the further development to which it has given rise and in the development of all those features of the social structure which this institution of ownership touches. The possession of wealth confers honours: it is an invidious distinction.[30]
>
> (Veblen 1994, 17)

Veblen's understanding of the modern economy and his treatment of consumption, therefore, is founded on an understanding of the distortions produced by the overriding power of a pecuniary measure of value and the overwhelming force of economic rationality. He effectively sought to highlight what economic theory, as a 'restricted' economic abstraction, disavows.[31] Nonetheless, there are limitations to Veblen's account, relating particularly to his overreliance on the mechanism of pecuniary emulation.

Although Veblen's overemphasis of pecuniary emulation is central to Campbell's discontent with Veblen, Campbell's assessment, it has to be said, is rather peculiar. Whilst he is often on fairly firm ground, his judgement is frequently distorted by an apparent failure to grasp Veblen's underlying purpose. For example, Campbell (1995) maintains that Veblen's account of conspicuous consumption draws simplistic analogies with 'primitive' consumption practices. Whilst it is true that Veblen's account drew on an understanding of 'archaic' social formations akin to those developed by Mauss and Bataille, he was hardly intending to draw analogies. Far from seeking simply to equate 'archaic' and modern consumption practices, as Campbell assumes, Veblen was seeking to point up the extent to which modern society retains the memory of practices it claims to have superseded. His intent is to point out the persistence (in his terms, the 'conservation') of a logic that the market, aided and abetted by economic theory, has sought to occlude. Campbell's incredulity towards Veblen would finally seem to rest on an unacknowledged assumption that the 'rational' consumption of modern society is entirely distinct from the fundamentally 'irrational' practices of 'primitive' societies. Whilst Campbell (1987) usually manages to avoid taking modernity's self-representations at face value, he appears to fall into this trap in his assessment of Veblen.

Campbell is on firmer ground in targeting Veblen's preoccupation with pecuniary emulation. Veblen (1994, 52) explicitly accounts for the way in which 'the leisure-class scheme of life ... extends its coercive influence ... down through the social structure to the lowest strata' in such terms. Emulation, though, is a notoriously problematical concept. It is often presented in overly mechanistic terms, as though the trappings of social standing simply 'trickle down' the social hierarchy over time, as they are adopted by the aspirational masses subordinate to the leisure class. Emulation need not lead towards a simple convergence between ranks, however. It is just as likely to be animated by a perpetual dynamism of 'upward "chase and flight"' (McCracken 1990, 94), as different ranks strive to maintain their difference over time (Figure 1 below). Many historical changes in tastes and fashions are, of course, broadly suggestive of such processes. Muckerji's (1983) account of the widening patterns of consumption associated with the developing urban sphere of fifteenth- and sixteenth-century Europe,[32] for example, is consonant with the operation of an emulatory mechanism. Mennell's (1985) discussion of the reversal in the association of brown bread with poverty, when white bread became available to the working classes, suggests a more complex variation on the theme of 'chase and flight'. However, there are basic problems with all emulatory accounts, which invariably suppose that society amounts to a one-dimensional 'pecking order', consisting of 'nothing but individual striving to move up or to keep others down' (Douglas 1996, 57). It is in this light that Campbell (1987, 53) cautions against an incautious acceptance of Veblen – on the basis that 'individuals may gain success over their competitors through innovation rather than imitation ... and ... social groups (especially social classes) may actually be in conflict over the very question of criteria to be employed in defining status'. A substantial literature on 'subcultures' and 'neo-tribes' attests to the fact that 'remaining different from the Joneses' can be as important an influence on patterns of consumer culture as any form of emulation (Bennett 1999; Heath 2001; Hebdige 1979; Maffesoli 1996). Whilst Veblen was undoubtedly guilty of affording emulation undue emphasis, Campbell's dissatisfaction again stems from a conviction that Veblen's account is unconvincing with respect to *modern* consumption. However, if one alters the emphasis slightly, to suggest that Veblen stressed '*pecuniary* emulation' rather more than 'pecuniary *emulation*', the charge that Veblen's account is historically naive is somewhat tempered.

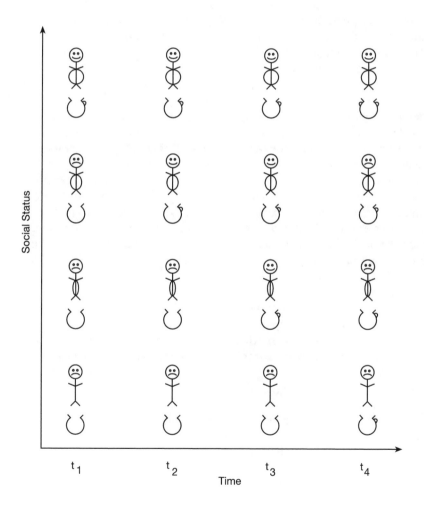

Figure 1 The dynamics of social emulation: handles on drinking vessels

Source: After Miller (1985)

Veblen's analysis actually points to the specifically modern forces unleashed by the gradual transcendence of a situation maintained by sumptuary law (Hunt 1996a; 1996b). Sumptuary law once served to regulate consumption in relation to social rank, in accordance with the image of the 'great chain of being' (Lovejoy 1961).[33] It codified and legitimated 'specific standards of decorum and decoration under the doctrine of "consumption by estates"' (Appleby 1993, 163), thus ensuring the maintenance of the moral order of society. This effectively precluded all possibility of emulation; hence the fears of social disintegration aroused by the early modern expansion of consumption.[34] Veblen's stress on *pecuniary* emulation – on the nature of the mechanisms that, by the end of the nineteenth century, appeared to have firmly established themselves as determining the nature of consumption in the absence of sumptuary law – was, undoubtedly, a considered emphasis, relating closely to Veblen's particular historical vantage-point. The nineteenth century was, as Bauman (2001b, 28) notes, 'the century of great dislocations, disencumbering, disembeddedness and uprooting (as well as of desperate attempts to re-encumber, re-embed and re-root)'. Whilst Veblen's work is not without its problems, therefore, the reaction it prompts from Campbell is rather akin to throwing out the baby with the bathwater. In fact, Campbell's dissatisfaction arguably stems less from Veblen *per se*, than from other high-profile attempts to account for the development of modern consumption in emulatory terms. Campbell is understandably dissatisfied, for instance, with the importance McKendrick *et al.* (1982) grant to emulation in accounting for the upward turn in consumer demand in late eighteenth-century England.[35] In the light of the limitations of McKendrick's account, it is unsurprising that Campbell (1987) should seek a different explanation. In the account he develops – which focuses on neither the utilitarian nor honorific role of consumption, but on its relation to pleasure – there is little need for emulation, and consequently no room for Veblen.[36] Although this fails to grant Veblen a fair hearing, Campbell (1995) has subsequently used this dismissal of Veblen in order to reject a whole host of other theorists who appear to follow in his footsteps.

Most of the recent uses to which Veblen's work have been put stand accused by Campbell of tearing it from context and extrapolating it to shore up notions of consumption as 'communicative display'. For example, the 'unqualified emphasis [Veblen lay] on the manifestation of "pecuniary strength"' has, Campbell (1995, 102) maintains, been diluted to the extent that 'a simplified and amended version featuring "taste"' figures in work seeming to inherit Veblen's mantel. However, Veblen's approach, stripped of its unnecessary preoccupation with pecuniary emulation, does, despite Campbell's negative pronouncements, offer considerable insight into the logic of social differentiation. Bourdieu's (1984) account of the interaction between economic capital and symbolic capital is arguably the most forceful example. As Bourdieu (1984, 483) points out, moreover, consumption 'need not be conspicuous in order to be symbolic'. In other words, it is not just in the consumption of 'status symbols' that the logic of social differentiation is manifest. All consumption practices, however mundane they may appear, serve to classify an individual's social status, and hence provide for the reproduction of the social order. Campbell, it seems, is failing to spot something in Veblen that others have been keen to press into service. We can begin to reassess Campbell's dismissal of Veblen by revisiting, with Baudrillard (1998a), the polarized conceptions of the consumer from which Campbell's adjudication ultimately derives.

Paralleling Campbell (1987), Baudrillard (1998a) also begins by noting that consumer demand is typically portrayed in polarized terms. It is regarded as *either* the innate property of the sovereign individual (which Galbraith [1967] termed the ' "accepted sequence", where the initiative is supposed to lie with the consumer and to impact back, through the market, on the manufacturers' [Baudrillard 1998a, 71–72]); *or* it is taken as being externally 'conditioned' (for instance, by advertising, in Galbraith's 'revised sequence'). For Baudrillard (1998a, 72), the 'revised sequence' at least 'destroys the fundamental myth of the "accepted sequence"…namely that it is the individual who exercises power in the economic system'. Nonetheless, the dichotomy between 'real' and 'false' needs that follows from the 'revised sequence' remains, for Baudrillard, wholly untenable. The very idea of 'false needs' relies on a mischaracterization of consumers as 'mere passive victims of the system', and rests on the infeasible assumption that, 'by producing particular goods and services, companies at the same time produce all the means of suggestion tailored to gaining acceptance for them' (*ibid.*, 74). Yet as Baudrillard plainly states, 'advertising is not all-powerful' (*ibid.*). A more adequate understanding of consumption must, therefore, avoid falling back on the myth of the sovereign consumer, whilst simultaneously moving beyond the equally inadequate notion that modern consumers are routinely inculcated with 'false' needs. Here, a significant departure from Campbell's (1987) line of reasoning becomes evident, for Baudrillard does not consider the possibility of a simple *mediating* position (to be supplied by Veblen or anyone else). Rather, he scrutinizes Galbraith's position further and discovers a basic problem with it. This problem happens to point us in the direction of Veblen. It does so, moreover, in a way that remains far truer to Veblen's work than Campbell's characterization achieves.

Baudrillard affirms that the 'accepted sequence' amounts to nothing other than an ideological justification of economic rationality. It rests on an idealized and naturalized version of society as a social contract between pre-existing individuals. In attempting to overturn this conception, however, Galbraith falls into an essentially similar trap. Rather than *subverting* the ideology of the market, he succeeds only in *inverting* it, offering an equally idealistic vision of a 'natural' state of affairs (which is, in fact, the precise mirror image of the one supplied by modern economic ideology). Galbraith's tack is to dismiss the idealistic vision of 'economic science' and replace it with a more 'realistic' one (one that better captures the grim realities of monopoly capitalism). In doing so, he envisions dominant business powers manipulating hapless consumers, which he sets against a utopian vision of what a less exploitative set of economic relations might entail. This vision, however, merely appeals to a mythical notion of ' "harmonious" individual satisfaction … limited in terms of the ideal norms of "nature" ' (Baudrillard 1998a, 74). Galbraith effectively sees consumerist desires as a distortion of 'basic' needs (such that a second home, for example, is artificially granted the same importance as basic shelter). But by remaining tied to 'idealist anthropological postulates' (*ibid.*, 73), Galbraith fails to see that it is the supposedly 'timeless postulate' (*ibid.*, 75) of 'need' itself that ought to be put into question. For 'need' is a construct of modern economic discourse that *necessarily presupposes* a formal equivalence between different demands. The nostalgic vision that Galbraith pits against an all-too-concrete reality is, then, nothing other than the mirror image reflected back from the mythical ideal issuing from economic discourse. It bares absolutely no resemblance to the principle uncovered by Mauss, Bataille, and Veblen, which reveals consumption to be 'an unlimited social activity' (*ibid.*), not a 'natural necessity'. For Baudrillard, it is this principle that is vital to any adequate understanding of consumption.

In the light of Galbraith's shortcomings, the significance Baudrillard finds in Veblen becomes more readily apparent. Baudrillard (1981, 76) regards Veblen as offering, 'in a way far superior to those who…have pretended to surpass him, the discovery of a principle of total social analysis, the basis of a radical logic, in the mechanisms of differentiation'. As we shall finally hammer home in the next section, it is in terms of this 'differential structure that establishes the social relation, and not the subject as such', that modern consumption must be understood (*ibid.*, 75). As Ritzer (1998, 6) notes, therefore, 'Baudrillard rejects Veblen's concern with imitation, the study of prestige at the "phenomenal" level, and his concern with superficial conscious social dynamics'. Baudrillard is not, in other words, simply a theorist of consumption as 'communicative display'. He is principally concerned, not unlike Bourdieu, to reveal the way in which consumption operates as a form of social classification (cf. Frow 1987; Holt 1997). Although Veblen's preoccupation with pecuniary emulation 'posited the logic of differentiation more in terms of individuals than of classes' (Baudrillard 1981, 75–76), Baudrillard nonetheless holds that Veblen brought out 'how the production of a social classification…is the fundamental law that arranges and subordinates all other logics' – including the logic of use-value and needs, in terms of which consumption is conventionally understood. Baudrillard does not conclude, therefore, with Campbell, that Veblen's account is inadequate in its grasp of the specificity of modern consumption. Rather, he sees Veblen as glimpsing the '*logic of sumptuary values*' lying behind the modern system of economic exchange (*ibid.*, 115). Baudrillard contends that this 'sumptuary logic' – the agonistic logic of the gift, the potlatch, and symbolic exchange – has only ever been *displaced*, not *replaced*, by the logic of economic equivalence. 'In consumption generally', Baudrillard (*ibid.*, 112) asserts, 'economic exchange value (money) is converted into sign exchange value (prestige, etc.); but this operation is still sustained by the alibi of use value'. Consumption, in other words, involves an operation that has nothing to do with needs and their satisfaction by the use-value of objects, which only serve to disguise its underlying social logic.

To illustrate as much, Baudrillard proposes the art auction as an exemplary site for glimpsing the sumptuary process lying behind economic exchange. Whilst this might appear an unlikely setting for the detection of a logic purportedly unconnected to economic value and the logic of equivalence, Baudrillard (*ibid.*, 122) insists that the art auction represents 'a sort of *nucleum* of the strategy of values'. It represents a situation in which consumption is not tied to needs, and permits us to decode 'the birth of the sign form in the same way that Marx was able to uncover the birth of the commodity form' (*ibid.*, 112). The art auction, as Baudrillard notes, is always a *singular* event. Its rules are 'arbitrary and fixed, yet one never knows what will take place, nor afterward exactly what has happened because it involves a dynamic of personal encounter ... as opposed to the economic operation where values are exchanged impersonally, arithmetically' (*ibid.*, 116). It develops according to its own internal rhythm: 'In the altercation and the out-bidding, each moment depends on the previous one and on the reciprocal relation of partners' (*ibid.*). Unlike the mercantile auction, therefore, it does not serve to equilibrate supply and demand. Rather, it departs from a strictly economic calculus, returning us to the agonistic or fatal situation of symbolic exchange. In the art auction, 'exchange value is no longer offered (in exchange for); it is wagered...it ceases to be an exchange value and is transferred out of the realm of the economic. It does not, however, cease to be an exchange' (*ibid.*):

> In the sumptuary act, money is nullified as a general equivalent, as form, and so as a specific (capitalist) social relation regulated by this form. The social relation instituted in this act by the auction is still one of aristocratic parity (among partners). Contrary to commercial operations, which institute a relation of economic *rivalry* between individuals on the footing of formal *equality*, with each one guiding his own calculation of individual appropriation, the auction, like the ... game, institutes a concrete community of exchange among peers. ... Competition of the aristocratic sort seals their *parity* (which has nothing to do with the formal equality of economic competition).
>
> (*ibid.*)

The '*reciprocal wager*' (*ibid.*, 116) of the art auction thus lays bare the logic of symbolic exchange. It is a process founded upon the rivalry of expenditure and involving parity between protagonists defined by their relations with others – rather than a formal situation of competition on an equal footing between abstract economic agents. The whole process reveals that, in 'expenditure, money changes meaning' (*ibid.*, 112). Thus, it 'is not the quantity of money that takes on value, as in the economic logic of equivalence, but rather money spent, sacrificed, eaten up according to a logic of difference and challenge' (*ibid.*, 112–113). In other words, a different logic – a logic of difference – is revealed as prior to (as lying behind) the logic of equivalence enshrined in money. As Butler (1999, 82) remarks, 'what we see in the auction ... is the moment *before* use value and exchange value, that exchange before them that makes each possible'. It also 'makes possible' or gives rise to *sign-value*, not only in the transmutation of expenditure into a sign of prestige (Veblen's principle), but also in the way in which a *symbolic* object (that is, a *particular* painting) is turned into a *sign* by the same process. As Baudrillard (1981, 112) puts it:

> The decisive action is one of a simultaneous double reduction – that of exchange value (money) and of symbolic value (the painting as an *oeuvre*) – and of their transmutation into sign value (the signed, appraised painting as a luxury value and rare object) by expenditure and agonistic competition.

The crucible of the art auction, therefore, turns objects into signs, even as it returns the conventional sign of wealth to the sumptuary logic of expenditure. Here, we have finally reached the stage where the problems posed by sign-value and its relation to modern consumption can be confronted.

Empire of the sign

> Instead of assuming that goods are primarily needed for subsistence plus competitive display, let us assume that they are needed for making visible and stable the categories of culture
>
> (Douglas and Isherwood 1996, 38)

Although Campbell's (1995) misunderstanding of Baudrillard rests on a misleading conflation of 'sign' and 'symbol', something far more serious than a semantic dispute is involved here. Campbell's suggestion that Baudrillard's emphasis on signs amounts to a theory of consumption as 'communicative action' is, in fact, entirely erroneous. Let us begin to set matters straight by going back to basics, hopefully laying some of Campbell's misgivings to rest in the process. Consider, for instance, Bauman's (1999a, 71) statement that, with relatively

> few exceptions (like the language of gestures and etiquette – it is not by accident that the word 'language' has been spontaneously applied to these phenomena) ... non-linguistic culture operates with material

which by itself is directly related to 'non-informative', in some way 'energetic', needs.

This would seem to sum up, in a nutshell, Campbell's frustration with Baudrillard's emphasis on (ethereal) signs at the expense of (material) goods. Yet as Bauman goes on to say, the 'energetic' dimension of certain cultural systems does not mean that they bare no comparison whatsoever to purely linguistic systems. On the contrary, we are 'justified in extrapolating (to the non-linguistic spheres of culture)...exactly those features, which characterize the linguistic interaction in its capacity as a case of a more inclusive class of self-regulating systems' (*ibid.*, 77). Bauman is suggesting that culture, whether in the linguistic or non-linguistic sphere, operates primarily as a form of *structuring* activity, aimed at increasing the orderliness, predictability, and understandability (or meaningfulness) of the world – even though that activity is invariably aporetic and ridden with ambivalence. For as Simmel (1968) notes, the task of culture is always a tragic one: order is always inevitably accomplished at the cost of further disorder; meaningful-ness is invariably achieved with an irreversible measure of ambivalence.[37] Culture, then, is a matter of rendering some events more likely than others, thus ensuring that 'knowing how to proceed' is not beset by the kind of disabling angst that would otherwise arise – even though it is destined to remain riven by anxiety. To see consumption as such a 'structuring system', which is precisely what Baudrillard's conception involves, does not, then, rely on some kind of unreasonable extrapolation from cases such as 'saying it with flowers'. Nor does it seek to deny that consumption has an 'energetic' component. What it does claim, however, is that this 'energetic' dimension is far from the whole truth of the matter, and that the broader 'cultural' dimension of consumption is vital for understanding our society as a consumer society. This 'cultural' dimension implies that consumption acts as a structuring device, operating in much the same way as language does, to lend consistency and coherence to the world – though always imperfectly and clouded by an irreversible ambivalence. Accordingly, if 'one admits that need is never so much the need for a particular object as the "need" for difference (*the desire for the* social *meaning*)', a very different conception of consumption emerges (Baudrillard 1998a, 77–78). Thus, as Douglas and Isherwood (1996, 49) emphasize, 'consumption goods are most definitely not mere messages; they constitute the very system itself'. To summarize, consumer goods constitute a *system*, one that is inherently *cultural* (that is, which is intrinsically related to the aporetic structuring of the social world).

On the basis of this conception, we might propose that the true signifi-cance of consumption lies in the distinction it marks (and the transition it effects) between a natural, energetic system, and a social or cultural system. Despite the all-too-common tendency to reduce it to its 'natural' basis, the most important factor to grasp with respect to consumption is that 'a bio-functional, bio-economic system of goods and products (the biological level of need and subsistence) is *supplanted* by a sociological system of signs (the level of consumption proper)' (Baudrillard 1998a, 79, emphasis added). Here, Baudrillard invokes the structural principle uncovered by Lévi-Strauss' study of kinship, where *natural* relations (relations of consanguinity) are supplemented by *social* relations, thus 'superimposing upon the natural links of the family the henceforth artificial rules ... of alliance' – rules that govern an *expanded* social body (Lévi-Strauss 1969, 595). This principle is more important in terms of its general *form* than its particular *content*, insofar as it 'introduces a *general* opposition between nature and culture' (Easthope 1999, 34). It involves the imposition of interdictions and taboos which, whatever their actual character, necessarily afford priority to culture over nature. Likewise,

> what confers on consumption its character of being a social fact is not what it apparently preserves of nature (satisfaction, enjoyment/pleasure), but the essential procedure by which it breaks with nature (what defines it as a code, an institution, a system of organization).
>
> (Baudrillard 1998a, 79)

This is why Baudrillard should insist on the cultural character of consumption against all naive appeals to its 'natural' or 'energetic' aspect (which Baudrillard terms its 'primary level'). Baudrillard's position also amounts, however, to an appeal for a broader recognition of the particular importance of consumption today, which is arguably the primary means by which the ineluctably cultural character of human relations is now expressed:

> What is sociologically significant to us, and marks out our age as an age of consumption, is precisely the generalized reorganization of this primary level [of consumption] into a system of signs which reveals itself to be one of the specific modes, and perhaps *the* specific mode, of transition from nature to culture in our era.
>
> (*ibid.*)

Thus, if consumption amounts to an institution with codified rules, its *structural* dimension (like that of language) is of far greater significance than any particular usage of consumer goods in everyday life – as status symbols, signs of subcultural dissent, or personal style (analogous with particular utterances or speech effects).

> The circulation, purchase, sale, appropriation of differentiated goods and sign/objects today constitutes our language, our code, the code by which the entire society *communicates* and converses. Such is the structure of consumption, its language [*langue*], by comparison with which individual needs and pleasures [*jouissances*] are merely speech effects.
>
> (*ibid.*, 79–80)

In thinking of consumption as an activity involving signs, therefore, we are well advised to acquaint ourselves with Calvino's (1972, 48) maxim: 'Signs form a language, but not the one you think you know'.

To pursue this further, we need to appreciate the specific sense in which a sign system (and, as we shall eventually see, the commodity system itself) is a structural system. A structure, according to Hjelmslev, may be defined as 'an autonomous entity of internal dependencies' (cited in Barthes 1990, 3). The distinguishing feature of a structure is that the existence of any particular element 'does not precede the existence of the whole, it comes neither before nor after it, for the parts do not determine the pattern, but the pattern determines the parts' (Perec 1988, 189). This means that 'knowledge of the pattern and its laws...could not possibly be derived from discrete knowledge of the elements that compose it' (*ibid.*). We shall shortly see the full implications this holds for consumption, in terms parallel to the distinction between language as structure (*langue*), on the one hand, and individual utterances (speech effects), on the other. The basic lesson is brought out most clearly, however, in Perec's inspired example of a jigsaw puzzle (cf. Lévi-Strauss 1960). The structural quality that defines the jigsaw 'means that you can look at a piece of a puzzle for three whole days, you can believe that you know all there is to know about its colouring and shape, and be no further on than where you started. The only thing that counts is the ability to link this piece to other pieces' (Perec 1988, 189).

In line with Perec's example, it is a commonplace of semiotics to note that the meaning of a sign – like that of a piece a jigsaw puzzle – derives entirely from its 'negative difference' from all other terms making up the system to which it belongs. To employ another analogy, the terms of a language are, like the holes of a net, strictly empty in themselves. They are defined not by any particular positive content, but by their boundaries with neighbouring terms. 'Their most precise characteristic is in being what the others are not' (Saussure 1959, 117). Hence Saussure's dictum that 'language is a form and not a substance'. The *differential logic* to which the sign accords applies both *phonetically* (at the material level of the signifier – for instance, the difference between 'cat', 'mat' and 'hat': each is a word of one syllable differing only in the letter or sound with which it begins) and *semantically* (at the conceptual level of the signified – for instance, the differences between 'rock', 'stone', 'cobble', 'boulder', 'brick', 'pebble', 'gravel', 'sand': all relate to a similar category, but derive their sense from their relative differences along such dimensions as size, smoothness or roughness, natural or cultural status, and separate or aggregate context). The terms constituting a particular system of language – such as English or French – are, therefore, like the pieces of a jigsaw puzzle, first and foremost *arbitrary*. The signifier (the material inscription on the page or its acoustic equivalent in spoken language) is dictated by convention: there is no necessary or natural relation between actual trees and the signifier 'tree' ('t-r-e-e'). Indeed, in French, 'arbre' ('a-r-b-r-e') is the corresponding signifier: different languages conventionally employ different (yet equally arbitrary) signifiers. The same principle is reflected in the arbitrary quality of the signified, which carries the implication that different languages 'carve up' the world in different ways. Take, for example, colours. Colours form a natural spectrum, determined by the infinite play of wavelength. Different languages divide up this continuum into *discrete* units in different (though equally arbitrary) ways. In other words, signs necessarily imply a degree of discretion (this becomes evident when we occasionally have to resort to saying something is 'reddish-brown' or 'yellowy-green'). And this discretion is wholly determined by convention (there is no guarantee that the particular degree of discretion will remain constant from place to place, period to period, or culture to culture). It is thus the case, as Baudrillard suggests, that a *code* governs any process of signification. If one is not attuned to the code, the meaningfulness of signs will be entirely absent. Hence: 'The origin of meaning ... is to be found ... in difference, systematizable in terms of a code (as opposed to private calculation) – a differential structure that establishes the social relation, and not the subject as such' (Baudrillard 1981, 64).

The necessity of such a code entails that language is an ineluctably social phenomenon. Language always already implies at least two: a private language is a strict impossibility – a contradiction in terms. Thus, as a structure 'whose material can be disassociated from the subjects speaking it' (Baudrillard 1981, 64), language necessarily *precedes* any meaningful individual utterance. When I speak or write, the sense I produce is enabled but simultaneously constrained by language. I draw upon the pre-existing structure of language to express myself, in terms that antecede me and can never finally belong to me. My meaning is spoken by language as much as it is spoken by me (Heidegger's formulation: 'language speaks us'). Language is, then, 'a social form in relation to which there can properly speaking be no individuals, since it is an exchange structure' (*ibid.*, 75). The structural property of language entails that humans are social before any possibility of their being characterized as individuals. This is not to say that the subject is wholly determined by language, however. Language exists in advance of the individual, 'not as an absolute, autonomous *system*, but as a structure of exchange contemporaneous with meaning itself, and on which is articulated the individual intention of speech' (Baudrillard 1981, 75). To this extent, it bears a striking resemblance to the structure of gift exchange described by Mauss. This, however, is precisely what modern consumption disavows. Consumption is conventionally regarded as an economic relation involving a 'private calculation'. For Baudrillard, however, 'a consumer is never isolated, any more than a speaker' (*ibid.*). Despite its usual characterization, 'consumption does not arise from an objective need of the consumer, a final intention of the subject towards the object' (*ibid.*), which would be analogous to explaining language in terms of an individual need to speak. The implication is that any individual use of goods (like an individual utterance) is subordinate to the system of objects (analogous to the system of language). Needs and objects do not, therefore, accord to an individual logic, but to a thoroughly social one. This is far from simply saying that goods are used as a form of 'communicative display' by consumers – that goods permit us to 'say something about ourselves'. It is, rather, that the distinctions marked out by the *system of objects* speak us: 'there is social production, in a system of exchange, of a material of differences, a code of significations and invidious (*statuaire*) values' (*ibid.*).

In order to understand the nature of the system of objects, which carries the direct implication that consumption is a social relation and not an individual relation to needs, it is necessary to develop the sense in which consumer goods act as 'sign-values' in more detail.[38] For Baudrillard, sign-value is as vital to the nature of the commodity as the more familiar properties dissected by Marx: use-value and exchange-value. This is not simply because such a conception permits a fuller understanding of the commodity, in relation to consumption as well as to production. It also brings to the fore the fundamental distinction between the commodity form (underpinned by a logic of *equivalence*) and the form of the gift (the *singularity* of the object in symbolic exchange). Let us initially pursue the sense in which sign-value is fundamental to the commodity, which will lead us to a clearer understanding of precisely what modern consumption involves (sign-value) and also disavows (symbolic exchange). We can begin to do so by considering the nature of the commodity in the light of our previous discussion on the nature of structural systems.

Unlike the gift, the commodity is characterized by alienability and fungibility. The commodity, one might say, is an eminently interchangeable object, marked by a kind of generalized equivalence that issues from its relation to the commodity system. The fact that it possesses exchange-value is, for Marx, a necessary feature of the commodity (though it is not a sufficient feature; a commodity must also possess a use-value). Because it is produced as an exchange-value, for sale on the open market, a commodity is necessarily a part of the system of exchange-value. The system of exchange-value is, moreover, a self-referential system (in the sense that its terms are defined against one another, such that £10 is ten times as valuable as £1, whilst £1 has a tenth of the value of £10). Marx set great store by this property of the commodity, because the principle of commensurability to which it accords underlies many of the most pernicious aspects of capitalism. However, whilst he held that exchange-value is defined by commensurability, Marx saw use-values as strictly incomparable. Thus, where it makes sense to compare £10 worth of cheese and £10 worth of chalk, it makes no sense to compare chalk with cheese in and of themselves. Yet as Baudrillard (1981) points out, there is something at fault in Marx's analysis here. Marx overlooks a particular kind of comparability, which is what the very notion of 'use-value' encapsulates. In stark contrast to the singularity of the gift (the symbolic object), the use-value of a given commodity – cheese, chalk, a tomato, etcetera – possesses, in the first place, a kind of equivalence to all other use-values of the same kind. One tomato on the supermarket shelf, for example, is near identical to the next. Chalk, cheese, and tomatoes also have in common that they are *sufficient to their purpose* as use-values. Thus chalk is useful as chalk; cheese is useful as cheese; a tomato is useful as a tomato. The comparability, equivalence and fungibility of commodities thus applies in terms of use-value, even though Marx held use-values to be incomparable in the obvious ('chalk and cheese') sense. All use-values, in other words, accord to a rational, functional common denominator. This effectively means that a *naturalized* conception of use-value slipped, unnoticed, into Marx's reasoning. We have already noted Marx's passing recognition of the fact that use-values are 'no part of' objects themselves; that they are projected onto the natural world. In neglecting to develop this recognition to any extent, Marx failed to see that use-values are cast off as the shadows of exchange-values – and that use-value thereby serves as the alibi of exchange-value. This throws into sharp relief the singularity of the gift. For unlike the gift, the commodity is marked by the logic of equivalence, through and through (and not only in terms of exchange-value, as Marx implied). To this extent, the commodity bears an uncanny resemblance to the sign: it 'has no more meaning than a phoneme has an absolute meaning in linguistics' (Baudrillard 1981, 64).

Rather than its meaning deriving from its use-value (functionality) in relation to the subject, 'the object finds meaning with other objects, in difference, according to a hierarchical code of significations' (*ibid.*). As such, the object is 'reified into a sign' (*ibid.*, 65). Accordingly, we should properly speak not of 'objects' but of 'sign-objects'. Where the conventional discourse of consumption affords primacy to the functional aspect of the object, as a use-value and a source of individual satisfaction, this is to deny the formal priority of the system of objects – to deny that 'sign-objects exchange among themselves' (*ibid.*, 66). The 'functionality of goods and individual needs only follows on this, adjusting itself to, rationalizing, and in the same stroke repressing these fundamental structural mechanisms' (*ibid.*, 75). Barthes (1990, 265) expresses the same proposition in the following terms:

> as soon as real objects are standardized (and is there any other kind today?), we must speak not of functions but of function-signs. Whence it is understood that the...object possesses, by its social nature, a sort of semantic vocation: in itself, the sign is quite ready to separate itself from the function and to operate freely on its own, the function being reduced to the rank of artifice or alibi.

The system of objects, therefore, coincides with the advent of a fully fledged consumer society. Accordingly, the sign-object bears properties that are specific to its nature and which mark a fundamental contrast with the symbolic object (the object of gift exchange). We can consider these properties, and the significance they hold for modern consumption, by returning to the situation described by Mauss.

Recalling the inalienability of gift and giver, Baudrillard (1981, 65) says that the gift-giver

> separates himself from it [the gift] in order to give it ... divests himself as if of a part of himself – an act which is significant in itself as the basis, simultaneously, of both the mutual presence of the terms of the relationship, and their mutual absence (their distance).

The *ambivalent* nature of this exchange both issues from and affirms the fact that 'the gift is a medium of relation *and* distance; it is always love and aggression' (*ibid.*). The implication this holds is that gifts are 'not autonomous, [and] hence not codifiable as signs'; hence, the gift is 'not amenable to systematization as commodities and exchange value' (*ibid.*). In contrast, the object of consumption, as a sign-object, no longer symbolizes but occludes this (ambivalent) relation. The sign-object 'is no longer the mobile signifier of a lack between two beings, it is "of" and "from" the reified relation (as is the commodity at another level, in relation to reified labor power)' (*ibid.*). The reification of labour-power in the commodity is thus paralleled by the reification of social relations in the object of consumption. The alienation achieved by the commodity in the sphere of production is mirrored in the consumptive sphere. This issues directly from the way in which modern consumption deals with a virtual and necessarily ambivalent desire by recasting it as a need or a want that can be satisfied by the 'reality' presented by the *profusion* of objects of consumption.[39] Thus, insofar as the singularity of the gift-object served to express the 'transparency of social relations in a dual or integrated group relationship' (*ibid.*), the advent of the sign-object coincides with their opacity. To say that sign-objects, as the terms of a structural system, exchange amongst themselves, is to highlight that the sign-object 'is neither given nor exchanged: it is appropriated, withheld and manipulated by individual subjects as a sign, that is, as a coded difference' (*ibid.*). Where gift exchange symbolizes a social relation, the reified sign-object of commodity exchange 'only refers to the absence of relation itself, and to isolated individual subjects' (*ibid.*). On this basis, Baudrillard achieves a Lacanian reformulation of the Maussian notion of obligation. The social relation made manifest in the *symbolic* object is, as Baudrillard (*ibid.*) says, 'a total relationship (ambivalent, and total because it is ambivalent) of desire'. In this sense, 'the symbol refers to lack (to absence) as a virtual relation of desire' (*ibid.*). In its singularity, the symbolic object marks what Lacan characterizes in terms of the *constitutive* lack of the subject; a lack that is necessarily occasioned by the accession of the subject into the symbolic order of language. A constitutive lack, by definition, can never be made good. Nonetheless, it compels the subject to action, in an interminable search for completion that is as inevitable as it is impossible.[40] What Baudrillard draws attention to, here, is the historical – and, indeed, geographical – specificity of the way in which this lack is figured, explicating the way in which the modern disavowal of this lack (the belief that it can be readily satisfied) connects with the imperatives of capitalism. As a necessarily expansive system, capitalism feeds off the insatiability of a desire that is no longer manifest in symbolic terms, but which is disavowed by the hollow promises of total satisfaction made by consumer capitalism – promises on which it can never deliver and which, if it could, would sound its own death-knell (cf. Scitovsky 1976).

The political-economic definition of consumption, as Baudrillard (1981, 113) notes, is the 'reconversion of economic exchange value into use value, as a moment of the production cycle'. However, as we are now in a position to see, consumption also entails 'the conversion of economic exchange value into sign exchange value' (*ibid.*). The role of sign exchange-value in modern consumption implies the 'production of sign exchange value in the same way and in the same movement as the production of material goods and of economic exchange value' (*ibid.*, 114). Sign exchange-value is not, then, autonomous from the production of the commodity. 'The difference – a major one – between the aristocratic potlatch and consumption is that today differences are produced industrially, they are bureaucratically controlled in the form of collective models' (*ibid.*, 119).[41] To identify the specificity of modern consumption in terms of the conversion of economic exchange-value into sign exchange-value is not, therefore, to claim that a given material object functions, by itself, as a transparent medium for the transmittal of information, messages or meaning (about social status or anything else). Nor is it to suggest that this has somehow become more important than an object's material use-value in 'postmodern' times. It is simply to bring out, as forcefully and rigorously as possible, the role of consumption in the redefinition of capitalism and in the figuration of capitalist society as consumer society.

Accordingly, where Marx saw capitalism as gestating 'an abstract, systematized productive force, radically different from concrete labour and … traditional "workmanship"' (Baudrillard 1998a, 75), Baudrillard sees capitalism as proceeding to accomplish precisely the same transformation in the sphere of consumption. Consumption has become '*the most advanced form of the rational systematization of productive forces at the individual level*, where "consumption" takes over logically and necessarily from production' (*ibid.*). Consumption is, in effect, the continuation by other means of the system Marx considered under historical conditions where production assumed primacy. In a manner radically different from the logic of symbolic exchange, modern consumption assumes the form of an abstract and systematized 'consumptive force', which is thoroughly imbricated within the structure of capitalism, and which performs a direct role in its survival. For example, even 'the technology of final consumption must keep pace with the requirement to absorb the increasing quantities of commodities produced' under the capitalist mode of production (Harvey 1982, 122).[42] Hence the insistence behind Baudrillard's (1981, 31) contention that the 'fundamental theorem of consumption' is 'that the latter has nothing to do with personal enjoyment ... but rather that it is a restrictive *social institution* that determines behavior before even being considered in the consciousness of the social actors'. This institution is hidden behind the economic ideology of consumption, which presents consumption as a matter of individual need and its satisfaction, to be found in the functional use-values of objects, just as it is hidden behind the ideology of consumption as unbridled pleasure. Behind this naturalized mythology and behind the constant invocations to indulge oneself, consumption plays an ever more important systemic role.

From consumption to the city

> Beyond a certain frequency, need knows absolutely no limit.
>
> (Burroughs 1959, viii)

We have covered some considerable ground in attempting to provide an account of consumption adequate to the task of elucidating its connections to the city. Before finally heading off in this direction, however, it is worth restating the most important implications, and also considering the deficiencies, of Baudrillard's conceptualization of consumption. The single most important point to be taken from the foregoing discussion is that consumption cannot satisfactorily be considered in the individualized terms that the consumer society itself seeks to promulgate. This point is crucial to the conception of consumption deployed throughout this book. The *systemic* role played by consumption – which is definitive of the consumer society as such – requires constant vigilance. We must, at all costs, avoid taking at face value the naturalized, individualized self-representation that the consumer society itself promotes – which would amount to taking 'the ideology of consumption for consumption itself' (Baudrillard 1981, 62). The derivation of this basic point has involved a detailed, at times technical, discussion. The significance I have claimed for it remains to be seen, as we turn to consider the relations between consumption and the city. The basic point it encapsulates, however, is a simple one and, as such, does not present any inherent difficulty of comprehension.

The limits of the theoretical framework established here are at least twofold. The first concerns the sense in which Baudrillard's focus on the structural/systemic dimension of consumption fails to adequately capture the qualities of the unruly consumer (cf. Bianchi 1998; Gabriel and Lang 1995). This points up one limitation of Baudrillard's analysis: however much it is true that consumption represents a systemic imperative, it is also a form of practice, and on questions of practice, Baudrillard's account remains underdeveloped. Whilst Baudrillard is hardly the representative of the manipulationist school he is often taken to be, it has been left to de Certeau (1984, xi) to demonstrate that consumers are never simply 'passive and guided by established rules'; that they are always active and discriminating (cf. de Certeau *et al.* 1998; Genosko 1992; Buchanan 2000). Consumption is never a ventriloquized acting out of rules; it is always an act of enunciation (we should note in passing that Baudrillard clearly recognizes as much). The tendency for the truth of this observation to slide into a romanticized image of the consumer, heroically resisting the powerful currents of consumerism, should not, however, go unchallenged. As Roberts (1999, 27) notes, all too often 'the activity of the cultural critic is mistaken for the critical activity of the consumer' (we should note that de Certeau scrupulously avoids this slippage, merely demonstrating how consumers, quite literally, *take liberties*). Baudrillard's argument is particularly important in highlighting that the consumer society is programmatically geared towards 'deflecting' the possibility of resistance. As Bauman (1996a, 34) notes, offering an apposite analogy, we are encouraged 'not [to] ruminate on the way to rectify commodities displayed on the supermarket shelves – if we find them unsatisfactory, we pass them by, with our trust in the supermarket system unscathed, in the hope that products answering our interests will be found on the next shelf or in the next shop'. By extension, the consumer society, more than any form of society preceding it, seems eminently capable of embracing the resistance it encounters, drawing would-be resistance into the very system it pitches itself against. Baudrillard's emphasis on consumption as a *structuring* structure, as opposed to a *structured* structure (to deploy Bourdieu's [1977] terminology), is, then, very much a considered emphasis.

Whilst it remains a perennial problem of social analysis, the recursive nature of social structures as both structured and structuring is not, in principle at least, an insoluble issue. Indeed, it is a false conception of the problem that perceives them as mutually exclusive, or as the basis of strictly opposed approaches (Bauman 1972). There is, however, a second and potentially more problematical issue evident in Baudrillard's account of consumption, which relates specifically to the notion of symbolic exchange. This concept has often been charged with embodying a certain nostalgic primitivism (Lyotard 1993b). It is not difficult, for instance, to see the relationship between symbolic exchange and the sign as a parallel instance of the relationship between use-value and exchange-value in Marx, where the former term represents little more than the shadow cast by the latter, which effectively serves as its alibi. This is, clearly, a contentious issue. Rather than address it in any detail here, however, let us simply note that the structure of symbolic exchange is not directly comparable to that of the sign, unlike the case of use-value and exchange-value, and also that Baudrillard does not employ it in a particularly nostalgic vein (cf. Hefner 1977; Turner 1987).

Having put these expansive but necessary remarks on consumption in place, it is time to turn our attention to the relations between consumption and the city. Let us end this chapter by drawing attention to Baudrillard's (1998a, 64) remarks that it is in the city that consumption makes its presence most palpably felt:

In a small group, needs, like competition, can doubtless stabilize. There is less of an escalation of signifiers of status and the stuff of distinction. We can see this in traditional societies or micro-groups. But, in a society of industrial and urban concentration such as our own, where people are crowded together at much greater levels of density, the demand for differentiation grows even more quickly than material production. When the whole social world becomes urbanized…'needs' grow exponentially – not from the growth of *appetite*, but from *competition*.

It is in the urban environment that the logic of consumption reveals itself most clearly: 'the differential "chain reaction" … has … urban space as its locus' (*ibid.*, 65). Whilst the expansion of 'needs' in the city parallels the expansion of production, it also sets up a tension between the two, as two different logics interfere and interact:

> Urban concentration (and hence differentiation) [initially] outstrips productivity. That is the basis of urban alienation. A neurotic equilibrium does, however, establish itself in the end, somewhat to the advantage of a more coherent order of production – the proliferation of needs washing back over the order of products, and becoming integrated into it after a fashion.
>
> (*ibid.*)

The establishment of this 'neurotic equilibrium' has proved to be a particularly complex affair in political-economic terms, as Castells' (1977) work on 'collective consumption' testifies (Chapter 4). The significance of collective consumption has, nonetheless, rapidly faded from view. It effectively assumed the role of a vanishing mediator, paving the way towards an increasingly individualized, market-oriented form of consumption. Today, as Baudrillard suggests, 'the *language of cities* is competition itself' (*ibid.*). In consequence, the powerful aspirations for inclusion in the seductive world of consumption assume an ever more important role in shaping the social topography of the city. And, as we shall see, nowhere else are those whom the process excludes more prevalent.

Part II
Consumption and the city

3   Consumption and the city, modern and postmodern

The literature of modernity and the themes of modernism were not concerned with shopping.

(Wolff 1990, 58)

Nor is it the case that shopping was 'invisible' in the literature of modernity. Quite the contrary.

(Wilson 1995, 68)

The 'form' hypermarket can thus help us understand what is meant by the end of modernity.

(Baudrillard 1994a, 77)

Introduction: consumption and (post)modernity

The same process of rationalization of productive forces which took place in the nineteenth century in the sector of *production* reaches its culmination in the twentieth in that of *consumption*.

(Baudrillard 1998a, 81–82)

Disciplinary society is succeeded by the sophisticated city.

(Alliez and Feher 1989, 54)

Much of the current reassessment of consumption – of both its historical and contemporary significance – can be related to a retrospective acknowledgement that things have not turned out in the manner anticipated or projected by earlier modern thought. It is only now, it might be ventured, that we find ourselves in a position capable of recognizing that modernity has been, above all, an *exceptional* state of affairs: not, as it represented itself, the final solution to the enigma of the world, but a temporary state under which that particular self-representation was propagated, promulgated and vaunted.[1] Whilst, for many, this coming-to-an-end of modernity gives cause for nostalgia, lamentation or alarm, a more considered approach would recognize such responses as thoroughly modern ones, founded on all the false hopes that modernity offered. The task in hand, therefore, is to uncover a different history, to document the process of disappearance to which modernity has subjected itself, and to assess the processes that become evident when history is brushed against the grain. All of this can be associated with the idea of the postmodern. This is, however, a term that must be read with care. As we have previously emphasized, the appellation 'postmodern' should not be read as implying an epochal conception of history, 'a diachronic sequence of periods in which each one is clearly identifiable' (Lyotard 1992, 90). The very term signals the end of any such teleological metanarrative as 'history'. The 'idea of a linear chronology is itself perfectly "modern"', says Lyotard (*ibid.*). Still less does the term imply the beginnings of some kind of 'postcapitalist' situation – the 'end of history' as the triumph of market liberalism (Fukuyama 1992).[2] To the contrary, the contention to be developed here is that capitalist society has, through changes intimately related to consumption, witnessed the inauguration of a 'new mode of domination', which 'distinguishes itself by the substitution of seduction for repression, public relations for policy, advertising for authority, needs-creation for norm-imposition' (Bauman 1987, 167–168). Precisely how this transformation has taken shape, and the place of the city in all of this, is the subject of this chapter.

Human encounter with the world has always been a profoundly enigmatic affair. Traditionally, human societies attained a measure of ontological security from social arrangements that accepted the fundamental ambivalence of the world (Giddens 1990; 1994). Because traditional modes of existence were symbolically tied to the world of appearances, such societies were able to employ well established, ritualized ways and means of being-in-the-world. The continuity of such an existence was assured by the experience of tradition alone, and legitimated by forces beyond human powers. 'Nature ... in the infinite detail of its illusory manifestations ... was conceived above all as the work of hidden wills' (Bloch 1962, 83). This world of illusion amounted, in other words, to an acceptance of the *illusion of the world* (the term 'illusion' is apt 'not in the sense of its power to fool you, but in its power to put something into play, to create something: scene, space, a game, a rule of the game – to invent, in fact, the mode of appearance of things' (Baudrillard 1993b, 59–60)). Modernity, in stark contrast, was founded firmly on the disavowal of the fact that ambivalence is inevitable; that appearances are intrinsically deceptive. Modernity was founded on a commitment to the *reality of the world* (Bauman 1991).[3] Modernity thus held out the dream of an attainable order. And order, as the promise of the removal of ambivalence and contingency from the world, necessarily cast contingency as a threat, and demonized ambivalence. This was, of course, a complex and multifaceted affair. It involved, for instance, a transformation of time, as time became 'the property of man' (Le Goff 1980, 51). Above all, however, the inauguration of modernity saw the force of *reason* as uniquely capable of removing the dead weight of tradition, such that the human capacity for reconsidering – collectively or individually – society's ways and means of proceeding could be given its release.[4] To this extent, modernity may be characterized as instating a 'canonical morality of change' (Baudrillard 1987b, 63). As Bauman (1992a, x) notes, the 'war against mystery and magic was for modernity the war of liberation leading to the declaration of reason's independence'. Such a declaration, however, internalized an ultimately insurmountable problem, insofar as it entailed that 'the world had to be *de-spiritualized*, de-animated: denied the capacity of the *subject*' (*ibid.*). This capacity was to be reserved for the human will alone. Thus modernity instated an opposition between the human and the natural world, where previously the world as a whole was conceived as a great artifice, as a purposeful grand design, in which – as, for example, in the Doctrine of Signatures – God's hand could be read (Foucault 1974).

If reason renders problematic any aspect of the world that appears, in its innermost nature, antinomical to the modern idea of order *as such*, such enigmatic qualities are in need of rectification. To this extent, the process of disenchantment amounts to the 'forced realization of the world' (Baudrillard 1993b, 45): the de-spiritualization of the world of appearances demands nothing less than the annihilation of ambivalence. This is, however, a demand for the impossible. It is an impossible demand insofar as modernity's attempt to rid the world of its now foreboding contingency inevitably produces its own sworn enemy; for reason can carve out a niche for itself – assert its existence, as it were – only in opposition to its subordinate other. Reason, that is to say, can appear as coherent and complete only against some kind of deluded, illusory, and irredeemably irrational 'unreason'. And yet, whilst reason would seek to do away with its subordinate double – which necessarily appears as a flawed version of itself – it is, nonetheless, dependent on it. Its 'other' serves as its own alibi and accomplice. But, in spite of this, the successive instatement of modernity proceeds on the basis that '*[a]ll ambivalence is reduced by equivalence*' (Baudrillard 1981, 135).[5] It is in this sense that modernity can be characterized as 'the immense process of the destruction of appearances … in the service of meaning … the disenchantment of the world and its abandonment to the violence of interpretation and history' (Baudrillard 1994a, 160).[6]

If the reduction of ambivalence by a calculus of equivalence entails the self-assertion of reason in opposition to its other, it nonetheless receives its justification from elsewhere. Insofar as modern reason relates to the promise of ordering the world, there is an obvious need for some manner of assessing the relative value of the manifold different ways by which social existence, freed from the ties of tradition, might be taken forward. In effect, reason receives its power only insofar as it is capable of being harnessed in the name of *progress*, and the promise of progress requires some means of assessing its degree of realization. It demands, in short, precisely the kind of accountability that is enshrined in the principle of equivalence. Accordingly, over the course of its history, modernity engendered an immense series of abstract reductions and formalizations, which began with the reduction of the multiplicitous appearances, senses and qualities of the world to discrete signs, and would ultimately include the reduction of nature to the functional object-form; of heterogeneous concrete human labours to an homogenous abstract labour value; of multiple temporalities to clock time; of multiple spatialities to a mappable Euclidean geometry; and so forth. Modern reason, and the promise of progress towards an ordered world it held out were, in other words, mutually reinforcing principles.[7]

Progress, however, amounts to a considerably more circular and convoluted process than modernity itself ever cared to admit – all of which relates to the central contradiction embodied in reason. Modern reason, it might be said, only ever amounted to a *propensity to rationalize*: as Freud recognized – his own thought remaining to some extent contaminated despite his insight – the peculiarly modern faith in reason was never anything but a faith. This is nowhere more evident than in the specific case of progress: 'Rather than deriving its own self-confidence from its belief in progress, the educated elite [of Western Europe] forged the idea of progress from the untarnished experience of its own superiority' (Bauman 1987, 110). Progress was always already defined on the basis of an unprecedented modern self-confidence in the superiority of its own mode of existence. The idea of progress stemmed, above all, from 'a novelty in the experience of objective time' (*ibid.*). For whereas, 'for most of the history of Christian Europe, time-reckoning was organized around a fixed point in the slowly receding past', the belief in universal reason 'set the reference point of objective time in motion, attaching it firmly to [modernity's] own thrust towards colonizing the future in the same way as it had colonized the surrounding space' (*ibid.*). The faith placed in modernity's self-proclaimed historical centrality was such that it automatically registered as wanting any alternative mode of existence, which necessarily belonged to the 'wrong side' of history (Heffernan 1994).[8]

Progress was, therefore, far from identical with its modern representation as a rational path towards a final, glorious order. Insofar as the problems the modern reduction of the ambivalence of the world sought to overcome were always only ever displaced, rather than solved once and for all, this kind of displacement lay at the heart of the trajectory of modernity. The modern requirement of a means of evaluation of the pace of progress inevitably threw up new problems, problems that would eventually undermine the very idea of progress itself. For as soon as a departure is made from traditional, unreflexive modes of social existence, a series of aporetic differences between modes of evaluation makes itself evident. 'Through most of human history', writes Bauman (1993b, 4), 'little difference was seen or made between now strictly separated standards of human conduct, such as "usefulness", "truth", "beauty", "propriety"'. With the birth of modernity, however, such a process of differentiation necessarily asserted itself:

'The once unitary and indivisible "right way" begins to split into "economically sensible", "aesthetically pleasing", "morally proper"' (*ibid.*, 5). Whilst modernity can defend its own integrity by subordinating particular modes of evaluation for a time – most commonly, today, by elevating the 'economically reasonable' above all other principles – it will eventually reach a situation where its limits become recognizable, and where faith in modernity itself becomes subject to widespread disenchantment. Modernity has, from the start, been destined to proceed, in asymptotic fashion, towards its own inherent limits. Approaching its liminal condition, modernity becomes susceptible to a renegade form of reversibility – not the cyclical form of symbolic exchange but the inevitable return of everything that modernity sought to repress (Baudrillard 1990a).

The effective cancellation of modernity's progress is accomplished from within – the end result of the ultimate impossibility enshrined in modernity's conditions of possibility. In the terms Baudrillard (1993a; 1994a) employs, this eventually leads towards a situation of *simulation*. The initial disenchantment of the pre-modern world relied on the dialectical power of representation to stand in for, and to exchange with, the real. But the Renaissance emancipation of the sign from its pre-modern, hierarchical status saw its eventual replacement by the serial (re)production of the sign – marking a gradual shift away from the possibility of the *counterfeit* and its succession by the *copy* (cf. Benjamin 1975b). At such a point, where the pejorative implication of the copy as a (mere) counterfeit no longer holds, the *real* no longer equates to the *original*. But this shift itself clears the way for a third order of 'pure simulacra', wherein the status of the real is, finally, fully undermined. The pure simulacrum does not – indeed, cannot – (mis)represent or denigrate reality but rather entails an order of hyperreality, 'a real without origin or reality' (Baudrillard 1994a, 1), where 'the very definition of the real' becomes '*that of which it is possible to provide an equivalent reproduction*' (Baudrillard 1993a, 73). And whereas 'representation attempts to absorb simulation by interpreting it as a false representation, simulation envelops the whole edifice of representation itself as a simulacrum' (Baudrillard 1994a, 6). Once simulation assumes this pre-eminent status, it usurps the real: the real is, so to speak, no longer what it used to be. Today the real can no longer – (as) if it ever could – be distinguished as more real 'than the thing which feigns it' (Bauman 1992a, 151).[9] In line with such a situation, the postmodern can be characterized as 'the immense process of the destruction of meaning, equal to the earlier destruction of appearances' (Baudrillard 1994a, 161).[10]

With an eye towards the philosophical tradition of sophistry and its links to the notion of the simulacrum, Alliez and Feher (1989, 54) suggest that, in 'a quasi-mathematical way, the function of power appears to change in order to fit a posturban agglomeration involving a sophistication that gradually combines with and eventually replaces the former surveillance apparatus, just as the latter had replaced methods of confinement'. This 'sophistication' equates to the metamorphosis of modernity into postmodernity, which, in this respect at least, is closely associated with consumption (Jameson 1983). What ought to be stressed, in the light of the previous chapter, is that consumption needs to be understood, here, in terms of the increasing importance of its systemic dimension. 'The truth of consumption is that it is not a function of enjoyment, but a *function of production* and, hence, like all material production, not an individual function, but *an immediately and totally collective one*' (Baudrillard 1998a, 78). Certainly, consumption appears to be – and undoubtedly, for many, is – a highly pleasurable activity, characterized by a strong sense of freedom of choice. But consumption, as such, is necessarily tied to a *social* logic, over and above its connection with *individual* pleasure. Indeed, for Baudrillard (*ibid.*), as 'a social logic, the system of consumption is established on the basis of the denial of pleasure. Pleasure no longer appears as an objective, as a rational end, but as the individual rationalization of a process whose objective lies elsewhere'. In short, the equation of consumption with pleasure and freedom of choice is a profoundly reductive equation. The role consumption has come to play is increasingly significant to the reproduction of capitalism. As Baudrillard (1995b, 100) has put it,

> it was a vital necessity for capital to have workers and producers transformed into active consumers, and even direct stockholders in the capitalist economy (this doesn't change anything ... the strategy being, as always, to remove the tablecloth without changing the organization of the table).

In the company of strangers

> The city is a place of mismeetings.
>
> (Bauman 1993b, 157)

It was, perhaps, the nineteenth century that was witness to the full flowering of a series of changes initiated in the seventeenth century, when the notion of modernity first rose to dominance. For it was in the nineteenth century that the immense and varied changes wrought by modernity centred inexorably on the city (Berman 1983; Gilloch 1996).[11] As Charney and Schwartz (1995, 3) put it, 'the experience of the city set the terms for the experience of the other elements of modernity', representing the zenith of this process of acceleration and intensification. Hence, from Baudelaire (1970), to Simmel (1950b), to Engels (1973), contemporary writings on the nineteenth-century city expressed astonishment, perplexity, and often a pronounced concern over the developing conditions of modern urban life (Blanchard 1985). More often than not, such reactions owed more than a little to a powerful contrast – whether implicit, explicit, or (as in Engels' case) polemically overstated – with a purportedly more 'natural' prior mode of existence (Williams 1973).[12] Modern city life brought with it 'the rapid crowding of changing images, the sharp discontinuity in the grasp of a single glance, and the unexpectedness of onrushing impressions' (Simmel 1950b, 410), and must have seemed to represent nothing less than a fundamental and unnatural mutation of the human species. Such astonishment, however – and, to a lesser extent, such concern – showed a marked tendency to diminish as the features of city life became more commonplace. This occurred not simply as a result of an increased accumulation of material wealth in the city (responsible, for instance, for enabling the better functioning of the city) but was also the product of an accumulated wealth of everyday urban experience. The latter involved the generation of an heteronomous series of urban discourses – a veritable 'techno-cosmopolitanism' (Gregory 1994, 138) – that provided the conceptual language appropriate both to speaking of and to acting upon (in accordance with pre-specified ends) particular aspects of the city and city life. The diminution of such dispositions as astonishment and perplexity, and their replacement by Simmel's (1950b) celebrated 'blasé attitude' was itself, therefore, a process issuing from the new social relations engendered by the modern city, as urbanism became – in Louis Wirth's (1938) memorable phrase – a 'way of life'.

The kind of questions initially raised in writings on the city during the nineteenth century were undoubtedly the kind for which no answers are available.[13] Accordingly, the continued development of modern urbanism brought about an impressive ability not only to displace or defer all such questions, but also to disavow that any such displacement or deferral was underway. Modernity as a whole functioned in a manner that served to render inaudible those questions most uncomfortable to its own continued existence: 'Modernity had the uncanny capacity for thwarting self-examination … it wrapped [its] mechanisms of self-reproduction with a veil of illusions without which those mechanisms, being what they were, could not function properly' (Bauman 1993b, 3). To the extent that they were represented at all, those questions modernity would rather were not posed were portrayed as, in principle, answerable and, in practice, already in hand; about to be taken care of once and for all (typically by a higher authority, in possession of the appropriate knowledge, skills and expertise). But in spite of all this, modernity remained haunted by the spectre of ambivalence. This haunting manifested itself, most of all, in the city. The ambivalence thrown up on the reflex of modernity's own (impossible) desire to overcome ambivalence *as such* made its spectral presence felt in the fleeting, fragmentary experience of modern city life.

At the heart of Simmel's (1950b) suitably impressionistic observations on the nineteenth-century city – on its pace, rhythm and transformation of pre-existing temporalities; on its novel forms of social relation (effected and coordinated by the cognitive force of the money-form); and on its transformation of social and physical distance and space – is an insistent reference to the fleeting, fragmentary quality of modern metropolitan life. For here, in the modern city, the full solidity of presence as such seemed to give way to a kind of spectral presence – true to the sense of a haunting by something that has ostensibly been laid to rest. The ambivalence manifest in the fragmentary experience of city life came to be personified in the fleeting figure of the stranger (Simmel 1950c).[14] Hence Sennett's (1977, 39n) definition of a city as 'a human settlement in which strangers are likely to meet'. The stranger came to populate the modern city in almost direct proportion to the extent to which modernity as a whole sought to disavow the inevitability of ambivalence. For, with the new forms of distantiated social relations ushered in by modernity, the stranger's occupation of modern space was, on the one hand, entirely necessary, yet on the other hand, necessarily fraught with anxiety.[15] As Robins (1995, 45) notes, the city 'is an ambivalent object: an object of desire and of fear'.

'There are friends and enemies', writes Bauman (1990b, 143). 'And there are *strangers*' (*ibid.*). The stranger is, precisely, a 'category' defined by its imprecision: neither friend nor enemy; neither neighbour nor alien – breaching the barrier set up between the two. The allotropic stranger *necessarily* defies the parameters of modernity's ordering principle, in a manner that might appropriately be characterized as 'viscous' or 'slimy':[16] spreading across both sides of the opposition, invading both sides of the divide, straddling the boundary, and being, improperly, neither fully present on, nor fully absent from, either side. Yet, such a 'sliminess' or 'viscosity' simultaneously possesses a glutinous quality – necessary for sticking or gluing those terms modernity sought to hold together in their very opposition (Doel 1996). The possibility of the stranger is, therefore, a direct effect of modernity's vain attempt to annihilate ambivalence. It is the displacement and welling up of that selfsame ambivalence its own ordering functions sought to erase. Moreover, this displacement issued specifically from the *reconfiguration of social and physical space* intrinsic to modernity.

If the spatial constitution of earlier societies was such that physical and social distance were intimately correlated, the social form of the modern city – and the appearance of the stranger – points to the breakdown of any such simple correlation. In pre-modern societies, those whose lives were lived in close proximity in physical space were, in general, correspondingly close in social space. Likewise, those geographically distant were socially distant. Whilst such a situation could never guarantee a clear-cut division between near-by friends and neighbours, and far-off aliens and enemies, it nonetheless afforded a kind of certainty with respect to social relations (enemies close to home, for example, were at least recognizable as such: they did not bear the unpredictability of the *potential* threat necessarily borne by aliens). In the context of the collapse of the correlation between social and physical space, wrought by the forces of modern urbanization as 'remoteness changes into proximity' (Schuetz 1944, 503), the true significance of the predominance of the stranger may be discerned. For, in defiance of the boundaries that traditionally aligned physical and social space, the stranger is – seemingly unnaturally – 'socially distant yet physically close' (Bauman 1993b, 153). Thus was the ambivalence characterizing the spaces of the modern city made incarnate in the figure of the stranger.

The stranger was, as Simmel (1950c; 1978) showed, vital to the new conditions of social life modernity sought to impose, yet destined to arouse anxiety. The paragon of such a situation is to be found in the modern money system, which was both prerequisite to *and* sustained by precisely the kind of relation embodied by the stranger: 'The desirable party for financial transactions – in which, as it has been said quite correctly, business is business – is the person completely indifferent to us, engaged neither for nor against us' (Simmel 1978, 227). Equally, however, encounters with strangers inevitably take the form of anxiety-ridden 'mismeetings'; casual, fleeting, episodic encounters both born of and necessary to the modern political economy, yet irrevocably detached from any moral economy, where social and physical space are reconcilable and likely to coincide. Thus it might be said that the modern urban world was, at one and the same time, the world experienced by the stranger and the experience of a world of strangers – a world in which a universal strangehood had come to predominate (Lofland 1973).[17]

For those concerned to maintain the power of defining and sustaining order, the presence of the stranger functioned as a sign of the continued necessity of *imposing* order – seemingly the only way possible of confronting and rectifying the ambivalence that continued to pollute modern society. For if modernity heralded a situation whereby the appropriate standards for the evaluation of social life were subject to a process of differentiation, modernity's response to this – the only response consonant with its own self-representation as unified, coherent and cohesive – lay in the codifying power of the law. In practice, this led the way to an ardent ordering zeal, which sought to impose a set of norms towards which all socialization must aim – norms which, above all, abhorred the lack of order implicit in ambivalence of any form. The 'unnatural', boundary-straddling figure of the stranger posed the most serious of (*potential*) threats to the modern molar order. And hence, for modernity to maintain its (increasingly institutionalized and bureaucratized) concerns, the reinforcement of its defences against any potential breaching of its boundaries was of the utmost imperative. Social categorization, enumeration, legislation, surveillance, policing, and so on, represented the multifarious dimensions of the ordering zeal that arose in direct consequence of the need to secure the balustrades upon which modernity's continued functioning and survival depended (Roberts 1988; Ogborn 1993).[18] The process of etatization taking place, in territorial terms, at the national level, was fundamentally implicated in engendering such a range of processes: processes dedicated entirely, in the first instance, to the policing and defence of abstractly defined boundaries and abstractly erected barriers, both territorial and non-territorial. Modernity, it might be said, concerned itself with 'the harsh law of spacing' (Derrida 1976, 200), with constructing 'intellectually, by acquisition and distribution of knowledge' a clearly ordered, bounded and mappable 'cognitive space' (Bauman 1993b, 146). As Corrigan and Sayer (1981, 33) put it: 'the law is a moral topography, a mapping of the social world which normalises its preferred contours – and, equally importantly, suppresses or at best marginalizes other ways of seeing and being'.

The city provided, in microcosm, the most significant site for the encounter between the threat of ambivalence and the forces of law and order. Modernity's self-proclaimed ability to define, establish and maintain the boundaries necessary to the preservation of order was, of course, always intrinsically ideological, based upon unequal relations of power (Sibley 1981; 1995), such that the 'strong depend on the certainty of mapping' (Creswell 1997, 362), whereas the 'weak ... are left with furtive movement to contest the territorialization of urban space' (*ibid.*; cf. de Certeau 1984). In the old world at least, in the modern European city, moments of revolution, of barricades set up against the prevailing order, scatter the histories of the street. The class dimension of the modern city is, in other words, immanent to the urban order as such. Indeed, there can be no underestimation of the extent to which modern urbanization was thoroughly imbricated with the development of capitalism (Harvey 1985a; 1985b). The modern individual, prerequisite to and promoted by capitalism – initially as the 'free worker' (in the characteristic class form of the wage-slave, and thus intimately connected to the expansion of the money-form) – was one in the same figure as the stranger subject to surveillance on the modern city streets. And if capitalism was, on the one hand, an overarching system capable of fuelling the urbanization necessary to providing itself with a ready supply of labourers (housed in conditions concordant with, at least, the basic capacity to work and the reproduction of labour power) it was also, on the other hand, a system predicated on the necessity of segregating economic rationality from other, moral concerns (a process Bauman (1993b) frames in terms of *adiaphorization*: the social production of moral indifference). Thus, if capitalism was capable of generating an 'industrial reserve army' composed of masterless men who, potentially at least, posed the most significant of threats to the established order by their mere presence in the city, this was a concern beyond the economic affairs of capitalism, regardless of its directly economic causation. The amorphous, inchoate urban mass of strangers, which became the foremost object of surveillance for the modern state, was thus determined by capital, whilst being directly displaced into the arms of the state. The structural relation between state and capital was, therefore, sometimes contrapuntal, but ultimately none the less harmonious for that. Even as the state was given over to the direct repression of certain elements of modern urban society, however, the city itself was already beginning to gestate a profoundly different set of circumstances – the significance of which only fully began to realize itself a full century or so later.

The *flâneur*'s dislocation, or lost in (aesthetic) space

> To the *flâneur* his city – even if he was born in it, like Baudelaire – is no
> longer home. For him it represents a showplace.
>
> (Benjamin, cited in Frisby 1994, 94)

At the centre of the modern concern with law and order stood the stranger,
the personification of the specifically modern anxiety Bauman (1993b, 164)
names as *proteophobia*: 'The term refers to the apprehension aroused by the
presence of multiform, allotropic phenomena which stubbornly defy clarity-
addicted knowledge, elide assignment and sap classificatory grids'. That the
modern molar order remained under threat from that which it disavowed, no
matter to what extent it resolved to manage and police the cognitive spaces
of the city, serves only too well to illustrate that modernity's underlying
ordering principle was destined to generate as many, if not more, problems
than it solved. The stranger – the slimy, ectoplasmic apparition engendered
by the 'harsh law of spacing' – was the most immediate and apparent by-
product of modernity's overriding concern with boundary maintenance. If
the proliferation of the stranger was, however, for those smitten by proteophobia, the strongest possible confirmation that modernity required a
redoubling of its efforts to attain a rigidly codified social order, from an
altogether different angle, the personification of another mode of coping
with the modern city strolls into view – in the figure of the *flâneur*
(Benjamin 1973; Tester 1994).

As Shields (1994a, 61) states, 'Benjamin's *flâneur* is the inversion of
George Simmel's *The Stranger*'. The *flâneur* embodies precisely the opposite reaction to the disorientating turbulence of the modern city to those
concerns institutionalized in the law: a veritable *proteophilia*, which dis-
locates the cognitive space modernity sought to impose, tracing across it,
by way of a ludic, transversal peregrination,[19] the heterotopic contours of
an aesthetic space. As Benjamin (cited in Vidler 1993, 55) put it, 'space
directs winks at the *flâneur*.' Thus, one 'may say that if proteophobia is the
driving force of cognitive spacing – *proteophilia* prompts the efforts of
aesthetic spacing' (Bauman 1993b, 168). Accordingly, the *flâneur* – who
appears as a detached observer, a pleasure-seeking stroller on the streets of
the nineteenth-century city; as a loiterer, frittering away time – has
emerged as a 'key figure in the critical literature of modernity and urbanization' (Wilson 1995, 61): a figure embodying a disposition 'ultimately
more truthful than the zeal of the [urban] reformer' (*ibid.*, 76).[20] If the
proliferation of the stranger had much to do with modernity's preoccupation with order and ordering – with the often paranoiac insistence on the
necessity of *socialization* in accordance with the law; on the necessity of
definitive, molar social norms – the *flâneur* represents a manifestation
formulated in mirror-image terms: as a molecular counter-structure; an
aleatory and irruptive *sociality*.

Where the stranger was most commonly defined, by reformers of both left and right, but above all by the state, in relation to *work*, specifically in terms of his or her capacity or willingness to work – to which the criminalization of the urban poor under the vagrancy laws most clearly attests (Davis 1989; Roberts 1988) – the other face of the stranger, as *flâneur*, owed more to the realm beyond work: to *play*. And yet, despite the class dimension evident in the distinction, it should be noted that 'play' is defined, not on capitalism's own terms, as the simple opposite of work; it contains an excess that is irreducible to such a simple opposite, to 'leisure' (Lefebvre 1996). As Bauman (1993b) emphasizes, play is necessarily gratuitous or excessive (serving no 'sensible' or 'serious' purpose); free (in the sense that one cannot genuinely be forced or, indeed, ordered to play); and underdetermined (taking its cue from the 'unreality' or phantasy of the pleasure principle, rather than the overdetermination of the reality principle). It is not necessarily unconstrained, insofar as it accords to rules, but it is unconstrained by reality; its constraints are arbitrary and designed to allow things to happen, not to prevent them from happening. Above all, '*flâner* means to play the game of playing' (Bauman 1993b, 172):

> This play is conscious of itself as play. Its enjoyment is mature and pure. It is pure because the *aesthetic proximity* [of the *flâneur* and those caught within his gaze] does not interfere with *social distance*; the city stroller can go on drawing the strangers around him into his private theatre without fear that those drawn inside will claim the rights of … insiders.
>
> (*ibid.*, emphasis added)

As Buck-Morss (1989, 346) suggests, the *flâneur*'s guiding principle can be stated, in plain and simple terms, as 'look, but don't touch'.

Strolling through the public spaces of the nineteenth-century metropolis, the *flâneur* has frequently been regarded as embodying a visual mastery of urban space, condensed in a scopophilic gaze that is overcoded by its gender implications. The practice of *flânerie* would thus seem to exclude, categorically, the possibility of the *flâneuse* (Wolff 1990). As Wilson (1995) notes, however, the *flâneur* actually carved out a far more indeterminate existence, his apparently masterful gaze being intrinsically less sure or secure than is usually assumed. One cannot, and should not, minimize the centrality of gender to *flânerie* but, equally, the complexity of the relations between patriarchy, modernity and capitalism demand particularly careful scrutiny. Examination of these relations suggests that the gaze of the *flâneur* was not, in fact, the ultimate, indefatigable masculine gaze but, rather, a decidedly emasculated gaze. To conceive of the *flâneur* primarily in terms of the male gaze is, 'to overlook the desperation which motivates the wanderings of the *flâneur*' (*ibid.*, 73). That a number of women did – in admittedly exceptional and famous cases such as that of George Sand – engage in *flânerie*, signals something of the complexity of the *flâneur–flâneuse*. The interpretation of the *flâneur* as a voyeur, wandering the crowded urban public spaces from which the majority of women, being confined to the rigidly privatized domestic space of the home, were excluded, both misspecifies the complexity of the gendering of modern urban space, and universalizes a domesticity experienced predominantly by women of the middle classes (a misspecification that is, as Saegert (1980) notes, remarkably persistent). Moreover, as Falk (1997a, 180) points out, 'urban space offers a whole range of natural opportunities for shorter and longer halts involving different types of scopic regimes'.[21]

The public spaces of the nineteenth-century city were, of course, subject to a series of fundamental transformations, linked, as both Benjamin (1973; 1999) and Kracauer (1937) record, to the overwhelming dominance of the commodification process. The erstwhile public spaces of the city were witness to a blurring of their status by the growing prevalence of private commerce, generating such new cultural experiences as that offered, most famously, by the Parisian arcades (Benjamin 1999; Buck-Morss 1989; McRobbie 1992). In this case, as Green (1990, 25) notes, 'their spatial organization stimulated a series of links between parading on the boulevards, looking and acquiring'. Certain of these newly commodified spaces – perhaps most notably the department store (Benson 1986; Bowlby 1985; 1987; 2001; Chaney 1983; Corrigan 1997, 50–65; Laermans 1993; Lancaster 1995; Lawrence 1992; Leach 1984; Lears 1989; Miller 1981; Nava 1997; Pasdermadjian 1954; Reekie 1993; Williams 1982) were, in fact, targeted specifically towards *women* of the middle and upper classes (Benson 1994; Blumin 1989; Cross 1993; Crossick and Jaumain 1999; Domosh 1996; Rappaport 1995; Richards 1991). In these spaces, such women could themselves assume a role akin to that of the *flâneur*.[22] Moreover, as Lancaster (1995, 182) notes, men entering these establishments were likely to be ridiculed as 'molly husbands' by 'both women consumers and workers defending their "space"'. Accordingly, whilst the city offered women of the appropriate means distinctive new freedoms – and McRobbie (1997, 74) is insistent on this class dimension; women 'did not flock equally to consume' – the necessity of such separate spaces provides a clear indication that these women's experience of the city was of an overtly male-dominated environment (Blomley 1996). Equally, however, working-class women cannot be said to have been excluded from modern urban public space – notwithstanding the unequal and uneasy existence they led in relation to men, of all classes, occupying such territory. As Wilson (1995, 70) remarks with reference to working-class women, 'to read the journalism of the mid- and late nineteenth century is to be struck by their *presence* rather than their absence'. This is, perhaps, most evident in the overriding reformist concern with the presence of female prostitution (Acton 1968; Buck-Morss 1986; Walkowitz 1977; 1980; 1992). The upshot is that the modern gendering of the public and private *spheres* (Davidoff and Hall, 1987; Wolff 1990) did not map at all neatly or simply onto public and private urban *space*. To the contrary, the increasingly commodified urban arena was charged with a tense and equivocal sexualization, underpinned by a profound ambivalence: 'consumer culture emerges as a kind of liminal space which forged new links between public and private; ... it bridged the nineteenth-century separation of these spheres to situate consumption more precisely in the moral geography of the middle classes' (Tiersten 1993, 138–139). The lack of any clear, unambiguous *spatial* demarcation between public and private spheres was wont to heighten the dangers of the city for women, particularly those for whom the city was unfamiliar, as the following extract from the *Saturday Review* of 1862 suitably illustrates:

A 'Paterfamilias from the Country' makes his appearance in London, and sends his two daughters, young, lovely and guileless, on a shopping walk. Unacquainted with the moral geography of the West End, they innocently trip down the tabooed side of Regent Street. The natural consequence follows. A young gentleman of amorous disposition, seeing them there, upon the equivocal ground, solitary, sauntering and attractive, comes to the conclusion that they had rather be looked at than not, and begins to ogle them accordingly.

(Anon. 1862, 125)

Thus it was that a pronounced moralizing subtext to the ordering zeal of the urban reformer concerned itself with the presence of certain 'types' of women in the public spaces of the nineteenth-century city (Ferguson 1993). This contradictory situation reflected the hypocrisy of bourgeois male Victorian sexual attitudes – insofar as such men's concerns to render the public spaces of the city safe for their own wives and daughters were wholly incompatible with the frequent exploitation of working-class women by many such men – which took place within the domestic spaces of the home as well as the surrounding public spaces of the city (Barret-Ducrocq 1991; Trudgill 1973). The overriding reformist concern with prostitution can, however, be read as revealing, obliquely, the true status of the *flâneur*.

The presence of the prostitute in the nineteenth-century city came to be cast, in very clearly patriarchal terms, as a metaphor for the loss of nature; for the widespread breakdown of the 'natural order' that the modern city was seen as both manifesting and further accentuating. But, whilst the characterization of the prostitute as an 'unnatural type' signals the degree to which the presence of women in the city was constituted, by the bourgeois masculine discourse of the urban reformer, in terms of a retrospective naturalization of a more rigidly patriarchal prior social order, the *flâneur*'s experience of the city was equally framed in relation to such perceptions. Ryan's (1994, 49) remark, that whilst 'the city was the birthplace of a new set of values constituted within modernity and capitalism, it was also the arena for the reformulation of a pre-capitalist patriarchal culture', applies across the entire spectrum of men forging their lives and identities in the nineteenth-century city – but the *flâneur* specifically is marked by a far more contradictory nature than Ryan's assessment allows.

Of particular significance to the *flâneur*'s disposition is the fact that he occupied a paradoxical class position. As Wilson (1995, 63) notes: 'He is a gentleman … yet he is subtly *déclassé*, and above all he stands wholly outside production'. And whilst the role of the *flâneur* 'was open to one narrow segment of the population only, educated men … it … often led to poverty and obscurity' (*ibid.*, 72). Many of those *flâneurs* indulging in the conspicuous consumption synonymous with the pleasure-zones of the modern city carved out an uneasy existence as artists, writers and journalists, possessing 'the attitude of men who have been bought: while critical of and opposed to the philistinism of bourgeois society, they were paid to entertain it' (*ibid.*, 64). Their position was one which 'rejected conventional society, yet … [was] financially dependent on it, and as a result their attitude towards society was cynical or ironic rather than passionately and committedly oppositional' (*ibid.*). The *flâneur*'s experience of the modern city – which, in common with other nineteenth-century urban experiences, amounted to a fragmentary and episodic affair – resulted, for the *flâneur* as much as for the reformer, in a profound nostalgia; in a sense of loss for the *epic* quality of life retrospectively attributed to an earlier mode of existence. But, for the *flâneur*, the vertiginous turmoil of the modern city was marked by an overbearing sense of impotence.

The aestheticized, ludic existence pursued by the *flâneur* represented, above all, a reaction to, and a mode of coping with, the otherwise disorientating, agoraphobic spaces of the city. Indeed, the impotence experienced by the *flâneur* was etched directly into his peregrinations. According to Benjamin (1985, 40), the 'path of someone shy of arrival at a goal easily takes the form of a labyrinth'. So it was that the aesthetic spacing produced by the wanderings of the *flâneur* took on a labyrinthine form. The *flâneur* perpetually *lost himself* in the aesthetic spaces of the city, his peregrinations interminably deferring his satisfaction and contributing to the melancholy of his existence. As Wilson (1995, 75) suggests, in 'the labyrinth the *flâneur* effaces himself, becomes passive, feminine'. Such is the *flâneur*'s predicament in relation to the threat to the established patriarchal order posed by modern urbanization.

In a sense, therefore, the *physical* proximity of the *flâneur* to the prostitute in the public spaces of the modern city suggests a kind of uneasy (and undoubtedly unequal) parallel between their respective positions in *social* space – a parallel occluded in theorizations of the male gaze that simply accept its self-asserted mastery on its own terms. In the light of the virulent commodification beginning to saturate the public spaces of the city, prostitution had also come to serve as a metaphor for the omniscient presence of the commodity-form and the resultant commercialization of urban space.

Hence, to the *flâneur*, it seemed as though the 'whole of society was engaged in a sort of gigantic prostitution; everything was for sale and the writer was one of the most prostituted of all since he prostituted his art' (Wilson 1995, 73). Whilst there is, clearly, an immanent danger of romanticization with respect to portrayals of the *flâneur* as a wholly tragic figure, the respective existences of *flâneur* and prostitute were both ultimately stoical ones; both pinning their hopes not, self-evidently, in terms of the discourses of urban reform – but not in strict opposition to them, either. Rather, their lives took the form of a continued daily survival against the odds that modernity set.

So it was that the animated public spaces of the modern city, populated by those strangers objectified by the surveillance of the agencies of the state, came to represent a dramatized playground to the *flâneur*, a figure who exercised his ultimately defeated, emasculated gaze over the urban spectacle, cathecting the strangers around him in accordance with his own aestheticized vision of the world. The corollary of this dislocated, aestheticized space was a profoundly disjunctured, aleatory experience of life, which promoted a nostalgic longing for times past, a sense of melancholy for a lost world. The *flâneur* was, therefore, as Wilson (1995) insists, never more than a fleeting, spectral figure – ultimately an impossible, invisible, phantasmagorical figure – stoically inventing himself (and occasionally herself), and forging a seemingly heroic existence in the face of a vertiginous and disorientating urban world. As Buck-Morss (1989, 346) has suggested, however, the *flâneur* is perhaps 'more visible in his after-life than in his flourishing', his living on by other means ensuring that a myriad of subsequent cultural forms, 'no matter how new they appear, continue to bear his traces, as ur-form'. In effect, the proteophilia embraced by the *flâneur* provided for a new kind of aesthetic spacing – the contours of which were to become as intimately connected with the postmodern city as they were with the modern metropolis.

Seduction, repression, and consumption

> The system can do without the industrial, productive city, the space-time of the commodity and market-based social relations. ... It cannot, however, do without the urban as the space-time of the code and reproduction.
>
> (Baudrillard 1993a, 78)

As the primary locus of production, the modern city represented a world constituted in accordance with the law. To this extent, it relates to the original sense of production: 'not in fact that of material manufacture; rather, it means to render visible, to cause to appear and to be made to appear: *pro-ducere*' (Baudrillard 1987c, 21). However, within the space-time framework of modernity, for something to be made to appear, something must be forced to disappear. Accordingly, the modern city was, simultaneously, the primary locus of a foreboding anomie; of conflict, crisis, violence, madness, revolution and transgression; of everything that departs from the jurisdiction and margin of the law. Hence the systemic necessity of the active surveillance and repression characteristic of the modern city. In a dimension transversal to all of this, however, the *flâneur*'s peripatetic existence was predicated not in terms of any margin that might be transgressed but in a tangential relation to the molar order. The *flâneur*'s world was always already heterotopic with respect to modernity's cognitive social space. Indeed, the *flâneur* spontaneously redefined the institution of the law into the rules of an unremittingly reflexive game. For the *rule* is that which does away with both the law *and* its opposition/transgression – that which moves to a register beyond the law. As Baudrillard (1990b, 131–132) notes, whilst the law might be defined as that which it is possible to transgress, 'it makes no sense to 'transgress' a game's rules; within a cycle's recurrence, there is no line one can jump (instead, one simply leaves the game)'.

However, at a certain juncture in space-time, though doubtlessly irregularly and unevenly, 'the *flâneur*, through seduction, was transformed into the consumer' (Bauman 1993b, 173). The gaze of the *flâneur* – perhaps, prototypically, of the *flâneuse* (or is this merely another version of the myth of Eve or Pandora?)[23] – was, in effect, captured; or, more precisely, bought; by the spectacular display offered in the commodified spaces of the city. And herein lies the ultimate truth of the *flâneur*'s gaze. For the emergent captains of consumerism, the 'right to look gratuitously was to be the *flâneur*'s, tomorrow's *customer*'s, reward' (Bauman 1993b, 173). There was money to be made in the *flâneur/flâneuse*'s gaze, and it was this commodified specularity that came to provide the model that would henceforth irradiate social space: in 'the new consumer society, we are observing the (re)birth of *homo aestheticus*', writes Maffesoli (1997, 23). With the transmogrification of the world of the *flâneur* into the world of consumption, 'the miraculous avatar of the commodity into the shopper is accomplished. … It is no more clear what (who) is the object of consumption, who (what) is the consumer' (Bauman 1993b, 173–174). Consider, for example, Céline's (1988, 188–189) sudden recognition of such a situation in a 1920s New York diner:

> if they showered the customers with so much light, if they lifted us for a moment from the habitual darkness of our condition, there was method in their madness. The owner was up to something. … [W]e were being watched through the window by the queue of people we had just left in the street. They were waiting for us to finish eating so they could come and take our tables. Actually that was the reason, to keep up their appetite, why we were so well lit and displayed so prominently; we were living advertisements.

In such a world, 'the same homogeneous space, without mediation, brings together men [*sic.*] and things – a space of direct manipulation. But who manipulates whom?' (Baudrillard 1994a, 76).[24]

If the seduction of the city as a game was the *flâneur*'s remittance from the cognitive ordering of the law, the particular transformation the system has since undergone accomplishes nothing less than the extension of the role of *flâneur* to the whole of society. The postmodern city has witnessed the re-appropriation of the *flâneur*'s earlier ex-appropriation of the spaces of modernity, transforming it into a managed playground (Bauman 1993b). The situation accords to the logic of *seduction* (from the Latin '*se-ducere*: to take aside, to divert from one's path' [Baudrillard 1990b, 22]), rather than the forced realization implied by production and the law. Modernity has taken its own ironic revenge, distending the law and adsorbing it onto the surficial curvature of an asymptotic norm (Clarke and Doel 1995; Doel and Clarke 1998). As Baudrillard (1990c, 164) notes: 'The law is an *instance*,[25] whereas the norm is a curve; the law is a transcendental, whereas the norm is a mean'. In the transition from production to seduction, *anomie* becomes reinscribed and dispersed within the molecular, statistical field of *anomaly* – 'a field so normalized that abnormality no longer has a place in it' (Baudrillard 1990a, 26). Accordingly, today, there is no line one can jump, no law to transgress. In contradistinction to the *flâneur*'s earlier, pure, ludic existence, however, the play the system's mutation has come to instigate is, above all, controlled, in a cybernetic sense: it is operational in accordance with the *code*.[26]

Repression, then, is no longer the dominant form of social control, as the result of a levelling process that constitutes subjects first and foremost as consumers. We henceforth find ourselves in a situation regulated by the aleatory play of the code, where every 'sign and every message ... is presented to us as a question/answer' (Baudrillard 1993a, 62); where everything takes the form of a perpetual testing. This is the general form administered by the market, which is increasingly marked by its capacity to provide commodified 'solutions' to the problems of everyday life (even its capacity to gestate the 'problems' for which 'solutions' are necessary is no longer presented as a secret, waiting to be uncovered). Whilst all of this ensures for the continuing realization of value on the part of capital, it also, ultimately more importantly, marks the imposition of a distinctive new mode – or code – of social integration. We have entered the phase of the 'lottery society', governed by the seductive fascination of an indeterminate and indifferent play of aleatory forces. 'The whole system is a kind of realization of indifference' (Baudrillard 1995a, 91).

This process involves the adsorption of the preceding order, and its transposition to the realms of hyperreality. Consequently, the law is not simply done away with. It does not effect an end of a mortal kind but lives on, in a kind of ghostly existence or second coming – in accordance with a *fractal* rather than a *fatal* mode of disappearance.[27] 'Even repression is integrated as a sign in this universe of simulation. Repression become deterrence is nothing but an extra sign in the universe of persuasion' (Baudrillard 1994a, 76). None of this is to say, therefore, that the ghosts of class do not continue to stalk the postmodern world. Nor is it to say that a radical process of individuation has served to dissolve, willy-nilly, the existence of supra-individual groupings or collectivities *per se*. Class is itself subject to a metastatic proliferation, which sees the progressive disappearance of the possibility of any *authentic* meaning for class or class struggle (particularly in the proliferation of 'staged authenticity'). Class continues to proliferate along new axes, defined, increasingly, in relation to cultural as well as economic capital (Bourdieu 1984) – whilst the burgeoning (and classically impure) middle classes attest to the extent to which the social order has become imbricated with consumption. Likewise, the irruption of an effervescent, molecular counter-structural sociality – the defining characteristic of Maffesoli's (1996) 'tribes' – is similarly the consequence of the metastasis attacking the oppositional clarity of the law. The transition from a situation requiring an active, centralized force of repression to a situation governed, in accordance with a statistical norm, by a passive and decentralized force of seduction is, by implication, the transition to an era of 'instant collectivities', which possess the transitory effervescence of the urban crowd without the necessitating the co-presence of its members in physical space (Shields 1992). The neo-tribe is a phenomenon that short-circuits and scrambles the differences between physical, cognitive, aesthetic, and moral spaces in a play of differences that is itself indeterminate and indifferent.

It is in this sense that the structural revolution accomplished by the system pulls off a feat of adsorption rather than eradication with respect to the modern social order. The seduced of society have been granted a kind of freedom, no longer being coerced into compliance with the law, on the implicit understanding that it will be in their own self-interest to play by the rules of the consumer society; that they will, however cynically, accept and internalize the belief that this society is suited to them, and therefore they to it. For, indeed, the market does possess – for those who are not dispossessed – a certain 'democratic' character, unlike all other social systems that have attempted to tackle the same basic problem. As Baudrillard (1995a, 92) puts it:

> Everyone is equal before the rule. We are not equal before the law, of course. The law is a principle of fundamental inequality. But everyone is equal before the rule because it is arbitrary. So there we find the foundations of a true democracy ... though not by any means of the usual political type.

The market is a system in principle – if only in principle – open to everyone. It is a system of sheer indeterminacy, based on neither productive meritocracy nor any anterior mark of exclusivity (such as inherited wealth or status). As if to reinforce the latter point, success in this system can buy into the nominal trappings of inherited status at the level of simulation: the market in landed titles, for instance, is both popular and profitable. Thus, when Baudrillard (1993c, 19) suggests that, today, 'the market ... is beyond good and evil', he is not writing as an apologist of the market (as some have presumed). Rather, he is pointing to potentially the most fundamental transformation in the long history of the market, the distant goal toward which it has always been aimed.

To all intents and purposes, accepting the freedoms offered by the consumer society amounts to an acceptance of the rationality, efficiency and providence of the market – which entails an acceptance of the importance of money, and all that obtaining it involves: above all, earning it; though obtaining it speculatively, inheriting it, accepting it in the form of credit, and so on, are more than acceptable. And, of course, attaining it illegitimately does not necessarily taint the notoriously colourless money thus obtained. The acceptance of this freedom, and the concomitant acceptance of the emphasis that needs to be placed on money and its acquisition, are mutually reinforcing self-orientations. For money, too, is a commodity, and it would therefore follow that the problem of attaining money is, precisely, a problem also best (that is, most 'democratically') dealt with by the market (specifically, by the labour market: a very particular market, to be sure, but a market nonetheless).

All of this follows directly from the widespread acceptance that the present system is the one most capable of guaranteeing personal freedom (Bauman 1988b). Faith in consumerism, the market, and the pursuit of money amounts in principle – and again it is the principle that is important – to an acceptance that this provides the only possible freedom from repression (even if this is a decidedly 'majoritarian' solution): 'Seduction versus terror. Such is the wager, since no other exists' (Baudrillard 1993a, 183). Seduction amounts to a 'package deal', the only available freedom from the residual 'second world' of repression – even if, for the silent majority, it amounts to the acceptance of money as a necessary evil, incurring the cost of trading one's labour-power on the market. 'What makes money so terribly attractive and prompts people to try so hard to obtain it, is exactly the possibility of buying oneself out of this second world' (Bauman 1987, 169). Failure to accede to the requirements of the consumer society constitutes the circumstances deemed necessary for the continued repressive enforcement of the law: as Bauman (1995, 204) puts it, 'yesterday's underdogs were non-producers, while today's underdogs are non-consumers'. The parallel worlds of seduction and repression that together constitute the consumer society are entirely mutually dependent: 'The two parts can only exist together – and only together can they be eliminated' (Bauman 1987, 168). They are the double face of the selfsame structuring principle.

Today, for the inhabitants of the consumer society, dependence on the market is more pronounced than ever. Our faith in consumption has reached the point where our own fragmented identities – cast adrift from both pre-modern communal and modern class-related positions – are themselves subject to the provision of a 'market solution' (which is not, of course, to say subject to solution by the market: quite the contrary). As Bauman (1990a, 102) puts it: 'Through the market one can put together various elements of the complete identikit of a DIY, customized self'. This is the direct consequence of a structural revolution that has set in motion the aleatory play of the forces of simulation. And if the dispossessed can readily perceive themselves as the victims into which they have been made, this is only to the advantage of the system. The spiralling rate of property-related crime (or at least media fascination with rising crime-levels), instigated broadly against 'the seduced' (or at least the trappings of consumerism: videos, television sets, mobile phones, etc.), indicates an intimate, sporadic, and aleatory connection between the two worlds on the level of social practice, which, whilst driven by (and attesting to) the seductive powers of consumerism, provides the vestiges of the law with perhaps its most sharply defined legitimation ever – as a public relations machine. And yet it should also be noted that the fundamental indeterminacy of the system has simultaneously given rise to an increase in (concern over) middle-class shopping disorders, which, notably, also find their ancestry in the nineteenth-century emergence of urban consumer culture (particularly in the medicalization rather than criminalization of conditions such as kleptomania: Abelson 1989; Camhi 1993; Schwartz 1989).

The equation of consumption with freedom is, then, a thoroughly reductive equation. Whilst it appears to pertain directly to the pleasure and well-being of society and its members, it is, in fact, a mode of ensuring for the reproduction of its own conditions of possibility. To repeat: 'As a social logic, the consumption system establishes itself on the basis of a denial of enjoyment. Enjoyment no longer appears there at all as finality, as rational end, but as the individual rationalization of a process whose ends lie elsewhere' (Baudrillard 1998a, 78).

Conclusion: towards the posturban question

> Little by little social consciousness ceased to refer to production and to focus on everyday life and consumption. With 'suburbanization' a process is set in motion which decentres the city.
>
> (Lefebvre 1996, 77)

> From thirty kilometers all around, the arrows point you toward these large triage centers that are the hypermarkets, toward this hyperspace of the commodity where in many regards a whole new sociality is elaborated.
>
> (Baudrillard 1994a, 75)

Modern social life was born of the city and its reconfiguration of social and physical space. The postmodern city amounts to its posthumous continuation, its fractal form. The postmodern city is 'less an identifiable city than a group of concepts – census tracts, special purpose bond-issue districts, shopping nuclei' (Pynchon 1996, 14). Among the models of the 'posturban' space this entails is the hypermarket. As Baudrillard (1994a, 77) suggests, 'the role of the hypermarket goes far beyond "consumption", and the objects no longer have a specific reality there: what is primary is their serial, circular, spectacular arrangement – the future model of social relations'. The specificity of the object thus becomes generalized and reinscribed across postmodern society as a whole:

> objects are no longer commodities: they are no longer even signs whose meaning and message one could decipher and appropriate for oneself, they are *tests*, they are the ones that interrogate us, and we are summoned to answer them, and the answer is included in the question.
>
> (*ibid.*, 75)

Henceforth, we will have been subject to a perpetual testing by the system, and whilst market research, electoral polls and so on are quite candidly nothing but this (Clarke and Doel 1995), the entire sphere of consumption possesses the character of an indefinite referendum, constituting the 'space-time of a whole operational simulation of social life' (Baudrillard 1994a, 76).

> We live in a dark and fearful time, a time of polls and ratings and market research. ... They take polls, do research. They find out what people want, and give it to them. *This doesn't work*, since when people find their alleged ideas thrown back at them ... these ideas are somehow *changed, diluted*, not what they meant at all. They already *know* the stuff, it's the same stuff they told the market researcher, *it's boring* ... *they're finding out what they already know*.
>
> (Heimel n.d., unpaginated)

In such a situation the role of the hypermarket is revealed as a model of the end of the social, of the 'reinscription of contradictory fluxes in terms of integrated circuits' (Baudrillard 1994a, 76). We have been witness to a morphogenesis of a cybernetic kind. Whereas, once, the space of the city was subject to the mechanical control of an earlier social order – workers 'absorbed and ejected at fixed times by their work place', regulated 'by a continuous rational constraint' (*ibid.*, 75–76) – today this takes place in accordance with 'scenarios of molecular control that are those of the genetic code' (*ibid.*, 77). Such codes are, moreover, spontaneously reintegrated into the system by the computer modelling of shopping centres and their indeterminate, fuzzy 'functional' regions (Clarke 1995).

Whereas, in relation to the initial phase of the development of consumerism, with the department store at its cultural and commercial zenith, 'cities remained cities', 'the new cities are *satellized* by the hypermarket or *shopping center* ... and cease being cities to become metropolitan areas' (Baudrillard 1994a, 77). This nuclear, satellitic form irradiates social space: for example, 'the hypermarket ... establishes an orbit along which suburbanization moves' (*ibid.*). Such a morphogenesis reveals the hyperreal status of such models. For the function of the hypermarket has nothing whatsoever to do with the modern functionalism of retailing. Such 'forms' are of the order of 'functional nuclei that are no longer at all functional' (*ibid.*, 78). They are dedicated to creating a space of control that has taken its departure from the classical regime of surveillance which, in line with an aesthetic of disappearance, has itself become integrated, ever so lightly, into the interior design of the shopping centre; the closed-circuit TV camera becomes part of the 'decor of simulacra' (*ibid.*, 76).

Such '[n]egative satellites of the city ... translate the end of the city, even of the modern city, as a determined, qualitative space, as an original synthesis of society' (*ibid.*, 78). The modern city, the 'city historically constructed is no longer lived and is no longer understood practically. It is only an object of cultural consumption for tourists' (Lefebvre 1996, 148).[28] The experience of the postmodern, posturban world results from an experience of an aleatory transformation of space-time, which emerges as presence mutates into telepresence.

> From the town, as theatre of human activity with its church square and market place bustling with so many *present* actors and spectators, to CINECITTA and then TELECITTA, bustling with *absent* televiewers, it was just a short step through that venerable urban invention, *the shop-window*.
>
> (Virilio 1994, 65; cf. Clarke 1997b; Friedberg 1993)

It is this capacity for the system to mutate in response to its own contradictory conditions of possibility, and to reinscribe space-time in its own simulacral image, which entails that all future understandings of the city must proceed to ex-appropriate it as an impossible object. The 'city as the haunting ground of the *flâneur* turns into the *telecity*' (Bauman 1993b, 177). Towns become 'window-towns ... that have the paradoxical power of *bringing individuals together long distance*, around standardised opinions and behaviour' (Virilio 1994, 65). 'Strangers might now be gazed at openly, without fear. ... They are infinitely close as objects; but doomed to remain, happily, infinitely remote as subjects of action' (Bauman 1993b, 178). We have truly witnessed the sophistication of the city; its transposition to the realms of hyperreality.

4   Seduced and repressed

Collective consumption revisited

with Michael G. Bradford

Introduction

For a time, in the 1970s and 1980s, so-called 'collective consumption' – the non-market provision of goods and services to correct certain persistent and widespread market failures – became one of the central concerns of urban studies (Castells 1977). For many, the evident need for collective provision emanating from the heart of the Western city exposed the limits to capitalist urban development, holding out the hope of an entirely different form of sociopolitical organization – even if, at the same time, collective consumption could just as easily be thought of as a means of perpetuating capitalism, however much it benefited members of the working class. This entire situation, however, was to witness a dramatic reversal, as one state of affairs rapidly gave way to another. By the mid-1990s, Dowding and Dunleavy (1996, 43) were lamenting the fact that 'the collective consumption approach has not continued to develop in a unified way since the mid-1980s, and has shown some signs of overall loss of direction and vigour'. Even at the time, this was something of an understatement. With the benefit of hindsight, it seemed increasingly evident that collective consumption amounted to little more than a transitory phase on the road towards a fully fledged, market-driven form of individualized consumption (Saunders 1986a [1st edn 1981]). Ever since, the concerns raised by the debate over collective consumption have slipped from the forefront of academic and political attention. Of course, a basic concern with the urban-political processes shaping consumption opportunities within the city – and, with them, whole ways of life – has not entirely vanished from the urban research agenda. Still less have such issues vanished from the city itself. Nonetheless, these matters appear to be commanding less and less attention, as other avenues of academic enquiry have opened up, and as urban policy has adopted an increasingly neo-liberal stance.

In the academic sphere, a number of factors conspired to remove collective consumption from its formerly central position. To begin with, the promise of an overarching framework for urban theory that was widely read into the notion of collective consumption never quite materialized. The intense debate the concept inaugurated ultimately generated as much heat as light, with the detractors eventually gaining the upper hand in showing that the specificity of the 'urban' could not be convincingly defined in terms of collective consumption. Moreover, with the advent of the 'cultural turn' in the social sciences, other approaches to consumption forced their way onto the research agenda. Ultimately, though, it was not academic fad and fashion that caused the loss of momentum with respect to collective consumption. The single most important factor lying behind the shift away from the topic was the erosion of the distinction between the 'public' and the 'private' in the political sphere. Once common suggestions that the whole question of public and private consumption is amenable to rational consideration appear, today, both remarkably optimistic and incredibly anachronistic (Thompson 1977). In the course of a few short decades, the ideological valencies of the terms 'public' and 'private' have shifted almost beyond recognition, with numerous developments combining to give 'considerable force to the view that "collective" and "private" consumption are increasingly hard to distinguish' (Dowding and Dunleavy 1996, 37).[1] A parallel series of 'blurrings' has been witnessed in relation to public and private *space* (one thinks, for example, of the shopping mall); public and private *sectors* (particularly with respect to the notion of 'public/private partnerships'); and also, perhaps, in terms of the gendering of the public and private *spheres* (Glennie and Thrift 1996). Such a situation makes for an increasingly complex picture (Garmaniko and Purvis 1983). Yet despite this, the distinction between 'collective' and 'private' consumption arguably remains as pertinent, if also as problematic, as ever. As the work of Zukin (1990; 1995; 1998) attests, the basic tension between private economic interests and social justice – the selfsame concerns that underpinned debate over collective consumption – remains one of the most potent factors affecting the contemporary urban landscape. In this light, the present chapter seeks to explore the extent to which the dimensions of public (collective) and private (market) consumption might yet prove vital to conceptualizing the city today.

Past deliberation over public/private consumption, can be recognized, with the benefit of hindsight, as tracing the lines of a fundamental change in the figuration of society. This argument is developed, initially, through a reconsideration of the seminal contributions of Castells and Saunders on urban consumption. The divergent positions these authors marked out, we argue, might be seen not so much as alternative formulations of broadly the same set of issues, but as unwitting characterizations of distinctively different social conditions – both of which saw the city take on a central role. Castells' theorization can be seen, with hindsight, to have involved a conception of the city under peculiarly modern conditions; whilst Saunders' account, albeit obliquely, began to outline the conditions specific to the postmodern city. Accordingly, the second section of the chapter reconsiders the specificity of private and public consumption under putative postmodern conditions, rejecting Saunders' formulation, and accepting the broad thrust of the critique attracted by his work. It develops an alternative theoretical frame-work, drawing on the work of Bauman (1998a). In this light, the case of housing – which consistently formed a central aspect of the collective consumption debate, and came to represent 'the most examined area of consumption' (Pennance 1977, 145) in urban studies – is briefly revisited. The intention is not to use this particular example to attest to the veracity of the theoretical account. It represents a more modest attempt to illustrate the insights that such an account is capable of revealing. Taken together, these theoretical and empirical observations serve to demonstrate that the issues first brought to light by the collective consumption debate remain vital to urban theory. Without further ado, therefore, let us begin with a retrospective look at the urban theory of the 1970s and 1980s.

Genealogies of urban consumption: the 'new urban sociology' and the city

Castells' (1977) immensely influential neo-Marxian reformulation of the 'urban question' was intended as an intervention into a body of urban studies that Castells (1976a; 1976b) saw as profoundly ideological. Isolated from the subsequent development of the 'new urban sociology' that it almost single-handedly served to initiate, the theoretical edifice Castells (1977) constructed still houses what has since been distilled as its central claim – that the process of collective consumption is definitive to the shaping of the city under conditions of advanced capitalism – with an incontrovertible (if, by now, rather dated) style. There can be little doubt that Castells' specific conception of the late-capitalist city was subject to a fairly damning critique, yet there is also a sense in which Castells alighted upon a profoundly timely formulation.[2] Whilst the prominence of his argument might have been short-lived, it caught and reflected a telling glimpse of the city at a specific conjunction, thus setting the stage for a highly important strain of writing on the city.

'The "urban" seems to me', mused Castells (1977, 236), 'to connote directly the processes relating to labour-power other than in its direct appli-cation to the production process'. Provision by the state of 'means of consumption' that could not be assured by the operation of the market (since they commanded 'lower than average profit rates' [*ibid.*, 460]) served to ensure nothing less than the survival of capitalism. Collective consump-tion was strictly 'necessary to the reproduction of labour power and/or to the reproduction of social relations' (*ibid.*). Although some orthodox Marxists demurred (Lojkine 1976), Castells (1977, 461) held that the provi-sion of the means of collective consumption 'plays a fundamental role in the struggle of capital against the tendency for the rate of profit to fall'. Since ventures that individual capitalists would, in the absence of solvent demand, fail to take on were effectively redeemed by the welfare state, collective consumption helped 'to raise proportionately the rate of profit attributed to social capital as a whole' (*ibid.*). The overall balance of tendencies and counter-tendencies determining the trajectory of the rate of profit had become, for Castells (1977), intimately related to the development of the city itself. Needless to say, such a conception afforded the 'urban question' an unprecedented centrality:

> The urban units ... seem to be to the process of reproduction what the companies are to the production process, though of course they must not be regarded solely as *loci*, but as being the origin of specific effects on the social structure (in the same way, for example, as the characteristics of the company – production unit – affect the expression and forms of the class relations that are manifested in them).
>
> (*ibid.*, 237)

Castells' (1977) overarching vision served to revitalize urban theory as a whole. His particular formulation was frequently diluted, in the numerous repudiations of his argument; indeed, all manner of definitional problems were endlessly rehearsed in the literature. The important point, however, is that the relational logic of Castells' formulation cannot – without losing sight of his central point – be reduced to the kind of pronouncement on the characteristics of collective consumption that was contained in the majority of the attacks on the overall coherence of his conception. As Preteceille and Terrail (1985, 127–128) suggest, this whole issue represents

> a real difficulty, which cannot be resolved so long as the search continues for criteria that can be used to classify different concrete processes of consumption so that each one may be unequivocally described as less or more collective. ... [O]n the contrary ... what is needed is an analysis of the social relations

both directly involved and indirectly implicated in the 'collectivization' or 'socialization' of the means of consumption. Dowding and Dunleavy (1996, 49) raise much the same issue, which they extend to include the neo-classical economic concept of 'public goods':[3]

> The characterization of goods and services as collective or private is not one of strict necessity (true in all possible worlds) but is place- and time-specific. A given good, 'correctly' identified as collective in one possible world, may be 'correctly' identified as private in a different possible world because of features which are not strictly characteristics of the good itself.

Commenting on the 'non-excludability' characteristic of public goods, Dowding and Dunleavy (*ibid.*, 62n) stress that 'excludability is an "economic" rather than a merely "technological" phenomenon'. Thus, whether or not a park, for example, could hypothetically become a private rather than a public good 'depends upon the technical possibilities of monitoring entrants, average wage costs of employing park wardens, and the relative wealth of the local population', *inter alia* (*ibid.*, 49). To return to Castells' account, the crux of the matter relates to the 'possible world' of which his city formed a part.

For Castells, under conditions of advanced capitalism, the city attained its specificity as a result of the way in which the processes ensuring the expanded reproduction of labour-power operate through, and simultaneously structure, space: the production of the late-modern city is internally related to the dynamics of capitalism. The socialization of the means of consumption, one might say, possesses a kind of elective affinity with the city (a position not too far removed from Saunders' [1985] adjudication on the matter). The point of greatest import, however, concerns the question of state involvement in matters otherwise (at other times, and in other places) dealt with by the market (Dunleavy 1980; Preteceille 1986; Harrison 1986). For Castells, collective consumption amounted to a functional necessity for the reproduction of capitalism, however much it might have benefited individual members of the working class. Here, however, the disputes over the substance and definition of collective consumption reveal the problematic nature of the functionalism Castells' account embodies. For Lojkine (1976, 121), for instance, collective consumption 'refers ... to the totality of medical, sports, educational, cultural and public transport facilities' provided by the state. Conspicuous by its absence from this list is public housing, which amounted to one of its principal components for Castells. Lojkine excludes it on precisely the same basis that Castells includes it: its effect on the overall rate of profit (with Lojkine drawing precisely the opposite conclusions to Castells). The implication is that Castells effectively sought out a mechanism to fit his argument, which suggests the need for a far more engaged way of elucidating the actual mechanisms at work (Saunders 1979; 1986a; Dunleavy 1980; Mingione 1981). In effect, Castells was accused of deploying eloquent theoretical reasoning on a matter that called for something other than theoretical eloquence (Pahl 1978). Nonetheless, the notion of a relatively autonomous state, responding to systemic contradictions (Poulantzas 1973), forcefully underpinned the idea that the distinction between privately and collectively provided goods was not the result of any inherent qualities of particular goods or services themselves, but was achieved through historically and geographically specific incidences of social contestation in an increasingly politicized city (Theret 1982; Pickvance 1976; 1977). Once again, however, assumptions about collective interests, collective consciousness and collective action tended to lean on theoretical rather than empirical support.

Here, one might usefully consider the sociogenesis of the conditions Castells addressed. Much of the collectivization experienced in Western Europe followed directly from the experience of the two World Wars.

Notions such as 'homes fit for heroes' pointed, for the first time, to an explicit concern with social justice that had been largely absent in the preceding period of market capitalism. Such a shift led to the beginnings of a situation where the possibility of state intervention was increasingly accepted as a valid form of social and economic regulation. This was, of course, most pronounced, in Europe, in the planned recovery following the Second World War, and helped, at least in part, to establish the 'consumption norms' intrinsic to the development of a Fordist regime of accumulation (Aglietta 1979). For Aglietta (*ibid.*, 159), Fordism was largely 'governed by two commodities: the *standardized housing* that is the privileged site of consumption; and the *automobile* that is the means of transportation compatible with the separation of home and workplace'. Clearly, the US and Western European experiences differed starkly in terms of the collectivization of these means of consumption but, nonetheless, much of Castells' work can, in hindsight, be seen as referring to the mode of economic regulation specific to this period of capitalist history, specifically in relation to the welfare state. In retrospect, what was most significant about the expansion of collective consumption, and the role of the welfare state *per se*, was the ease with which it embodied evidently contradictory positions. To the obvious question: was the welfare state 'an agency of repression, or a system for … mitigating the rigours of the free-market economy? An aid to capital accumulation and profits or a social wage to be defended and enlarged … ? Capitalist fraud or working class victory?' (Gough 1979, 11), the paradoxical answer is, 'both'. The extent to which this overdetermination was destined to represent a fundamentally unstable resolution is far easier to appreciate with the benefit of hindsight than it was at the time. Even so, Castells did show an acute sensitivity to the way in which collective consumption involved the *displacement*, rather than the resolution, of the systemic contradictions made manifest in the capitalist city, offering a perceptive diagnosis of the way in which capitalism as a whole tends to operate (cf. Harvey 1982).[4] A case in point concerns the manner in which collective provision engendered the conditions that led to the fiscal crisis of the state (O'Connor 1973; Fine and Harris 1976; Gough 1975).

The general train of Castells' argument, therefore, captured and reflected an important, though conjuncturally specific, process, which saw significant levels of state provision make their mark upon the city. In parenthesis, although this anticipates an argument yet to be made, it is worth quoting Bauman's (1999b, 184) remarks on the subsequent collapse of this conjunction and the demise of the welfare state, since it makes clear the factors responsible for the conjunction in the first place:

> The welfare state was a product of a unique historical conjuncture and there is nothing to keep it afloat once the 'overdetermination' generated by that situation ceases to be. With the state no longer bent on the recommodification of capital and labour, and with productivity

and profitability finally emancipated from employment, the welfare state has lost a large part of its sociopolitical utility, and particularly that part which underpinned the cross-spectrum consensus [it once enjoyed].

In part, one might argue, the fundamental political and theoretical turn-about that remained unanticipated in Castells' work was hidden by the assumptions it made with respect to collective consciousness and collective action. Castells' initial formulation was, over time, transformed by refinements centring on issues such as the significance of urban social movements (Castells 1984; Lowe 1986; Smith and Tardanico 1987) – which effectively lost sight of the structural changes that led to the demise of the welfare state. Tellingly, a much-revised role has since been given to collective consumption (Castells *et al.* 1990). More pertinent to our present concerns, however, is the way in which the issues foregrounded by Castells' account of collective consumption were subsequently taken up and developed by those working within different traditions (as well as in different historical and geographical contexts). For, in addition to prompting further research within the tradition Castells himself represented (Preteceille 1981; 1986), his arguments resurfaced, to a greater or lesser extent shorn of their Marxist trappings, in the 'urban politics' of such British writers as Saunders (1979) and Dunleavy (1980). Whilst a number of authors began to explore the implications of public/private sectoral cleavages in defining the interest groups shaping urban politics (Dunleavy 1979; 1986),[5] it was, perhaps, Saunders' vigorous restatement of the significance of these themes that most dramatically transformed the concerns Castells had first established. Most importantly, and in a manner not dissimilar from the inflection given in our own characterization of Castells' work, Saunders (1984a, 211) suggested that:

> Collective consumption is proving to be not a permanent feature of advanced capitalism but a historically specific phenomenon ... the period of collective provision may come to be seen in retrospect as a temporary 'holding operation' or period of transition between the decline of the old market mode and the emergence of a new mode of private sector provision which has today become both possible and attractive for an increasingly large proportion of the population.

With the benefit of hindsight, it seems as if Castells' theorization of a city politicized by issues centring on collective consumption was pinned, to a large extent, on certain revolutionary hopes that have since been severely undermined. In fact, Saunders' work came to reflect the fundamentally changed political conditions of the 1980s, in much the same way that Castells' work had captured the situation of the 1970s (or, perhaps, the after-glow of the 1960s). In Savage and Warde's (1993, 157–158) words:

If Castells' work was an attempt to theorise urban politics from the perspective of the significance of the welfare state, that of Peter Saunders approached the subject via privatised consumption. What Castells was to the 1970s ... Saunders was to the 1980s.

Whilst Saunders' work is equally as difficult to characterize succinctly and without oversimplification as that of Castells, certain enduring motifs recur. Chief amongst these is the independent role attributed to consumption in the stratification of society (Saunders 1986b; 1988; 1989). Saunders' (1986a) 'dual theory' of the state emphasized the distinction between the class politics of production and the different sets of sectoral interests mobilized by the politics of consumption (which are held to determine the character of the *local* state, in particular: Saunders 1984b). As Warde (1990, 230) notes, Saunders has often cited Moorhouse's (1983) essay, which 'attacks the presumption in industrial sociology and social stratification theory that people's central life interests are concerned with their paid work'. For Saunders (1986b, 156), social research errs fundamentally in continuing to use

> essentially nineteenth-century theories of stratification (whether Marxist or Weberian) to analyse late-twentieth-century cleavages and relations ... which are now structured as much by the exercise of state power (as employer and as provider) as by the private ownership and control of these societies' productive resources.

Ostensibly, therefore, Saunders breaks decisively with a tradition of analysis he regards as (increasingly) anachronistic. The implications this carries are considerable. Saunders' genealogy of consumption sees Castells' city, for example, as little more than a temporary blip – as belonging to a transitional phase between the class-divided nineteenth-century city, characterized by a purely market-determined provision of goods and services, and a new social order, based upon a newly discovered, acceptable face of capitalism. Saunders' genealogy of different modes of consumption is explicitly not intended to represent an evolutionary 'stage' model. Nonetheless, collectivized provision seems, on Saunders' account, to have primarily served to lay the conditions for the bulk of the population to enter into a new situation characterized by a predominantly private 'mode of consumption'. This new situation distinguishes itself from nineteenth-century market provision to the extent that private consumption is buoyed up by the state in various ways (income transfers, discounts, subsidies, tax relief, etc.), but differs significantly from the earlier phase of collectivized provision in that it has shed the 'nannying' role the state had purportedly come to assume.

Underlying Saunders' argument, as Warde (1990) notes, is an ideal-typical distinction between public and private 'modes of consumption' (Table 1).[6] Clearly – and without entering into the detail of either Saunders'

Table 1   Warde's schematization of Saunders' thesis: privatized *vs.* collectivized modes of consumption

Mode	Privatized	Collectivized
Property rights	Ownership	Non-ownership
Access	Purchased	Allocated
Control	Consumer	Bureaucratic
Sector	Market	State
Quality/satisfaction	Good	Poor

Source: Warde (1990, 232)

argument or Warde's critique – the message conveyed by Saunders fundamentally valorized *private* over public consumption. In consequence, and somewhat reminiscent of the attention received by Castells' work, Saunders' thesis attracted considerable criticism, at least within the British literature. Whilst a number of specific themes were singled out for attention, much of the dissent Saunders' work attracted was concerned with countering his championing of the market. In essence, it sought to demonstrate that the new 'acceptable' face of capitalism was, in fact, a wholly cosmetic affair. As such, privatization – less euphemistically, 'decollectivization' – was seen as having far less favourable – or at least far more contradictory – implications in terms of social justice than Saunders allowed (Burrows and Butler 1989; Duke and Edgell 1984; Harloe and Paris 1984; Warde 1990). We restrict ourselves, here, to two particularly pertinent observations.

First, Saunders' conceptualization of the increased importance of *sectoral* divisions (consumption cleavages) was, for many, overly confident in its adjudication on what actually amounts to a far more complex state of affairs (Harloe 1984; Pratt 1986). Put in its most starkly analytical form, Saunders (1984a, 206) argued that:

> Consumption sectors, which are constituted through the division between owners and non-owners of crucial means of consumption such as housing, crosscut class boundaries, are grounded in non-class-based material interests and represent an increasingly significant form of social cleavage which may in certain circumstances come to outweigh class membership in their economic and political effects.

For Saunders, therefore, social class, defined in relation to production, 'is often a poor guide to a household's consumption location' (1986b, 158). Whilst this may well be (increasingly) true, it does not, in and of itself, imply either that class and consumption are unrelated, or even that individual cross-class consumption possibilities translate into aggregate patterns (Hamnett 1989). Nor does it deal adequately with the question of the differential likelihood of collective consciousness and collective action in relation to class, on the one hand, and consumption cleavages on the other hand.[7]

Thus, whilst Saunders provided a contribution to this issue, he failed to engage explicitly with more challenging alternative interpretations of the same sort of concern (Bourdieu 1984; Savage *et al.* 1992), or even with a sufficiently diverse range of empirical sources. Crompton (1996), for instance, has subjected Saunders' thesis to critical examination, marshalling a good deal of evidence to suggest that class relations remain crucial to advanced capitalist society, but that employment-related changes, which diminish the possibility of collective identifications focused on the work-place, increasingly serve to mask such relations. To sum up, many have found in Saunders' work an ideological fixation on the benefits of private property rights over and above any negative consequences that private consumption and market provision might generate.

This leads on to a second observation. One neglected aspect of Saunders' work is the concept of the 'mode of consumption' itself. Terminological appearances notwithstanding, the extent to which this parallels the Marxian notion of 'mode of production' is questionable. In Saunders' hands, 'mode of consumption' arguably pertains to a quite different level of abstraction than does 'mode of production'. The main purpose of such a conception is linked, as noted above, to Saunders' (1984a, 209–213) genealogy of different 'modes of consumption'. The point it seems to underlie relates to the notion that advanced capitalist society is, and will continue to be, increasingly polarized between a residualized minority, reliant on a dwindling public mode of provision – those 'cast adrift on the waterlogged raft of what remains of the welfare state', as Saunders (1986a, 318) put it – and a majority who are increasingly integrated by their common experience of private consumption (whatever their class position). As is well known, it was housing tenure, and specifically the various benefits of home-ownership, upon which Saunders' argument primarily turned. Saunders drew attention to the economic conse-quences of owner-occupation in terms of the wealth-bearing potential of home-ownership (a generationally self-reinforcing tendency – in the absence of negative equity); and to the 'ontological security' afforded to homeowners (though purportedly denied to those in public [or private] rented accommo-dation). Saunders (1990, 83, emphasis added) even went as far as to suggest that 'a widespread desire for owner-occupation is likely to be fuelled by certain *natural* dispositions as well as by economic and cultural factors'. Needless to say, such statements received considerable criticism. As Savage and Warde (1993, 161) note, 'Saunders conflates the control which people have when purchasing their houses on the market with the control people have when purchasing specific services on the market'. Private medical care, education, and home ownership, however, arguably involve fundamentally different modalities of consumption behaviour. It is worth noting that Saunders' preoccupation with housing tenure has meant that his work has commanded less attention within North American urban studies. Despite his appeal to certain supposedly 'natural' predispositions with respect to tenure, his generalizations drew, for the most part, in an all-encompassing manner upon a peculiarly English situation (particularly in relation to 'ontological security', but also with respect to the financial benefits of home ownership, which related closely to the British taxation system of the time). Whilst a good deal of his argument might have been applicable to home ownership within England and Wales (Scotland exhibits a notably different tenure structure), the public/private dimensions to education and health tend to diverge markedly from the generalizations Saunders presented as nigh-on universal. Saunders also tended to generalize from the complexity of the housing market in a manner that suited his own ends (compare Ball 1983; 1985; Forrest *et al.* 1990, particularly table 8.1, which provides a sober assessment of the various generalizations made in work on housing tenure; Forrest and Murie 1988). Despite the analytical tone characterizing Saunders' work, therefore, as it gathered steam, his argument increasingly reflected the surety of the New Right that the market had finally overcome what looked like increasingly dated leftist concerns with socialization (let alone socialism). In particular, it accepted and underscored the monetarist philosophy that people only value what they, as individual consumers, pay for. This is not to imply that general theoretical statements cannot be advanced on such issues as Saunders raised. It is, however, to suggest that Saunders failed to conceptualize adequately the issues he succeeded in high-lighting as being of increasing importance. If we change tack, at this point, we can develop a rather different theorization of the issues Saunders brought to the fore.

Public/private consumption, and the emancipation of capital from labour

Whilst Castells undoubtedly provided a more sophisticated theorization of public and private consumption and the city than Saunders, Saunders nonetheless managed to isolate a series of profoundly important social changes, the very possibility of which was effectively foreclosed in Castells' account. *Contra* Saunders, however, Castells' insistence that the city needs to be understood in relation to capitalism still retains its importance. Even so, Saunders' (1986b, 162) provocative remark – that, as academics,

> [w]e cannot afford to allow our own positions as public-sector workers and (as is normally the case) as socialists committed to collectivistic values to obscure our understanding of a fundamental development which is going on under our very noses and which is restructuring social relations in some most profound and significant ways

– needs to be taken seriously. How, then, given that Saunders' work carries the danger of affirming certain elements of a pernicious New Right ideology, are we to square this circle? The answer, which we sketch initially in purely theoretical terms, requires recognition of the limits of Marx's own account of capitalism. Hindsight affords a position from which to appreciate the extent to which Marx's theory of the capitalist mode of production reflected the specificity of *modern* capitalism. Such an appreciation permits a better understanding of the extent to which capitalism has departed from its modern form, to take on a distinctively postmodern, consumerist guise.

A defining hallmark of modernity was the belief that, through the application of reason, the world as it was could be improved upon. As the preceding chapter has detailed, a key watchword of modernity was 'progress': a notion that employed the power of reason in the belief that the world could, with sustained and concerted effort, be 'made to order'. Where a pristine and untamed nature diverged from what was deemed desirable from the modern vantage point, it was the task of humankind to *impose* order; to take the world to task and knock it into shape. As Bauman (1997a) suggests, Freud (1955) captured the essence of this attitude, disclosing the extent to which modern society as a whole was affected by the selfsame principles that were stamped upon the more esoteric concerns of science in the abstract. Science and knowledge sought understanding in order to predict and control; to overcome convention and superstition in the name of progress. Freud pointed to the extent to which modern civilization rested on the basis that certain 'natural' human impulses were incompatible with its own (rational) ends. The brave new world encapsulated in the intrinsically modern notion of 'culture' – a term deriving its strength and meaning from its opposition to 'nature' – required the repression, and indeed the correction, of certain unfortunate, supposedly innate, human tendencies. It necessitated, in Freud's terms, the subjugation of the 'pleasure principle' to the 'reality principle'. As Freud wrote: 'Civilized man has exchanged a portion of his possibilities of happiness for a portion of security' (cited in Bauman 1997a, 2). In a truly modern civilization, certain human traits were to be sacrificed in order to overcome the unbearable uncertainty of the premodern world. This was the cost paid for a more certain, predictable, and thus less insecure mode of existence. Were such supposedly desirable conditions to prevail, human beings needed to be moulded to fit the template of the new social order. Certain ways of being were incompatible with this desired end and could not, on any rational basis, but demand eradication.

The main obstacle to reaching the goals modernity set itself 'was the basic human inclination to do no more than satisfy one's needs' (Bauman 1997b, 24). It was the notorious 'work ethic' that was initially enlisted to overcome this obstacle. Thus, as Foucault made clear, amongst a whole host of modern institutions – including prisons, asylums, military barracks, schools – the factories that would eventually fuel the growth of industrial capitalism were, first and foremost, factories for the production of order. Modern industrial capitalism, in accordance with Freud's observations on modern civilization, established itself on the basis of the repression of certain previously accepted human behaviours; on the basis that appropriate conduct needs to be induced – that humans cannot be relied upon to behave spontaneously or autonomously in the 'right way'. For modern capitalism to reproduce itself, people had to become the cogs vital to its overall functioning, which entailed the necessity of surveillance and monitoring, of training and corrective action, in order to ensure that people behaved accordingly. As Bauman (1988b, 71) notes, in individual terms, hard work now became the thing that 'made the difference between affluence and indigence, autonomy and dependence, high or low social status, presence or absence of self-esteem'. Such terms internalized and reflected a mode of social integration based on repression and devised to ensure the attainment of the central systemic necessity (the reproduction of capital). The survival of capitalism as a whole, therefore, depended on the engagement of the majority of the people as *workers*.[8] This, however, is where Marx's dissection of capitalism stopped short. Marx was, inevitably, in no position to foresee the possibility that modernity would eventually create, on a global scale, 'expanding enclaves of "post-modern" existence in which people are consumers first – and workers only a very distant second' (Bauman 1997b, 24). Indeed, such an hypothesis still sparks controversy, confirming Baudrillard's (1998a, 81) point that it 'is difficult to grasp the extent to which the current training in systematic, organized consumption is *the equivalent and extension, in the twentieth century, of the great nineteenth-century-long process of the training of rural populations for industrial work*'. As Bauman (1983, 41) insists, 'consumerism does not mark any significant departure from the kind of society which emerged in Western Europe with the advent of industrialism. On the contrary, it seems to signify the fullest deployment to-date of the techniques of power which brought industrialism into existence'. The transition is between modes of domination, not between kinds of society.

What retrospectively turned out to be the preconditions for a fully developed system of consumerism were lain over many centuries (a process considered in detail in the next chapter). The development of a systematized, rationalized regime of accumulation of the kind described by Aglietta (1979), however, which instated mass consumption as an integral part of capitalism, is of relatively recent vintage. The development of Fordism involved a shift from an initial 'extensive' regime (which transformed the labour process in line with Taylorist principles, but did not yet 'recompose' daily life in accordance with a utilitarian functionalist logic) to an 'intensive' regime (which finally transformed the reproduction of labour-power in such a way, thus fully integrating the mode of consumption with the mode of production). This related specifically to the shift, at the beginning of the twentieth century, from the production of Department I (capital) goods to Department II (consumer) goods. Under these new conditions, 'the needs and satisfactions of consumers [became] productive forces that have now been constrained and rationalized like the others (labour power, etc.)' (Baudrillard 1998a, 82). As we have argued in detail in previous chapters, capitalism has, since Marx's dissection of its structure and operation, gestated an abstract and systematized 'consumptive force', constituting the subject as consumer in a manner directly parallel to the earlier rationalization of the worker.

Such a conception reveals the extent to which the consumer society increasingly ties the consumer into a network of dependencies. 'First new commodities make the necessary chores that much easier, and then the chores become too difficult to do unaided. ... Now what is necessary cannot be distinguished from what is unnecessary but which one can no longer do without' (Haug 1986, 53). For example, the 'private car, together with the running down of public transport, carves up the towns no less effectively than saturation bombing, and creates distances that can no longer be crossed without a car' (*ibid.*, 54). Reiterating the argument of Chapter 2, this does not imply any form of manipulation of the consumer, nor does it assert the generation of 'false' or 'artificial' needs. Rather, it entails that:

> new commodities ... create their own necessity – which sometimes has been expressed ... as the ability of the market to create 'artificial' needs (it would be much better to express the phenomenon in terms of the capacity of the market to render new needs practicably indistinguishable from the 'natural' ones; given the plan of most contemporary American cities, and the contemporary space relation between residence, work and leisure, it would be futile indeed to argue that the need for a car, or any other vehicle of personal mobility, is an 'artificial', or worse still – a 'false' need).
>
> (Bauman 1987, 164)

The twentieth century saw both the imposition and subsequent fragmentation of the system of mass consumption. Having reached a point where productive capacity exceeded consumptive capacity – where goods 'became harder to sell than to manufacture' (Galbraith, cited in Baudrillard 1998a, 72) – capital turned to market research in an attempt to anticipate consumer preferences prior to production, and advertising in order to influence them subsequently. Once again, this need not be read as implying the manipulation of the consumer (though both corporations *and* their critics tended to assume such, prematurely regarding attempts as accomplishments). Its real importance lies in the fact that it instilled within firms a new 'marketing orientation', which was ultimately to bring about the fragmentation of mass consumption into a series of increasingly segmented 'niche markets': a process associated with a parallel move towards 'flexible production' (Lee 1992). Such a transition finally undermines the class logic of mass consumption (where the possession of particular objects was determined for the most part by income levels), imposing an ethos of endless individual choice that belies the systemic role consumption has come to play. These circumstances are, undoubtedly, characteristic of only certain, perhaps expanding, areas of the world – a factor to which we shall shortly return. The crucial point, however, is that the 'inhabitants of these enclaves are kept in place not by coercion but by seduction, by the creation of new desires rather than by normative regulation' (Bauman 1997b, 24). This amounts to nothing less than 'a substantive change in the mode of domination central to social integration' (Bauman 1987, 167).

In this light, the repression that Freud saw as central to modern civilization has proved itself no longer vital to the maintenance of social order or the reproduction of capital. Indeed, circumstances have conspired to ensure that repression has been 'turned from an irritating necessity into an unwarranted assault launched against individual freedom' (Bauman 1997b, 2). As Freud pointed out, any gain is made at a cost – hence the propensity for civilization to produce its own discontents. Whilst this principle still holds true, today, 'the gains and the losses have changed places: *postmodern men and women exchanged a portion of their possibilities of security for a portion of happiness*' (*ibid.*, 3). In short, capitalism has left the premises of modernity far behind, having detached itself from the reality principle to embrace the pleasure principle as its principal means of perpetuation. 'The work ethic has been replaced by a consumer ethic; the savings-book culture of delayed gratification has been replaced by the credit-card culture that "takes the waiting out of wanting"' (Bauman 1997a, 24).

It is in this light that the notion of the consumer society takes on its full meaning and, *pace* Saunders, that the concept of a 'mode of consumption' may be employed at a level of abstraction parallel to Marx. In a postmodern society,

> capital does not engage the rest of society in the role of productive labour; more precisely, the number of people it does so engage becomes even smaller and less significant. Instead, capital engages the rest of society in the role of consumers.
>
> (Bauman 1987, 179)

We have, in other words, witnessed the emancipation of capital from labour; more precisely, we have witnessed such a process – colloquially referred to as 'downsizing' – within certain 'expanding enclaves' of the world, and chiefly the Western world. Such a situation involves a complex set of spatial transformations. It is, at least in part, dependent on the continued expansion of capitalist industrialization within the Third World, where a cheap and largely unorganized labour force makes the emancipation of capital from labour elsewhere more feasible and more practicable. In Bauman's (1998a, 64) words:

> The present-day corporations do not need more workers to increase their profits, and if they *do* need more workers they can easily find them

elsewhere and on better terms than those attainable locally, even if this leads to the further impoverishment of the local poor.

It also involves, in a reversal of the earlier geography of capitalism, a new ruralization of labour in the First World. Transformations such as the latter are, however, riven with further spatial complexities; for example, First World rural industrialization is juxtaposed with the commodification of the countryside as a space of leisure. We shall return to consider the city in such terms in due course. Initially, however, it is worth outlining the full consequences of the emancipation of capital from labour, and the resultant transition from a situation dominated by repression to one where seduction becomes ever more central, as a means of grasping the significance this carries for the public/private distinction under postmodern conditions.

Whilst the active pursuit of repression proved vital, under modern conditions, for maintaining the majority of the population as actual or potential labourers – with only the very few privileged and rich being exempt – the chances of a greater number purchasing such exemption increased as capitalism advanced on the road towards the emancipation of capital from labour. Today, within the expanding enclaves of the consumer society, a sizeable majority has been granted permission to buy their way out of the modern, repression-dominated world. These represent, in Bauman's terms, society's *seduced*. The displacement of the repression-centred mode of social integration by new, seduction-centred mechanisms comes about in direct proportion to the extent to which systemic reproduction is better achieved by letting social integration be guided by 'needs' rather than ensuring it be constrained by norms: increasingly, members of the consumer society are 'efficiently and effectively integrated (in a way resonant with their actual role in the reproduction of capital) through a new cluster of mechanisms – seduction, public relations, advertising, growing needs' (*ibid.*, 180). In such a situation, mechanisms centred on repression become increasingly irrelevant – at least, that is, for the majority. Even the proliferation of CCTV cameras surveying the city streets do not represent a panoptic gaze but a *selective* gaze, focused on those deemed to be 'out of place' and 'out of time' (Norris and Armstrong 1999; Williams and Johnstone 2000; Williams *et al.* 2000).

Whilst an expanding proportion of the population defines the visible contours of the consumer society, another, almost separate world is created. Such is the world of the *repressed*; a part of the consumer society that 'may be easily overlooked ... as it is precisely the domination of the market which makes it irrelevant, marginal, and theoretically "alien", "residual" or "not-yet-eliminated". ... However, this other part of society ... is as equally inevitably produced by the market' as the part that conforms to the image the consumer society itself promotes (Bauman 1987, 168). Moreover, if the reproduction of capital increasingly depends on the consumer, and to a declining extent on the labourer, a fundamentally different role is now afforded to those marginalized in the process. The consumer society no longer grants them a valid place in society, rendering their very identity as members of society problematic.

> Before [the] emancipation of capital from labour, the poor were first and foremost, 'the reserve army of labour'; they kept capital growth options open, and helped to keep the capital–labour conflict off the limit where they would jeopardize the reproduction of the system. The poor were, therefore, not just an unavoidable, but an indispensable part of the system – in no way an alien body. After the emancipation of capital from labour, the poor could play a similar 'inner-systemic' role only if they could be seriously considered as 'the reserve army of consumption'.
>
> (*ibid.*, 180)

But, of course, such a role is simply unnecessary under the new economic conditions: those on the margins of the consumer society are 'structurally redundant'. As social integration and systemic reproduction no longer require the active repression of the majority to ensure the availability of a supply of labour – as people became more important as the source of demand (a body of consumers) than as a source of productive potential (a body of labourers) – the poor count less and less.

> It is one thing to be poor in a society of producers and universal employment; it is quite a different thing to be poor in a society of consumers, in which life-projects are built around consumer choice rather than work, professional skills, or jobs.
>
> (Bauman 1998a, 1)

It is the lot of the 'new poor' to remain constrained by the forces of repression in a world increasingly saturated by the images of seduction. The emancipation of capital from labour has not, therefore, sounded the death knell for the mechanisms of repression once borne by all but the most privileged. Such mechanisms live a ghostly afterlife, serving to constrain, if not define, the 'residual' lives of those who fail to live up to the eligibility criterion of the consumer society – those who are not in a position to engage in market consumption (to all intents and purposes, those without money; or, rather, without access to credit [Hallsworth 1992]). In such a world, where the consumption ethic has replaced the work ethic, the repressed – once the majority, now a minority – find themselves 'denied the opportunity to follow the rules of the work ethic in a world in which the only access to the resources needed to exercise one's freedom is still through the door marked "work"' (Bauman 1997b, 24).

The seduced of society, therefore, need no longer fear the stick of repression, insofar as they willingly accept the carrot of seduction. The threat of repression remains, however, to shore up the decentralized, deregulated, privatized form of auto-surveillance that has been delegated to society's seduced by the formerly centralized powers of the state. Such a situation is thoroughly divisive:

> We have two worlds, at opposite poles, which are increasingly out of touch with each other – much as the no-go areas of contemporary cities are carefully fenced off and bypassed by the traffic lines used for the mobility of the well-off residents.
>
> (*ibid.*, 25)[9]

It is the presence of this foreboding, second world, however – increasingly visible in an ever more divided city – that ensures the perpetuation of this 'new improved' brand of capitalism. Though one might be fortunate enough to find oneself a member of the consumer society, one can never become a fully paid-up member, freed, once and for all, from the foreboding threat of renewed repression. Consequently, one must work hard to earn the right to money, credit, and freedom, in order to ensure that one does not put oneself in the position where one might become lost to the other world. Hence the pervasive sense of ambient fear that increasingly saturates the city (Doel and Clarke 1998). The division between the 'seduced' and the 'repressed' is vital to the new systemic arrangement. It is, perhaps, *the* postmodern divide. Nonetheless, some have mistaken it as merely a perfunctory distinction. Warde (1994, 60), for instance, characterizes Bauman's distinction as a division between 'a majority of happy shoppers and a minority subject to state surveillance and control' that directly equates to Saunders' public and private modes of consumption. However, as we have sought to show, Bauman's conceptualization incorporates more than an empirically derived generalization of the kind that Saunders advanced, providing for an incisive understanding of the structural underpinnings of the dismantling of the welfare state. In the next section, attention is focused on the way in which the distinction between public and private consumption has been blurred by the altered relations between the market and the state – and how this serves, in contradictory fashion, to exacerbate yet at the same time mask the division endemic to the consumer society.

Market freedom and social polarization: public/private consumption today

The 'restructuring' of the welfare state that began, in the wake of an increasingly neo-liberal form of global political economy, in the closing decades of the twentieth century, saw the pattern of public and private provision change beyond all recognition (Burrows and Loader 1994; Pinch 1996). Public provision today provides, at best, an increasingly fragile safety-net, with the former ideals of collectivization a distant, fast-fading memory. Concurrently, state subsidies for various forms of market consumption have increased to fill the vacuum left by the receding boundaries of the welfare state. Such a process, of course, relates to global economic change; specifically, to the emancipation of capital from labour. Minimizing state intervention (or, more precisely, state expenditure), is firmly tied to the need to ensure that, within a given national space, capital is not taxed in such a way as to discourage its locating or remaining within that national territory (Jessop 1994).[10] These new imperatives fully accord with the recasting of the poor as a drain on resources. Rarely, if ever, are the poor any longer regarded as a potentially valuable labour force. In consequence, the welfare state has transmuted into the 'workfare' state – a thoroughly paradoxical state of affairs, since there 'is simply not enough paid employment any more to support the model of full-time jobs for life' (Bauman 1997b, 24). 'Workfare' has, of course, a sinister underlying purpose. The necessity of ensuring one is 'available for work', and increasingly of 'earning' what was formerly a right to welfare – typically via an endless succession of often useless training schemes (run, of course, by private agencies) – creates an atmosphere in which the poor are, subtly or not-so-subtly, reproached for their fate and stigmatized for their social location. Such are the means of minimizing the numbers left reliant on state provision, and often, to put it bluntly, of punishing them for their fate. Such means have, perhaps, reached their most extreme in the United States, where a time-limit has been imposed on need: people 'will be allowed to stay out of work for no more than two years during their entire life' (*ibid.*, 25) – despite the lack of demand for their labour-power. A further consequence of this stigmatization, notable on both sides of the Atlantic, is the way in which the decline of the welfare state has coincided with an increasing prison population. Only partly the result of an increase in crime, to some extent triggered by the gap opened up between the seduced and the repressed, this reflects a new attitude towards those dispossessed by the consumer society; an attitude held with increasing passion – of the kind unleashed by ontological uncertainty – by those who need to feel safe within its invisible walls (Young 1999). The consequence is a penal system that is far less oriented towards social rehabilitation, and far more towards the generation of private profit (Shichar 1995).

The effort to castigate as 'spongers' those still reliant on collective provision runs in parallel with the state's increasing role in buoying up market transactions: through the provision of subsidies of various kinds; through the contracting out to the private sector the delivery of nominally collectively provided services; and through the grafting of certain market-related forms onto the remaining core of collectively delivered services. In short, the state has become directly involved in the promotion of private consumption, both practically and – as if this were somehow less practical – in promulgating a consumerist ethos of 'choice'. The blurring of public/private consumption this de-differentiation of state and market entails is one way in which the division between the seduced and repressed cannot be said to be simply stamped out in terms of public and private consumption (in other words, these two sets of distinctions cannot be directly read off from one another). Furthermore, the increasingly blurred division between public and private provision serves to mask the fact that a good deal of public money is now used not directly, to help the most needy, but to subsidize market-provision for the not-quite-so-needy – as well as to improve the urban fabric, and the 'image' of the city, in order to attract capital – with the supposed aim of improving the lot of everyone in the long run. Unfortunately, the widespread failure of property-led development to generate such 'trickle-down' effects means that a more prosperous urban economy far from guarantees a more equitable and socially just city (Robson *et al.* 1994). At the same time, policy changes, such as those exemplified by City Challenge and the Single Regeneration Budget in the United Kingdom, have instated a new competitive situation amongst the needy. This amounts to a competition for increasingly scarce resources: the rhetoric of a more 'efficient' use of such resources, supposedly achievable via market allocation, has served to hide the overall erosion of total funding levels. The blurring of public and private consumption has, in effect, widened the gap between the seduced and the repressed whilst concealing the division it supports and accentuates in the process.

Over the last few decades, the shift from collectivized forms of provision towards individualized, market provision has firmly taken hold. Certain goods and services of vital importance to the city, notably housing and transport, had already seen a shift in this direction by the 1970s, in both the United Kingdom and the United States. This trend has continued, with ever-growing proportions of the urban population owning not only the homes they occupy but also one or more cars. Many other collectively provided services, such as education, health and pensions – which, in the UK at least, had formerly been collectively provided – have seen dramatic shifts towards private provision and delivery. Though it is clearly impossible to document the full range of changes here, they are amply illustrated by the changed circumstances surrounding housing provision in the UK (Williams 1992; Goodwin 1997).

The case of housing, which, thanks to Saunders, has been central to debates over collective consumption, exhibits some clear underlying trends but also a good deal of complexity. Even in the rest of Western Europe, where owner-occupation is generally lower than in England, Wales, and the Republic of Ireland, and where other tenures are much more widely acceptable, there has been a notable trend towards owner-occupation. In the UK, however, this trend has proved to be anything but straightforward. Saunders' research output peaked at a time when owner-occupation was booming, particularly in south-east England, where the economy suffered far less from the recession of the early 1980s, where the expansion of business and financial services was generating high incomes, and where the deregulation of the financial sector had led to a highly competitive situation in terms of mortgage supply. Saunders' argument undoubtedly reflected the time and place in which he was working. The housing slump experienced in the late 1980s, which turned out to be far more protracted than the usual downturns exhibited periodically by the housing market, may have since revised Saunders' views. Owner-occupation is no longer seen as an unequivocally sound financial investment, in the wake of the negative equity experienced by many households, especially in the Southeast, during the housing slump of the late 1980s. In its wake, the wave of building-society repossessions experienced across the country reinforced the point (even though they revealed, to a large extent, the way in which the competition generated by financial deregulation led directly to irresponsible lending policies). Tax advantages to owner-occupiers have also been substantially reduced, with tax relief on mortgage interest being considerably lower, throughout the 1990s, than it was in the 1970s and even in the mid-1980s, and eventually abolished. Similar housing-market slumps have occurred in the USA – where owner-occupation has also been subsidized through tax relief on mortgages – serving as a sharp reminder to existing and would-be owner-occupiers of the negative as well as positive effects of property ownership. Despite this, however, a marked contrast between owner-occupiers and the socially excluded has arisen in terms of the advantages each group receives. At a time when the former could automatically enjoy tax relief, the latter became increasingly internally differentiated with respect to welfare entitlements – particularly in the face of the kind of urban policies noted above, where allocations based on criteria of need have given way to competitive allocation mechanisms. Nonetheless, it is undoubtedly true that an increased economic uncertainty has come to characterize home ownership since Saunders propounded its advantages.

The push towards increased home ownership in the UK was based on far more than economic considerations alone. It involved a particularly strong political drive, eventually supported by all the main parties. In the face of declining electoral popularity, the Labour Party came to accept the subsidized sale of local authority rented housing, a policy which had helped to propel the Conservatives into office in the 1979 election and to maintain their momentum in their landslide electoral victory of 1983. This policy, as the 'right to buy' slogan suggested, represented an ideologically driven attempt to break up the so-called 'monopoly position' occupied by local authorities, and to destroy Labour's traditional urban electoral strongholds. It amounted to an attempt to graft onto sections of the urban working classes the kind of concerns with property ownership that had traditionally translated into Conservative votes. The 'right to buy' legislation was intended, in effect, to tip the balance from collective to private consumption on the assumption that this would, in and of itself, alter voting predilections. If the policy involved, on the one hand, the carrot of belonging to Margaret Thatcher's vision of a 'property-owning democracy', on the other hand, it was reinforced by making public housing less and less attractive. This took the form, primarily, of a reduction in the subsidy granted to local authority rented accommodation. Consequently, by the mid- to late 1980s, tenants in housing associations (non-profit-making organizations), rather than those in local authority housing, were paying the lowest rents. The stock of urban local-authority housing, the mainstay of collective housing provision in Britain, had effectively been reduced by a massive sell-off, which represented the largest of the Conservative governments' series of privatizations. The other motive behind this policy was fiscal, pertaining to the monetarist preoccupation with reducing the public sector borrowing requirement.

The pattern of council-house sales did, however, exhibit considerable geographical variation. The 'right to buy' was taken up with most gusto in the towns of the Southeast, where the overheating of the housing market had put owner-occupation beyond the reach of many lower income groups, in contrast to comparable groups elsewhere in the country. The stock that remained in local-authority hands comprised, unsurprisingly, less desirable property: medium- and high-rise flats, maisonettes, and houses on difficult-to-let estates. Many of these were located in the inner city, or on peripheral estates (typically the areas of greatest social disadvantage). Given the kind of stock, its location, and government attacks on the kinds of people who had priority access to it (for example, young lone mothers), it was unsurprising that the tenure, stock and tenants were increasingly marginalized, residualized and stigmatized. In short, therefore, a major reversal in attitudes towards – and provision of – the major form of collectivized housing went hand-in-hand with strong positive support for private housing – with continued (if relatively lower) tax breaks that, for a time, still tended to exceed subsidies for public rented housing.

This, clearly, amounted to a situation giving rise to a new form of social polarization, with the political will to tackle it on grounds of social justice becoming ever more conspicuous by its absence. It was, at the same time, an attempt to shift the boundaries marking out the established class structure of the city. Put bluntly, the intention was to foster an expanded membership of the consumer society through home ownership, with little regard for those the process left high and dry (to invert Saunders' 'waterlogged raft' analogy). The actual consequences, however, were considerably more complex than those intended. Local authority housing was sold off, at a subsidized price, to sitting tenants, or in some cases to private developers for rehabilitation and subsequent private sale. Some of these homes went to those on low incomes; others were bought by people who did not need the subsidy, some buying them as second homes. Of those on low incomes, some did benefit from this means of access to low-cost home ownership (others in the same position benefited from what became known as 'build-for-sale', whereby private developers were sold public land at reduced prices, with payment due only after the sale of the houses built). Equally, however, many found the cost of 100 per cent mortgages – and, additionally, service charges if they had purchased flats – simply beyond their means. This was one way in which a new group of marginal homeowners was created. In times of high interest rates and job insecurity, the benefits of home ownership became, for many, very limited. Moreover, the profits generated by the sales of local authority housing had to be used in such a way as to contribute to savings in public expenditure. Legislation prevented any channelling of funds towards the provision of replacement public housing. As a result, nearly two decades of successive Conservative governments saw a reduction in the proportion of the population occupying publicly rented housing, and a massive drop in the development of new dwellings for public rental. Once set in sway, the process was destined to continue under New Labour, despite the lip service paid to concerns with 'social exclusion'.

The institutions enlisted to supply 'affordable housing' in place of the local authorities were the housing associations, funded through one of the largest quangos, the Housing Corporation. In the late 1970s, these voluntary-sector organizations were regarded as the 'third arm' of the housing system, but they rapidly began to replace the withered second arm of the local authorities, to form a part of what Wolch (1990) has aptly termed the 'shadow state'. Successive governments effectively colonized the housing association movement, pushing them first into rehabilitation (in the 1960s and 1970s) and, later – in the Conservative drive to owner-occupation and private finance enacted by the 1988 Housing Act – into various forms of low-cost home ownership. The new requirement that housing associations acquire private finance, rather than relying on central government grants, meant that the low rents associated with the pre-1988 'fair rents' provision disappeared, to be replaced by 'market rents' and 'assured tenancies'. Whereas fair rent provision previously allowed for various socially excluded groups to be accommodated, the new market rents acted to push out considerable numbers of those previously catered for by the housing associations. Typically, only those on benefits, or earning considerably more than the lowest low incomes, could afford the new levels to which rents rose. This had the effect of creating a new poverty trap for those on benefits: if a job opportunity were to be taken up, benefits would be lost and rents could no longer be afforded. Obtaining a job meant moving home.

In addition, the move to private finance created a requirement to obtain a return on the money invested. This led, quite simply, to the widespread abandonment of the provision of affordable housing for those in greatest need, and a reorientation towards a relatively wealthier market. Effectively, this meant building in areas where such a market could be found and solvent demand met. In consequence, the areas within which the activities of the housing associations had originally been zoned by local authorities were largely abandoned, at least by those associations taking up private finance to build new property for sale or rent. In short, therefore, the housing associations underwent a fundamental transformation, encompassing new forms of rent, new tenants, and new areas of operation, all of which meant that they were operating under conditions far removed from those to which they were accustomed. Given these circumstances, they did not always, initially at least, compete that well. Furthermore, they were, to some degree, hampered by the regulations imposed by the Housing Corporation. As a consequence, they failed to build the numbers of houses that were required to meet the 100,000 affordable houses per annum that were estimated, by a wide variety of experts, as necessary. Moreover, since their remit had effectively been oriented away from the most needy members of society, many of the houses that were built were not, in any case, that affordable. The upshot was that fewer houses were built at the end of the 1980s than at the start of the decade. In effect, the housing associations had been moved away from quasi-collective provision towards delivering a new form of private provision. Their activities and underlying rationale had witnessed a significant alter-ation, shifting them towards the centre of the triangle of private, public and voluntary housing provision shown in Figure 2.

Reinforcing this last point, the housing associations were also used by the government to bale out the private market, being given resources to buy unsold new houses from developers during the housing recessions of the early 1980s and early 1990s. Such property was often used to accommodate the kind of population that the associations had traditionally served. However, much of this property was built in the form of private estates with little, if any, community provision and where, in consequence, a car was essential. These estates, together with some of the estates the housing associ-ations took over from the local authorities – as part of the 'tenants' choice' provision of the 1988 Housing Act – became new, difficult-to-let 'problem estates'. The better management that was supposed to result, as the housing associations took over the role of landlord from the local authorities, was not, initially, automatically forthcoming – yet promises of better housing management under different forms of landlord continue to be repeated, under New Labour, as the attempt to reduce the level of reliance on public provision continues.

The overall result of the colonization of the voluntary sector by the state, and of central government restriction on local authority activity, has been a marked reduction in the provision of affordable housing and a considerably worse deal for the most marginal members of society. Simultaneously, a new group of owner-occupiers has been created who are marginal to that tenure. It has already been noted that considerable evidence has stacked up against Saunders' views of owner-occupation. The attitudes of these marginal home owners, and those who borrowed at the top of the housing boom, especially in the Southeast – later to face negative equity, higher interest rates, job inse-curity, and, in the extreme, repossession – can only reinforce the point. The overall net effect has been, however, that collective provision has been dramatically reduced as a portion of the total housing stock, has seen considerable institutional transformation (with agencies other than local authorities as landlords), and has been stigmatized – whilst private owner-ship has been extended and promoted as the goal to which every household

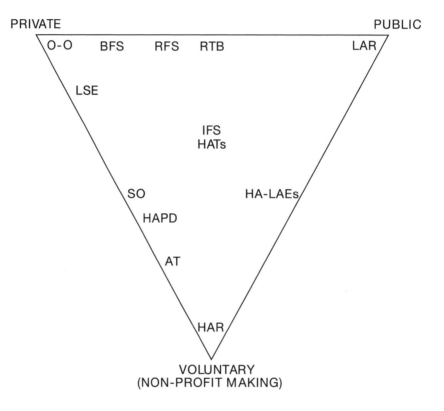

PRIVATE   PUBLIC

O-O   BFS   RFS   RTB   LAR

LSE

IFS
HATs

SO   HA-LAEs

HAPD

AT

HAR

VOLUNTARY
(NON-PROFIT MAKING)

IFS	Improvement-For-Sale: housing associations buying up property for improvement and private sale	AT	Housing Associations' Assured Tenancies with market rents and private finance (post 1988)
LAR	Local Authority Renting (traditional)	BFS	Build-For-Sale (subsidy to developers through land)
LSE	Leasehold Schemes for the Elderly: housing association retains 30%, private owner 70% of value of sheltered accommodation apartments	HA-LAEs	Housing Associations as landlords of old Local Authority Estates (post 1988)
O-O	Owner-Occupation through private finance (traditional)	HAPD	Housing Associations used to bale out Private Developers by buying up new estates in early 1980s and 1990s recessions
RFS	Rehabilitation For Sale: local authority stock improved by private developer for sale with public subsidies		
RTB	Right To Buy: subsidised sale of local authority stock to sitting tenants	HAR	Remnant Housing Associations Renting under Fair Rents scheme (pre 1988 norm)
SO	Shared Ownership: part renting from housing association, part buying through mortgage from private finance	HATs	Housing Action Trusts: old council estates divided up in various public/private/voluntary arrangements

Figure 2 The blurring of housing provision in the UK in the 1990s

should aspire. The spatial consequences of many of these changes have included the break-up of large expanses of housing of the same tenure or type. Local authority estates have become more internally differentiated, both visually and in terms of tenure. The process of gentrification has seen parts of the inner city and city centres reinhabited by those belonging to higher income brackets, in either 'new build' or converted property, especially in waterfront locations. Such developments have, however, tended to produce a new form of social segregation at the micro-scale, rather than a more genuine social mix. One classic example is the fenced-off private housing characterizing London's Docklands redevelopment. In parts of the US, where affordable housing has been built into private developments, a truer social mix has sometimes been achieved. In some cases, such as Boston's South End, such developments have been the anchor for urban revitalization. In the UK, however, such ideas have been rare outside certain rural developments, with a requirement that only 10 per cent of housing be earmarked as 'affordable'.

The case of housing reveals a particularly dramatic attempt by the state to promote an individualized, privatized form of consumption. It has proceeded regardless of the social costs involved, effectively running collectivized provision into the ground in many areas. There can be little doubt that there is a significant degree of correspondence between owner-occupation and those numbering amongst the seduced. Likewise, those reliant on public housing are likely to be those whom, in terms of a panoply of other state institutions, remain subject to repression – and for whom the freedom of the market is, and is set to remain, out of reach. But this does not amount to a strict correspondence; the relations between the two sets of terms are more complex than this. The increasing numbers of homeless; changes in the private-rental sector and the associated rise in multiple occupancy; the wide spectrum of housing qualities, social statuses and financing mechanisms captured under the single term 'owner-occupation' (Barlow and Duncan 1988), all speak of the fragmentation and, consequently, increasingly diverse nature of housing in the UK. The overall result, however, is relatively straightforward: whilst policies such as the 'right to buy' legislation, the colonization of the housing associations and so on, sought to open the doors of the consumer society more widely – thus granting increased numbers access to the freedoms offered by the market – the other side of the coin was, quite simply, a worse deal than ever for the very worst off. If a few were trampled in the rush or knocked off balance as the doors of the consumer society swung back in their faces, the fact that they amount to a small minority seems to ensure that they receive little sympathy or, indeed, practical help. They have had their chance; perhaps they simply weren't cut out for such a competitive world. Those who gamble well, it seems, are entitled to feel smug in the face of those who failed to play their hands successfully. The latter can look forward to some minimal form of safety-net provision – and all that goes with it. So it is that the poor are increasingly characterized as the undeserving poor, defined against the hardworking majority, who have won the right to enjoy the freedoms granted to them without harbouring feelings of guilt or responsibility. The kind of situation this represents is usefully summed up by Offe (1996, 47–48):

> If … as in the game-theoretic model of the 'prisoner's dilemma', the interacting partners are authoritatively *dis*associated, and if, moreover, they are prevented from spontaneously generating solidarity – because of the expectations and incentives built into the pay-off matrix – then we have reached the limiting case of an associative context that virtually precludes moral action.

In such a situation, lip service towards the socially excluded is perhaps the most one can expect.

This same underlying story has repeated itself for a host of other formerly collectivized goods and services. The basic plot line, as Bradford (1995) demonstrates for the case of education in the UK, is one of 'diversification and division'. Social polarization of the kind captured by the distinction between the seduced and the repressed is hidden behind a barrage of new mechanisms for more efficient, effective, market-orientated provision and delivery. Taken one by one, such changes might seem to involve moderate improvements, incorporating sensible enough, perhaps even laudable goals. They are also usually formulated with a keen eye on an increasingly important 'bottom line' – on the financial matters that were supposedly neglected in a less accountable era of collective provision (the implication being that it is difficult to challenge anything that self-evidently rational in economic terms). The net effect of such changes, however, which is simply inevitable under the conditions defined by the emancipation of capital from labour, is a further entrenchment of social polarization, to which even those politicians of formerly socialist inclinations have seemingly become accustomed. As we have already intimated, the electoral victory of New Labour in 1997, whilst perhaps tempering the unrelenting emphasis on individual property rights achieved during the so-called Conservative Restoration, failed to mark any significant change as far as issues of social justice are concerned. Indeed, the extent to which the transition towards a fully fledged consumer society is underpinned by the emancipation of capital from labour serves to ensure that political parties, of whatever hue, can do little more than live with the new division between those admitted to the world of consumption and those refused entry – between the new 'two nations' of the seduced and the repressed.

Conclusion

The dramatic shift from collective to market provision and its effects on the city has become, by now, a familiar story. Rather than limiting the issue to a consideration of empirical trends, however, which is too frequently seen as an end in its own right, the foregoing discussion has sought to relate such trends and tendencies to a far broader set of circumstances associated with the transition towards a fully fledged consumer society. The entrenched division between those increasingly tied to the market, and those systematically excluded from the world of consumption – despite being subjected to a constant barrage of images of consumer freedom that such a world projects – has figured amongst the most serious consequences of these changes. Numerous policy changes, moreover, have served to exacerbate this division, whilst purportedly purportedly extending the freedoms of the market to an ever-expanding proportion of the population.

By reconsidering the contours of consumption in the light of past engagements of urban theory with collective consumption, an account of urban consumption emerges that is very different from the increasingly prevalent focus on consumption as a force revitalizing city life; on the city, under the impress of consumerism, as a site of 'urban renaissance'. The city has, over recent years, undoubtedly witnessed a renewed vibrancy and vitality which, for a time, seemed to have disappeared almost entirely, especially from the central city. All too often, however, such terms have served as little more than modish buzzwords, detracting from the broader consequences of the city's unrelenting neo-liberal turn. Nonetheless, these contradictory aspects of the city should not be seen as entirely unconnected. They are twin born, and can only be understood together.

In order to elucidate the point, it is useful to end by rehearsing Bauman's (1993b) consideration of Turner's (1969) distinction between two fundamentally different forms of social togetherness: *societas* and *communitas*; respectively, 'structural' and 'counter-structural' modes of social being. The distinction differs from that drawn by Tönnies (1955) between *Gesellschaft* and *Gemeinschaft*, insofar as 'Turner intimates not a historical succession and temporal mutual exclusivity of the two forms, but their coexistence, interpenetration and alternation, and a perpetual and regular one at that' (Bauman 1993b, 116). Nonetheless, like Tönnies' distinction, the 'conditions of *societas* and *communitas* are mutually opposite in virtually every aspect. … If *societas* is characterized by its heterogeneity, inequality, differentiation of statuses, system of nomenclature, *communitas* is marked by homogeneity, equality, absence of status, anonymity' (*ibid.*, 117). For Turner, these forms of togetherness operate in relation to one another: 'the condition of *communitas* is dissipation or suspension or temporary cancellation of the structural arrangements which sustain at "normal times" the life of *societas*' (*ibid.*). What represents, from the point of view of *societas*, a 'social void' (*communitas*), serves as a fluid state within which rearrangements for the continuation of *societas* can crystallize out (Turner's examples centre on rights of passage). *Communitas* is, then, a mode of social being within which pre-existing orders can be (and are) dissolved, in a kind of levelling process that nonetheless allows and leads to the renegotiation of those aspects of

societas it first undoes. The two states are, for Turner, functionally indispensable; and herein lies a basic problem with Turner's particular formulation:
communitas becomes merely a functional necessity for *societas*, and 'the
commanding position of "structure" over "anti-structure" is reconfirmed,
obliquely, in the logic of explanation' (*ibid.*, 118). Rather than thinking in
such terms, Bauman suggests, we would be better 'to think of each as a
phenomenon in its own right and of its own, autotelic significance' (*ibid.*).
Likewise, rather than thinking of *societas* and *communitas* as separate *states*
of society, they are better thought of as two social *processes* – to wit, socialization and sociality:

> Socialization (at least in modern society) aims at creating an environ-
> ment of action made of choices amenable to be 'redeemed discursively',
> which boils down to the rational calculation of gains and losses.
> Sociality puts uniqueness above regularity and the sublime above the
> rational, being therefore generally inhospitable to rules, rendering the
> discursive redemption of rules problematic, and cancelling the instru-
> mental meaning of action.
>
> (*ibid.*, 119)

One senses, from this pithy rendition of 'sociality', a resonance with the
world of the *flâneur*. Indeed, the urban crowd, as first captured by Le Bon
(1982), serves both as an apt metaphor for, and an example of, the nature of
sociality. The crowd takes shape and disperses according to no particular
rule, carries people along for a while (though rarely against their own will),
may instantaneously coalesce into particular amorphous forms and then,
just as suddenly, change direction or fragment entirely. The crowd is saturated by the logic of ambivalence (cf. Bakhtin 1984).

Clearly, what much of the recent outpouring of work on the renewal of
city life has been groping towards is precisely the *reappearance* of such a
form of sociality, which connects to the distinction Bauman (1997a) draws
between, on the one hand, the situation of modern society uncovered by
Freud (1955), where security is attained at the cost of freedom, and on the
other hand the postmodern situation, where the trade-off is reversed; where
a renewed freedom comes at the cost of an increased level of insecurity.
Perhaps uniquely, however, the fluidity characteristic of sociality has, today,
become thoroughly imbricated with the machinations of capitalism.
Maffesoli (1996) frames the situation in terms of a new 'time of tribes'; the
irruption of a 'neo-tribalism'. The 'neo-tribes' are defined primarily in
aesthetic terms; in terms of the *experiences* 'membership' is likely to offer.[11]
In line with the counter-structural qualities of sociality, the boundaries of
the neo-tribes are infinitely permeable, and the tribes themselves infinitely
ephemeral. Neo-tribal 'membership' is assured by little more than the
possession of the appropriate paraphernalia (clothes, hairstyles, tastes, etc.).
It permits, therefore, the adoption of a *temporary* identity, appropriate to a

particular context today, yet easy enough to cast off, as life's circumstances change. In such a situation, the individual can reaffirm his or her existence without being overly concerned about the long-term consequences. Under such conditions, the rigid socialization required for the reproduction of the modern order is no longer a central systemic necessity. The system has found a way to thrive on a situation that is far more fluid, and has done so largely through the operation of the consumer market. This kind of social arrangement does not depend on any particular form of regulation. It is a process of spontaneous structuration – though each local structure so produced, being devoid of the means of its own perpetuation, is destined to live an ephemeral life. The sociality of the urban crowd is, in effect, no longer limited to the physical space of the city but plays itself out in the virtual spaces of the consumer society. The problem, however, is that this situation systematically excludes certain groups – those unable to live up to the role of the ideal consumer – from any legitimate sense of social belonging. Bauman (1993b) suggests that modernity represented a state of affairs where the moral impulse was, effectively, 'out-rationalized'. The modern effort of socialization rendered moral concerns a matter of indifference by removing moral self-sufficiency in favour of an heteronomous, externally imposed code of ethics (adiaphorization). The postmodern world, however, achieves much the same effect from a completely different starting point. Just as the *flâneur* kept moral sentiment at bay by means of – and as a means of permitting – an aesthetic engagement with the urban scene, so the sociality shaping the consumer society works to 'out-aestheticize' the moral impulse. The autonomous moral impulse is effectively quelled in both situations. Over recent decades, a powerful association has been forged between private freedoms and the market. All too frequently, in this context, the consumer society (the market) is seen as a universal panacea. It is therefore worth stating, as plainly and clearly as possible, that the ideological power of the term 'private', the present meaning of which has evolved over a far longer period of time (Williams 1976), serves to occlude the presence of those for whom the market fails. As Bauman (1987, 187) insists, in 'truly dialectical manner, [the] consumer society cannot cure the ills it generates except by taking them to its own grave'.

Part III

The consumer society and the postmodern city

5   The meaning of lifestyle

> Whether one uses the term lifestyles or even (the sociology of) everyday life,
> it is certain that this thematic can no longer be given cursory treatment.
>
> (Maffesoli 1996, 96)

Introduction

> The term 'lifestyle' is currently in vogue.
>
> (Featherstone 1991, 83)

On 18 July 2000, representatives of the marketing industry gathered in
Amsterdam for an event hosted by the Lifestyle Network (an organization
based in Slough in southern England). On the agenda was no less a matter
than 'Mapping the lifestyle genome'. In commercial quarters, as this somewhat fanciful title suggests, 'lifestyle' is widely regarded as offering an
unprecedented degree of 'marketing precision'.[1] In sociological circles,
however, the concept has seldom been viewed in such terms. Warde (1994, 69),
for example, has judged it 'a far from precise notion'. Indeed, most attempts
to define the term seem to resolve themselves in truism and tautology (Chaney
1996). Thus, for instance, Veal (1993, 247) concludes an exhaustive survey of
the concept with the striking banality that 'lifestyle' 'is the distinctive pattern
of personal and social behaviours characteristic of an individual or group'.
This underlying sense of meaninglessness is reinforced by the rampant expan-
sionism of the term, which has caused Sobel (1981, 1) to fear that it will
finally 'include everything and mean nothing'. Since 'lifestyle' is clearly such a
nebulous concept, is it possible to pin down its meaning at all? In a determined attempt to impose some clarity of thinking on the matter, Warde (1994,
69) has distilled five distinct senses in which 'lifestyle' is typically employed:

1   Market research uses the term to indicate market segments, that is
    groups of people with a higher than average probability of
    purchasing particular kinds or qualities of goods.
2   It may refer to the identifying characteristics of a status group, as
    was the case in the conceptual formulation of Weber (see Weber
    1978; also Turner 1988).

3   It can describe a set of practices – consumer behaviour, activities, body language, speech, etc. – which reflects a shared set of material circumstances and social class positions (for example, Bourdieu 1984).

4   The term may indicate a consumerist life-project, the pursuit by certain 'heroic consumers' of a coherent, comprehensive and highly stylized pattern of activities and possessions (see Featherstone 1991: 83–94).

5   Finally, it can imply pursuit of an ethical and principled way of life, irrespective of the purchase of commodities and shared class or demographic characteristics. This conception has affinity with the concept of neo-tribe, which has its origin in Maffesoli (1996).

Despite this effort of conceptual titration, the quintessence of 'lifestyle' remains infuriatingly elusive. Although a strong family resemblance runs through the list, it is almost as if mere accident attaches the same term to several distinguishable concepts.

Perhaps a clearer sense of the term's meaning emerges if one dispenses with the attempt to be comprehensive to consider, instead, the literal rendition of the term; as 'the stylization of life'. Although 'style' has been regarded as both a manifestation of consumer manipulation (Ewen 1990) and an opportunity for resistance (Hebdige 1979; Carter 1984), the term 'lifestyle' itself invariably implies that ways of life that were once stable correlates of social status are ever more 'free floating'. What was once a matter of class structure is, increasingly, a matter of individual expression and personal choice.[2] Hence Lowe's (1995, 67) pithy suggestion that '[l]ifestyle is the social relations of consumption in late capitalism, as distinct from class as the social relations of production'. Such a contention lays bare the connections between 'lifestyle' and consumerism. In present market-research usage, 'lifestyle' commonly refers to market segmentation and a renewed round of conspicuous consumption (Blackwell and Talarzyk 1983). However, a similar meaning was already apparent with respect to the standardization and conformity implied by mass consumption. One thinks, for example, of the 'other-directed' character-type isolated by Reisman *et al.* (1950), the 'consumership' associated with the rapid expansion of suburbanization and changes in modern city living (Bell 1958; Stone 1954), and the notion of 'inconspicuous consumption' (Katona 1964; Whyte 1954; 1960). Although the erstwhile correlation of 'lifestyle' with 'social class' (Myers and Gutman 1974) is precisely what current market-research usage seeks to undermine (O'Brien and Ford 1988), the common thread linking such different situations – mass consumption no less than its subsequent fragmentation – is the sense in which consumerism itself has become a 'way of life'. The extent to which previous ways of life appear to have broken free of their moorings, it would seem, is a consequence of this underlying change.

Giddens (1994, 90) has remarked that the 'notion of "lifestyle" has no meaning when applied to traditional contexts', whereas in 'modern societies, lifestyle choices are ... constitutive of daily life'. This is a suggestion that goes some way towards articulating the true meaning of the term. 'Lifestyle', one might say, is an oblique reference to the sort of situation that necessitates such a concept. The implication is that any adequate under-standing of 'lifestyle' needs to focus, first and foremost, on *the conditions under which the concept makes sense*. This is the purpose of the present chapter. From this point on, we will leave the definition of 'lifestyle' to take care of itself, and switch our attention to considering, first, the evolution and, second, the consequences of the kind of society within which the notion of 'lifestyle' emerges; namely, the consumer society. This will finally put us in a better position from which to appreciate the nature and meaning of 'lifestyle'. Let us initially turn, therefore, to the origins of consumerism.

The sociogenesis of consumerism

> The consumer orientation, first developed as a by-product, and an outlet, of the industrial pattern of control, has been finally prised from the original stem and transformed into a self-sustained and self-perpetuating pattern of life.
>
> (Bauman 1982, 179)

According to Sayer (1991, 55), our understanding of the development of modern capitalism requires 'a social psychology of consumption to comple-ment Marx's focus on production'.[3] In fact, introducing consumption into the picture complicates matters much more than Sayer implies (even if he is right to note that 'Marx ... can hardly be faulted for not foreseeing the Visa card' [*ibid.*]). The retrospective view of the development of capitalism we are afforded today throws up some particularly challenging questions with respect to Marx's account of proletarianization, industrialization, and class formation. Many of these issues have already been widely debated, particu-larly by social and labour historians, with most contributions to the debate serving to undermine the view – never quite Marx's own – that 'the prole-tariat was called into existence by the introduction of machinery' (Engels 1973, 61). Even so, the relations between production and consumption have been largely neglected in this literature, even if such a statement is less true today than it was a decade or so ago. There is still a marked tendency, for instance, for work on consumption to begin afresh on an entirely new plot – or, at least, to assume the character of a minor extension to existing accounts of the development of capitalism. Hence Sayer's (1991, 55) prefer-ence for thinking in terms of a 'complementary' account to sit alongside Marx: he merely mentions certain 'welcome moves in this direction', citing Campbell (1987) as an example. Campbell's work is, in fact, an ideal example of an approach to consumerism that remains detached from production in all but the most superficial ways. It offers no indication that our understanding of the development of consumerism might necessitate a substantial effort of reconstruction with respect to our understanding of the entire history of capitalism – the kind of major rebuilding programme that involves removing some of the original supporting walls and entirely transforming the shape of the original shell. Bauman (1982; 1983; 1987), in contrast, has managed to accomplish precisely such a reconstruction, bringing out its implications in terms of the development of consumerism. Although space restricts the extent to which we can do justice to his argument, it is worth retracing its outlines in detail, not least because of the implications it carries in terms of the *consequences* of consumerism – as we shall see in due course.

In the light of Bauman's account, the longstanding fixation of the modern intellectual on the working class becomes more readily recognizable for what it was. Once this is acknowledged, the way is opened for an interpretation of the history of resistance to the imposition of factory labour that frees it from its previous intellectual mythologization. Once freed, a clearer view of its role in the changing figuration of society becomes possible, along with a clearer view of the nature of consumerism itself. It is highly understandable why the version of events offered by Marx should have seemed so reasonable to the modern intellectual mind, in search of the 'subject of history'. As Bauman (1987, 174) notes, there were seemingly good grounds for 'fixing the search for "historical class" on the workers and proclaiming them the proletariat of the modern era':

> They showed signs of being conscious of the commonality of their fate, and of a determination to do something about it; they were stubborn, militant, they took to the streets, rioted, built barricades. In retrospect, we know that their militancy reached its peak in the vain attempt to arrest 'the progress of Reason', that is, the substitution of factory confinement for what memory held alive as the freedom of the petty producer. At the time, however, no such wisdom was available and it was easy to naturalize the historically occasioned militancy and impute to the restless and backward-looking factory hands the interests they did not possess. Violent resistance to being transformed into a disciplined and closely surveilled class of 'rational', capitalist society, could be taken as proof that 'class in itself' was already turning into 'class for itself'; the workers were accredited with a degree of 'settledness' in the 'rationalizing' society similar to that which came naturally to their intellectual mythologists.

As we shall shortly see, a candid acknowledgement of the way in which modern intellectual hopes were projected onto the working class puts us in a far better position to explain the evolution of consumerism. It reveals the way in which one, pre-modern, form of social arrangement (based on one form of power and control) gave way to another, modern, form of organization (based on a fundamentally different form of power and control). It also permits recognition of the way in which this latter arrangement has subsequently transformed itself again, into the postmodern form of the consumer society.

Bauman's account clearly owes much to Foucault, and embodies a distinctively Foulcauldian trait: a commitment to reinstating *the body* within social thought.[4] Before Foucault, as Bauman (1983, 32) notes, the major characters of virtually all accounts of the unfolding drama of modernity have been 'attitudes, motives, beliefs and expectations'. Social change has been one-sidedly conceptualized in terms of the 'transmission and internalization of values', with the social actor being reduced to a 'bundle of beliefs, motives and evaluative preferences' (*ibid.*, 33). This cannot but result in an 'understanding of modernity as, above all, a novel attitudinal syndrome' (*ibid.*, 37). Precisely this tendency is apparent in the vast majority of attempts to account for the origins of modern consumerism. It is evident, for instance in Taylor's (1996) account of the social attunement to 'comfort' in the development of 'ordinary modernity'. It is no less apparent in Campbell's (1987) treatise on the transformation of the work ethic, via the cultivation of individual sensibilities, into the cult of 'modern autonomous imaginative hedonism' (cf. Holbrook 1997; Campbell 1997b). Of course, it is unsurprising that Campbell should have adopted this stance, since the issue he addresses derives directly from Weber's (1976) thesis on the affinity between the protestant ethic and the spirit of capitalism. The enigma Campbell sets out to examine is how one set of attitudes (self-restraint, as embodied in the work ethic) could have transformed itself into another set of attitudes (unrestrained hedonism, embodied in the 'Romantic ethic' that purportedly paved the way towards modern consumerism). But although Campbell's account is underpinned by a basic recognition that the acceleration of capitalist production required a parallel acceleration of consumption, the version of events he recounts merely delivers a 'cultural resolution' to the problem of how this situation first arose. Bauman (1983, 42), in contrast, insists that the roots of consumerism 'do not lie in the cultural shift from the public to the private, from the puritan to the narcissistic attitude, nor in any other "cultural resolution" often constructed as an explanation of the apparently novel life patterns'. For Bauman (*ibid.*, 40), '[c]onsumerism was born as a twice removed offshoot of the frustrated resistance against disciplinary power which penetrated, and finally conquered the field of productive activity'.

The 'disciplinary power' to which Bauman refers first emerged in Europe in the seventeenth and eighteenth centuries. For Foucault, to whom we owe the notion, it heralded the birth of the modern social order. Its origins lie in the crisis afflicting traditional rural society at that time. Its ultimate consequences, however, were far removed from its original source. Disciplinary power, whilst not an essentially economic innovation, eventually gave rise to the widespread expansion of proletarianization, profoundly altering the relations between those who produced and those who benefited from their labours. This process simultaneously introduced a new kind of resistance, opening up new social fissures and sparking a conflict that was, in the end, never finally resolved but only ever displaced, or diverted, into an economic form. The initial diversion of the original power conflict into the distributive sphere was finally to see 'frustrated resistance' become caught up in the compensatory mechanism of consumption, at a second remove from the original conflict. This version of events explains why Baudrillard (1998a, 76) should insist, *pace* Campbell, that 'there has been no revolution of mores and the puritan ideology is still in force'; that 'it pervades all apparently hedonistic practices'. Let us consider these various changes in more detail, thus putting ourselves in a better position to appreciate their lasting significance.

Referring to the seventeenth and the eighteenth centuries, Foucault (1980, 104) identifies 'the emergence, or rather the invention, of a new mechanism of power ... which is ... quite incompatible with relations of sovereignty'. Its novelty lay in its 'capillary' form; in its diffuse, micro-scale deployment, which contrasted markedly with the previous form of 'sovereign power' – the kind of power that could be seized. The new form of 'disciplinary power' is, as Foucault (*ibid.*, 39) suggests, 'a synaptic regime of power, a regime of its exercise *within* the social body, rather than *from above* it'. It fundamentally recasts the relations between those wielding power and those subject to it. Prior to the advent of disciplinary power, for example, productive activity proceeded on the basis that 'the producer was in control of tools and materials which, when subject to the process of labour, resulted in the final product' (Bauman 1982, 12). Sovereign power was content merely 'to tax rather than organize production' (Deleuze 1992, 3), to extract a tithe from a process over which it had little or no direct control. The agencies of sovereign power 'did not consider the production of the surplus as their responsibility, fully satisfied as they were with the enforcement of its repartition' (Bauman 1983, 36). Where sovereign power was characterized by its *periodic* incursions into the lifeworld of the community, demanding an appropriation of the surplus product, the hallmark of disciplinary power was its *continuous* intrusion into the affairs of daily life. Its day-to-day operation consisted 'in rendering ... bodily behaviour routine, repetitive, mechanically predictable, subject to invariable rules amenable to codification and hence to "objective" scrutiny and assessment' (*ibid.*, 1983, 34). Little wonder, therefore, that the expansion of proletarianization – the divorce of producers from the means of production – would come to rely directly on the enforcement of disciplinary control. As Foucault (1980, 41) notes: 'once capitalism had physically entrusted wealth, in the form of raw materials and means of production, to popular hands, it became absolutely essential to protect this wealth'. Whereas the pre-industrial order saw the producer retaining ownership of the product up until the point at which a levy was exacted, proletarianization required 'that wealth be directly in the hands, not of the owners, but of those who labour' (*ibid.*). As Bauman (1982, 12) remarks: 'Never since the collapse of slave societies was such a formidable wealth entrusted to the care of people who had no stakes in its integrity'. However, although disciplinary power was eventually to become central to the regulation of a rapidly industrializing workforce, it was not initially ushered in to effect a change in productive arrangements. The new 'mechanisms of power emerged first for reasons of their own – but their availability made the emergence of the capitalist order plausible, as the means of control and surveillance which such an order required were already at hand' (*ibid.*, 43).[5]

In considering the origins of disciplinary power, Bauman questions Foucault's (1980, 104) specification of its emergence as an 'invention'. Was its appearance not, rather, a case of discovery? Bauman (1983, 34) notes that a 'strong case can be made for the assertion that power of exactly the kind described by Foucault was the major tool of social control, integration and reproduction of society well before the dawn of modernity'. Pre-modern forms of community were, by all accounts, characterized by an unrelenting degree of self-scrutiny and normative control that, in many respects, outflanked the disciplinary power pronounced as strikingly novel and distinctively modern by Foucault. Discovery, however, can mark as profound a change as invention, as it does in this particular case. However much it shared the attributes of later disciplinary power, the longstanding self-scrutiny practised by pre-modern communities was hardly recognized, or even recognizable, as a form of *power*. Against the highly visible form of sovereign power, which periodically descended on the pre-modern communities from the outside, the normative community controls 'dissolved in the totality of daily routine' (*ibid.*). Such locally embedded mechanisms remained more or less imperceptible, taken-for-granted aspects of everyday life. The dense pattern of sociability of the pre-modern communities meant that 'there was little chance for surveillance to emerge as an external and specialised agency responsible for the maintenance of the "pattern of conduct" as separate from the "way of life"' (Bauman 1982, 46). Only when the smooth running of the communities' self-reproduction mechanisms began to hit up against their 'carrying capacity' could such a possibility materialize. It was finally to materialize, in the face of the rapid demographic expansion of the mid- to late eighteenth century.

The unprecedented level of population growth experienced in the eighteenth century considerably exacerbated the 'vagrancy problem' that had plagued Europe through much of the early modern era (Febvre 1968; Beier 1985). This problem itself was in no small measure a consequence of the breakdown of customary tenurial relations and uprooting of the peasantry, sparked by the increasing market orientation of agriculture and the growing value of land. The rising tide of fear and insecurity associated with the 'dangerous classes' of 'masterless men', existing on the margins of society, signalled that the habitual, routinely practised community proto-form of 'surveillance' had finally outgrown its quotidian context. The situation prompted a rearticulation of the poor 'as a question of public order, rather than an issue of communal charity and individual salvation as before' (Bauman 1982, 114). The response this triggered finally permitted disciplinary power to emerge, for the first time, in its raw, unconcealed form. Within the confines of a rapidly growing number of 'spaces of enclosure' – the poorhouses, workhouses and other institutions of confinement designed as a means of 'separating order from disorder' (*ibid.*, 8) – the initial 'experiment of integral surveillance was carried out' (Foucault 1980, 71). It would finally sound the death-knell for 'the feudal principle of association between land-owning rights and administrative duties' (Bauman 1987, 25), already under pressure from those whose interests lay in the expansion of trade and commerce.

The 'proto-masses' to which the vagrancy problem had given rise threw the limits of sovereign power into sharp relief. It had become increasingly evident that sovereign power, 'armed only with coercive agencies specialized in preventing resistance against the expropriation of surplus, was organically unfit' for coping with the new situation (Bauman 1983, 35). The legitimacy of the rising demands the sovereign powers placed on the beleaguered communities could not but be undermined by the rising tide of disorder, against which it appeared singularly impotent.

> Time and again, more and more often the closer the dawn of the modern age, people had to defend their *Rechtsgewohnenhelten* against the rising appetites of their lords and princes. It was this contest which brought the phenomenon of power into relief and made its articulation, as a concept, possible.
>
> (*ibid.*, 35)

The bankruptcy of the pre-modern communities' self-reproducing mechanisms had rendered visible a form of power that was now in a position to be harnessed anew. Its conscious deployment, as a mechanism, simultaneously prompted a fundamental 'change in official forms of political power' (Foucault 1980, 39): 'the instituting of this new local, capillary form of power ... impelled society to eliminate certain elements such as the court and the king'.[6] The techniques deployed within the new spaces of enclosure gradually began 'to reveal their political usefulness ... and ... as a natural consequence ... came to be colonised and maintained by global mechanisms and the entire State system' (*ibid.*, 101). The way from the community-based proto-form of self-scrutiny to a fully fledged form of disciplinary power was paved by the centralization of the state, which took its cue from a new set of political interests (Corrigan and Sayer 1985). Despite its ostensible introduction as a means of coping with the 'excess capacity' that the pre-existing communities were unable to absorb, therefore, disciplinary power was ultimately to extend itself across the whole of the state territory, signalling a new 'politicization of space' (Foucault 1989, 336). The state took, as its model, the free towns that, from the later Middle Ages onwards, had been granted exemption from taxation. It was in the larger towns and cities that the dense sociability enjoyed by the small-scale communities had been subject to the most rapid disintegration; these 'were the very places where the old type of social power, with its instruments of control-through-space, revealed its limitations' most clearly (Bauman 1982, 46). From this point on, the 'cities were no longer islands beyond the common law. Instead, the cities, with the problems they raised, and the particular forms that they took, served as models for the governmental rationality that was to apply to the whole territory' (Foucault 1989, 336).

The profound discontinuity between sovereign power and disciplinary power owed as much to a change in the social imaginary as it did to changes in social practice. If sovereign power entailed one image of society, disciplinary power implied something altogether different. Sovereign power belonged to a society forged in the image of the 'great chain of being' (Lovejoy 1961). Lower and higher social ranks were essentially regarded as immutable, nonhomogeneous forms of human life. They were preordained as such, and tied together in a network of mutual obligation (the exchange of a portion of the surplus product for protection, both military and spiritual, for instance). With the collapse of sovereign power, however, a formerly unproblematic situation, in no need of 'solution', was no longer regarded as tenable. The advent of disciplinary power marked the *problematization* of the way of life of society's subordinate members; cast it as in need of corrective action.

> Being human, instead of remaining a natural condition enjoyed by all though in many alternative forms, became now a skill to be learned, an end to a tortuous effort, which everyone had the duty to undertake, but few were able to accomplish unassisted.
>
> (Bauman 1983, 36)

Thus, formerly inalienable ranks were recast as essentially malleable classes. Disciplinary power assumed its importance insofar as it was intrinsically geared towards 'the transformation of individuals' (Foucault 1980, 39). Its purpose was accomplished by means of the *regulation of the body*; the new spaces of enclosure were primarily a means of rendering the body *visible,* and thus amenable to strict control.[7] The routinization of behaviour this visibility permitted typically took the form of enforced labour, often on pointless tasks: 'work for work's sake' (*ibid.*, 42). The essential asymmetry between those wielding disciplinary power and those subject to its force contrasted starkly with the previous reciprocal relations. If its proto-form had its compensations, therefore, in its naked, unadorned form, disciplinary power was brutally one-sided. For the most part, this could only remain a burden to be suffered for those subject to its initial imposition. But circumstances conspired to ensure that it would not for long remain tied to its initial sphere of operation. Disciplinary power had ineluctably 'moved into the space of inter-class relations' (Bauman 1983, 35) and could only be regarded, by those on the receiving end, as 'an object of contest and as a negative aspect of the human condition which is not naturally balanced by tied benefits' (*ibid.*). It was thus destined to provoke fierce resistance. Whence this would arise, however, is crucial to the direction society would subsequently take. It could not, by and large, arise from amongst the most destitute, who found themselves in a virtually powerless situation. It arose from an altogether different quarter, as the consequence of disciplinary power began to unfold.

Let us first of all stress, however, that the routinized work-regime of the new spaces of enclosure was in no way, in its initial stages, 'subject to the logic of productive tasks' (Bauman 1982, 114). Indeed, the early workhouses and poorhouses, the madhouses and *hôpitaux*, were frequently filled with the sick, the maimed, the elderly, and others incapable of sustained productive labour. Any 'output – if expected at all – was a secondary consideration, almost a by-product of organisations designed with other objectives in mind' (*ibid.*). At the time, the new institutions were perceived as merely

> another variety of poor- or work-houses, and their owners as sui-generis agents of authorities, making the communal task of care for material and spiritual welfare of the poor their responsibility, and thereby simultaneously relieving the local tax-payer from an excessive financial burden and promising to secure the sought-after control of potential rebels as well as morally regenerating their souls.
>
> (Bauman 1983, 37)

The barely perceptible distinction between the parish-supported workhouses and those being run along lines that could, in effect, pay for themselves was nonetheless destined to establish itself more clearly. In the process, it was also to crystallize the class interests of the bourgeoisie. At first, the new power-masters preoccupied themselves with the social exclusion of the poor and the mad, with correcting the unrestrained tendencies of the morally bankrupt, and so on. The techniques and procedures employed, however, began to reveal their hithertofore unrecognized potential:

> what the bourgeoisie needed, or that in which its system discovered its real interests, was not the exclusion of the mad or the surveillance and prohibition of infantile masturbation [and other indications of moral turpitude] ... but rather, the techniques and procedures themselves of such an exclusion.
>
> (Foucault 1980, 101)

Exclusion, enclosure, surveillance and regulation began 'to become econom-ically advantageous and politically useful' (*ibid.*). For example, the closely scrutinized and regimented drill perfected in the workhouse made the use of new technological devices possible. It was their essential compatibility with a body of well disciplined, compliant subjects that prompted acceptance of the new technologies by those seeking to impose discipline and regulate behaviour patterns, however, and not a case of the pre-industrial poor being trained to cope with the regularities required by the new productive tech-nologies. Yet such practices inevitably began to 'lend themselves to economic profit' (*ibid.*, 101). The gains such a system delivered were not to remain a mere side effect of the imposition of discipline.

The techniques of discipline and surveillance had revealed themselves as those by which the bourgeoisie could maintain its position in the newly established hierarchical order, whilst productivity and profitability gradually turned into a self-propelling force. Where sovereign power had worked to ensure

> the upward flow of the surplus resulting from an essentially autonomous productive activity, the function of the new hierarchy was more than anything else to assure the reproduction of a form of life compatible with the continuation of [that] hierarchical order.
>
> (Bauman 1983, 36)

The means by which this would be accomplished led in the direction of widespread proletarianization, and would ultimately lay the foundations of industrial capitalism. The process did not, of course, follow a route that was clearly marked out in advance. Nor did the regime perfected in the work-house provide a model that could be unproblematically translated into a working version of the factory system. Nonetheless, as Ariés (1962) has convincingly demonstrated, work relentlessly became an increasingly impor-tant and time-consuming component of everyday life as the pace of modernization increased.

The course of industrial revolution was far more tortuous and convo-luted than the simple substitution of machine labour for hand labour. The pattern and form of mechanization differed greatly from one sector to another, from one region to another, and in a multitude of other ways (Samuel 1992). Well before 'machinofacture' took hold, discipline was already being exerted by a sustained process of deskilling, and by bringing into the labour market hitherto unexploited reserves of cheap, unskilled labour, particularly that of women and children (Rule 1987). The initial stages of industrialization called forth significant increases in the domestic system of production, as well as in much traditional craft production. Yet it 'was in the factory system that the concerns and the ambitions of the new type of power attained their fullest and most vivid manifestation' (Bauman 1982, 11). The increasing demand for labour as a *commodity* relied fundamentally on the enforcement of discipline, and the factory represented one of the most efficient means of imposing control on the body of workers. It provided a means of instilling a new kind of time-consciousness into the workers, for instance, of subjecting the work-force to the quickening pace of production (Thompson 1967).[8] Needless to say, the transformation from one state of affairs to another did not proceed smoothly. But the most significant dissent did not emerge from those who found themselves in the worst predicament, as we have already intimated. The initial blockage on the road to wholesale proletarianiza-tion took the form of a resistance movement rather than a revolutionary challenge.

The fiercest resistance came from a quarter that felt itself under siege: the artisans and petty producers of the traditional craft industries. The autonomy still enjoyed by many of the traditional industries bore more than a passing resemblance to the situation enjoyed more widely before the advent of disciplinary power. Needless to say, it would not be given up without a fight.

> If the labouring poor, ground in the mills of the Poor Law and anti-vagrancy regulations before they passed the factory gate, had no tradition of independence and self-management to recall as an alterna-tive to their present fate, the craftsmen were in an entirely different position. They had a fresh and rich memory of the guilds, kept alive before their eyes within the few crafts not only unscathed, but strength-ened by the initial stages of industrial revolution.
>
> (Bauman 1982, 115)

The voice of protest emanating from the craft tradition signalled much more than the fact that their livelihoods were under threat. The craft labourers, unused to accepting and unwilling to accept the imposition of discipline from outside their own carefully honed system of apprenticeship and control, reacted with a profound sense of moral indignation and 'outraged justice' (Moore Jr 1978). The knowledge and experience of collec-tive organization which the craft labourers 'derived from their memorized past' provided a ready-made means for the 'self-defence or pursuit of their claims, moreover, even if the initial objectives of the struggle proved unreal-istic and had to be adjusted to stubborn reality' (Bauman 1982, 115). 'Stubborn reality' would indeed ensure that the craft labourers' efforts fell 'short of attaining their backward-looking aims' (*ibid.*). But the particular structure of that reality was to channel events in a decisive way. The outcome of the struggle gave a seminal twist to the future trajectory of capi-talist development. It introduced into the capitalist system not only the possibility of an effective alternative to disciplinary control; it simultane-ously provided the means to ensure that the conditions on which capitalism thrived would remain in force.

Whatever 'productivity gains' mechanization was capable of delivering, the production process remained, in the early years of the industrial revolution, partial and patchy in its coverage. Together with the limits of the earliest technologies, the routinization imposed on those workers drawn from the massed ranks of the poor left little room for individual discretion. As a result, the production process as a whole remained reliant on the input of skilled labour, capable of exercising precisely the kind of autonomy that disciplinary power had elsewhere sought to crush. Thus, for 'several decades, until skilled machines sapped the indispensability of skilled labour', the cooperation of skilled labour was actively solicited by the factory owners (Bauman 1983, 38). In what hindsight reveals as an historically momentous exchange of rights and freedoms, the 'gradual surrender of one aspect of the craftsman's tradition – self-management – had to be obtained through the boosting of another aspect: market orientation and self-interest' (*ibid.*). The compliance of skilled labour was, in effect, bought. The ability of the craft labourers to control their own working conditions was traded-off against the ability to earn: an exchange of power for purchasing power (Gray 1981; Hobsbawm 1964; Moorhouse 1978).[9] This fateful process, dubbed the 'economization of power conflict' by Bauman (1983, 38), saw

> the conflict over *control*, triggered by the attempt to extend over the skilled part of factory labour the disciplinary forces developed in dealing with the unskilled part (or the fear of such an extension) ... displaced and shifted into the sphere of surplus distribution. Legitimation of the new structure of power and control was obtained through the delegitimation of the division of the surplus – the one thing which the 'sovereign power' in pre-industrial society sought to keep clear from contest.

The surrender of autonomy (reconciliation to heteronomy) in the productive sphere was undertaken on the unprecedented basis that the share of surplus value became open to negotiation. This was the price of ensuring the new work ethic reached this vital part of the labour force. The channelling of the initial power conflict into the sphere of distribution was based on the promise of greater levels of wealth, security, and living standards: everything, in fact, except the autonomy in the productive sphere relinquished in this momentous exchange. In its wake, money became 'a makeshift power substituted for the one surrendered in the sphere of production; while the experience of unfreedom generated by conditions in the workplace [was] re-projected upon the universe of commodities' (*ibid.*). Although this early compromise would hardly stem the subsequent development of working-class consciousness, it nonetheless established a set of conditions that were vital to the reproduction of capitalism, and which would eventually steer it towards its current, consumerist form.

Unlike the preceding 'static economy', in which sovereign power 'made a nonproductive consumption of the excess wealth', capitalism is dedicated to accumulation and the 'dynamic growth of the productive apparatus' (Bataille 1988, 116). As such, its reproduction relies on a parallel acceleration of consumption. The economization of the original power conflict assumes its significance not because of the (minor) expansion of the market achieved by adding a further component to the overall level of effective demand, of course. Its importance arises from the fact that it sowed the seeds of an *insatiable* demand. Having been blocked in the sphere of production and diverted into the sphere of consumption, the original desire for autonomy was destined to remain unfulfilled, however substantial the compensation might be. A compensatory mechanism is precisely that, and it remains highly 'unlikely that the emancipatory urge, originated and perpetually re-fuelled by the heteronomy of productive activity, will ever be quenched by a success, however spectacular, of its surrogate form' (Bauman 1983, 39). This originary diversion was thereby set to take on a self-perpetuating momentum of its own. The 'unsatisfied need for autonomy' was destined to put 'constant pressure on the consumer urge, as successively higher levels of consumption become disqualified and discredited for not bringing the hoped-for alleviation of stress' (*ibid.*). In fact, one could see the subsequent history of class struggle as the working out of a situation established at a relatively early stage in the development of industrial capitalism, which unwittingly turned out to be a self-serving force for the perpetuation of capitalism *per se*. Simplifying to the extreme, the compensatory mechanism of consumption made the notoriously unrevolutionary aim of 'a fair day's pay for a fair day's work' appear far more reasonable (and far more satisfactory) than it ever otherwise would have done. Whilst the nineteenth century saw continuing struggles to redress the exploitation of the masses, which secured modest reforms as well as promoting greater consciousness of the power of collective action, none of this could finally gainsay 'the annunciation of Acquisitive Man' (Thompson 1963, 832). This is, no doubt, to gloss over a considerable amount of complexity with respect to the interplay between production, consumption, and class consciousness. It would doubtless be impossible to specify a straightforward relationship between cultures of consumption and levels of working-class radicalism, for instance, and particular regimes of production frequently translated into highly specific cultures of consumption. Different cultures of consumption offered different levels of compensatory relief, prompted different motives for consumption, and led to differing degrees and forms of class consciousness (Stedman Jones 1983). Experiences of work conditions, pay and job security differed greatly by economic sector and geographical region, for example, and had a direct bearing on the extent to which it made sense to defer gratification or to indulge in the kind of immediate gratification that made life minimally bearable. In other situations, it made sense to marry consumption to investment, knowing that the pawnbroker would partially redeem the value of goods obtained when hard times returned. Such different cultures of production and consumption together inflected the degree and form of working-class radicalism. Nonetheless, insofar as consumption had become increasingly significant as the sustaining force of capitalist expansion, its compensatory role was set to become all important.

For this reason, consumption itself, for a time, became a battlefield (Furlough and Strikwerda 1999; Purvis 1998). If the workers' struggle for the right to 'the whole product of labour' was, as Bauman (1982, 13) has emphasized, not a 'phenomenon born of the distinctive character of the capitalist society' but an idea firmly established 'in the historical memory of the individual producer' since time immemorial, its conceptual development by the Ricardian socialists and by Marx permitted its translation into the arena of consumption. The workers' struggle could just as easily be conceived as the consumers' struggle. For example, an early advocate of the dividend on purchase found it possible to 'contend that consumption was the source from which all profit was derived and capital was unproductive without it; that therefore the consumer was entitled (as a consumer) to a return of [a share of] the profits derived on his purchase' (Anon. 1839). With the benefit of hindsight, consumer cooperation appears to have been engaged in a rearguard action. Although the fact was not yet evident at all levels of society, capitalism had already succeeded in severing the ties between consumption and need that the cooperative movement sought to re-establish and reaffirm. From the point at which it first assumed its compensatory role, offsetting the initial loss of productive autonomy, consumption had ceased to be the servant of need and become locked into a seemingly unstoppable vicious cycle fuelled by the dynamics of accumulation. Because of its status as a compensatory means of overcoming the persistent power asymmetry in the sphere of production, and because the attempt to meet the spiralling demand this asymmetry stimulates could not but drive the production process to ever-greater levels of intensity, the situation was destined, from the start, to be exacerbated at every turn. There is an inherent and apparently insoluble contradiction at the very heart of the system; a 'major structural fault' which is set to generate 'an ever increasing scale of contradictions' (Bauman 1983, 39). The ultimately unfulfilling effects of the compensatory drive to satisfy autonomy in the sphere of consumption makes industrial society 'the only one in human history which experiences stability as crisis rather than bliss' (*ibid.*).[10]

It is only our familiarity with this situation that disguises its fundamental absurdity. Indeed, the necessity of 'growth for growth's sake' was far from an accepted fact for the early political economists. The transition to a market economy was envisioned as a once-and-for-all shift, ultimately poised to improve the lot of all. Whatever inequalities were generated in the process might legitimately be subject to subsequent redistributive correction, providing that the mechanism of wealth-creation remained untouched by visible hands. The 'natural laws' of production were, for Adam Smith and the political economists following him, best left well alone ('a fair reflection of the "naturalization" of productive relations perpetrated by the growing heteronomy of producers' [Bauman 1983, 39]).[11] Likewise, redistributive interference (itself 'a fair anticipation of the needs arising from the "opening up" of the division of surplus to non-economic forces' [*ibid.*]), would inevitably be restricted by the total volume of goods that could be produced according to the working out of 'natural laws'. What went entirely unanticipated in the thought of the early political economists was the very possibility of interminable growth. A stationary-state, a finite equilibrium implied by the laws governing production, was expected to arrive imminently. It failed, of course, ever to materialize: 'incessant economic growth proved to be if not the undetatchable attribute of industrial society, then certainly a *sine qua non* condition of its survival' (*ibid.*). Accordingly, the economization of the original power conflict was ultimately to rebound in a 'politicization of economics':

> Being a condition of its survival, but not a feature guaranteed by the inner logic of industrial economy, economic growth – as a postulate more than as the reality of economic life – turned gradually into a major factor in shaping the system, its contradictions, and the way of coping with the contradictions, of a society based on an industrial mode of production.
>
> (*ibid.*)

In consequence, the state became decreasingly concerned with corrective redistribution, simultaneously assuming an increasingly functional role in the effort of stimulating production itself. Much later, the short-lived history of the welfare state was, as we saw in the previous chapter, to become the exception that proves the rule.

A further seminal twist arose from the politicization of economics and the inherent commitment to growth it entailed: consumption, already possessed of a significant *productive* role, finally began to assume a fully *reproductive* role. The initial 're-orientation of life interests from the reproduction of subsistence [in] its traditional form to the improvement of living standards, i.e., to … increasing consumption' (*ibid.*, 40), arose from the historic compromise reached with the labour aristocracy. The redrawing of the rules for the division of the surplus could easily have turned out to be a short-lived solution to an immediate problem, destined to fade away as quickly as it arose. Indeed, it might well have been envisaged that, in the long run, 'one could … dispose entirely with [the skilled workers'] strategic role in the factory production [process] through transferring their skills to new, more sophisticated machinery' (Bauman 1982, 115). But the essential compatibility of this makeshift means of obtaining compliance with the imperative of growth proved vital to the perpetuation of capitalism. It ultimately set the stage for a movement away from the original forces of repression and towards the seductive underpinnings of the consumer society.

Of course, the future consequences can hardly be seen as the explanation of the origins. At the time, it was simply the case that industry was faced with an unprecedented prospect of expansion, leading to a veritable explosion in the goods available on the market. The desire to consume was eagerly met by producers, only too willing to 'harness the energy of the power conflict, already channelled into the contest over distribution, to the commodity market' (Bauman 1983, 40). The aggressive advertising, which took off at the end of the nineteenth century, set about linking the urge to consume 'with specific goods the market had to offer; giving a tangible and purchasable content to compensatory dreams' (*ibid.*).[12] Ultimately, the temptations of the market would assume a force equal, if not superior, to disciplinary power: in terms of its ability to ensure a way of life compatible with the overarching requirements of the system. Consumerism, in other words, would come to take over where disciplinary power left off.

Where disciplinary society relied fundamentally on 'the organization of vast spaces of enclosure' (Deleuze 1992, 4) to mould compliant subjects by means of regimented bodily drill, compliance in the consumer society is assured without the need for confinement and regimentation. The very term 'compliance' seems out of place in the context of the consumer society, since its assurance is predicated on the *absence* of constraint – on the 'freedom of choice' offered to the consumer. Deleuze (1992, 4) prefers to speak in terms of 'controls': 'Enclosures are *moulds*, distinct castings, whereas controls are a *modulation*, like a self-deforming cast that will continuously change from one moment to the other'. Or, as Bauman (2002, 8) puts it, it is 'as if the moulds into which human relationships had been poured to acquire shape have now been thrown, themselves, into [the] melting pot'. Freedom of choice is, perhaps, the definitive feature of the consumer society; and thus, paradoxically, a strict necessity. The consumer society is a society in which 'we have no choice but to choose' (Giddens 1994, 75). It thereby puts the controls directly into the hands of the controlled. Just as industrial capitalism was erected 'on the formal emancipation of the labor force (and not on the concrete autonomy of work, which it abolishes)', the consumer society is 'only possible in the abstraction of a system based on the "liberty" of the consumer' (Baudrillard 1981, 82–83). Indeed,

> it is *necessary* that the individual user have a choice, and become through his choice free at last to enter as a productive force in a productive calculus, exactly as the capitalist system frees the laborer to sell, at last, his labor power.
>
> (*ibid.*)[13]

Little wonder, then, that consumption should come to figure 'not as a right or a pleasure, but as the *duty* of the citizen' of the consumer society, and as a *willing* duty, at that (Baudrillard 1998a, 80). Consumption assumes the force of a moral imperative, albeit that of a 'fun morality' (*ibid.*).

As the twice-removed offshoot of the original imposition of bodily discipline, consumerism could not but retain the traces of its origin: 'what was negated, could not but determine the substance and the form of the negation' (Bauman 1983, 40). As 'a compensatory reaction to *heteronomous* bodily drill', consumerism 'selects an *autonomous* bodily drill as its principal target': it is 'not about the emancipation of the body from control; it is about the joy of controlling the body of one's own will' (*ibid.*). One's 'body image', particularly for women (though increasingly for men), is subject to a 'beauty imperative' (Corbett 2000, 130). Hence Baudrillard's (1998a) suggestion that body itself might be regarded as 'the finest consumer object'. The imperative to consume demands, above all, that the body 'be *made fit* to absorb an ever growing number of sensations the commodities offer or promise' (Bauman 1983, 40). The objective is not simply 'to train the capacity to enjoy music' or whatever it may be; it is 'to make the body capable [of withstanding] a permanent exposure to the flow of sometimes deafening, sometimes barely discernible sounds' (*ibid.*, 40–41). Each and every possible sensation retains its attraction 'in as far as the elusive bodily sensation is translated into "objectified" indices of observable routine, and hence the relevance is shifted from the outcome to the bodily drill itself' (*ibid.*, 41). One thinks, for example, of *The Joy of Sex*, *The Bluffer's Guide to Wine*, and all such other manuals of self-instruction. The routines instilled are generally geared towards ensuring that the consumer might will and act to absorb *more* of the sensations on offer. Indeed, the consumer is forever 'haunted by the fear of "missing" something, some form of enjoyment or other. You never know whether a particular encounter, a particular experience ... will not elicit some "sensation"' (Baudrillard 1998a, 80). The contemporary relation to the body, therefore, is 'not so much a relation to one's own body as to the functional, "personalized" body. ... [I]t is mediated by an instrumental representation of the body' (*ibid.*). The inadequacies of the body, when set against this ideal, mean that it can never be left alone or unattended. It becomes a charge, in need of constant vigilance. Although it is now a question of *self*-discipline, the care of the body, as 'the crucial time- and money-consuming activity of the denizens of the consumer society' (Bauman 1983, 41), places it under a parallel kind of stress to that experienced by the body of the worker.[14] The consequences of this situation are legion, and intensely paradoxical. Let us begin to disclose their nature by reconsidering the way in which the puritan continues to cast his shadow over the consumer society.

The ghost of capitalism future-anterior

> The *flâneur* wanted to play his game at leisure; we are forced to do so. ... In Baudelaire's or Benjamin's view the dedication to mobile fantasy should lie on the shoulders of the *flâneur* like a light cloak, which can be thrown aside at any moment. But fate decreed that the cloak should become an iron cage.
> (Bauman 1994, 153 [with apologies to Weber])

We have already noted the enigmatic relationship between the puritan, whom Weber placed at the centre of capitalist development, and consumerism, the tenets of which appear to mark a fundamental discontinuity with the puritan ideology. For Campbell (1987), the transformation of industry and thrift into hedonism and consumerism rests on a fundamental attitudinal change, which nonetheless has its origins in certain distinctively puritan traits. Campbell points, for example, to the redirection of self-control towards the open display of emotional sensibilities, which finally promotes the kind of abstract appreciation of pleasure that can be harnessed to consumerism. Baudrillard (1998a, 82) similarly emphasizes the extent to which the 'themes of Spending, Enjoyment and Non-Calculation ("Buy now, pay later") have taken over from the "puritan" themes of Saving, Work and Heritage'. In line with the account developed above, however, Baudrillard stresses the sense in which consumerism is merely the continuation by other means of a system established on the basis of discipline and the work ethic. The apparently revolutionary transformation is, in fact,

> merely the semblance of a Human Revolution: in fact it is an internal substitution, within the framework of a general process and a system which remain in all essentials unchanged, of a new system of values for an old one which has become (relatively) ineffective.
>
> (*ibid.*)

A twofold error is involved in interpreting consumerism as a fundamental inversion or revolution, rather than an internal substitution of this kind. First, the attempt to conjure the 'modern hedonistic consumer' out of the inner-directed character of the puritan emerges from the kind of analysis that links 'Weber's discussion of Puritan ascetism to psychological repression' (Giddens 1994, 70). As we shall shortly see, there is a far more direct link between the puritan and consumerism than the one arising from a gradual transformation of self-restraint into the opposite attitude. Second, we need to recognize that the portrait of the puritan that has been handed down to us was constructed from a very particular perspective. A number of significant distortions become evident once the vantage-point is changed. The remainder of this section addresses these issues. The broader purpose, however, is to move towards a better appreciation of the *consequences* of consumerism and, in particular, their intensely contradictory nature.

'For many years now', writes Bauman (1987, 149), 'the "Puritan" has occupied a disproportionally large place among intellectual preoccupations'. The reason for the puritan's elevated position lay in the urge to see (and to forge) modernity as a fully rational society – the kind of society that surpasses the limitations of the past by severing the ties of tradition, whose future is assured by the application of reason. The fact is that 'Weber's moral tale armed the intellectuals with a most powerful aetiological myth of modernity' (*ibid.*). The pronouncement of an elective affinity between 'the frame of mind of a hard-working, profit-calculating industrialist and the prosaic severity of the reformed religion' (Bataille 1988, 115) served the imaginings of the modern mind well. Indeed, the mythic figure of the puritan was never so alive as he was in the modern mind. His 'passion for perfection, for a righteous life, for hard work, for the taming of instincts and emotions, for the delay of satisfaction, for a "lifework of virtue", for control over both body and fate', bore an uncanny resemblance to 'the central actor of a reason-guided society and the product of such a society' (Bauman 1987, 150). Little wonder, therefore, that the puritan came to occupy intellectual concerns more and more often as modern minds began to sense 'the first prodromal symptoms of the approaching end' (*ibid.*, 150). Carroll's (1977, 17) identification of the subsequent emergence of a 'remissive culture', populated by a hedonistic personality whose 'one conscious norm is to be anti-Puritan' reveals with absolute clarity the death of the puritan. It does much more than this, however. It simultaneously reveals that the 'puritan' stood, all along,

> as a shorthand for the acceptance of constraint and supra-individual authority, for the willing effort to repress emotional drives and subordinate them to the precepts of reason, for the belief in an ideal of perfection and objective grounds of moral, aesthetic and social superiority, for self-restraint and self-improvement.
>
> (Bauman 1987, 153)

And it reveals that all hopes for a world in which the puritan would reign supreme have since been irretrievably dashed. The irreversible transition from the modern to the postmodern world does not, however, entail that all its former features have vanished without trace. The spectre of the puritan continues to stalk the postmodern world: we

> [We] may assert that the puritan ethic, with all it implies, in terms of sublimation, transcending of self and repression (in a word, in terms of morality), *haunts* consumption and needs. It is that ethic that gives it this compulsive, unlimited character.
>
> (Baudrillard 1998a, 76)

This underlying compulsiveness is, as we shall see, fundamental to disclosing the fact that 'consumerism is not the dawn of a new historical era' but 'the last, most paradoxical and absurd, stage of the old one' (Bauman 1983, 42); its posthumous stage, one might say. Thus, the death of the puritan signals not only a departure from the modern scene (and hence an occasion for lamentation on the part of all those who never got to see the construction of the final, glorious, rational order they dreamt of); it also heralds the untimely arrival of a ghostly presence.

The shadowy presence of the puritan in the midst of the consumer society stems from the sense in which, for the puritan, orderliness was next to godliness. As Freud (1955, 30) first pointed out,

> order is a kind of compulsion to repeat which, when a regulation has been laid down once and for all, decides when, where and how a thing shall be done, so that in every similar circumstance one is spared hesitation and indecision.

This is very much apparent in Weber's (1976, 182) suggestion that his analysis of the spirit (*Geist*) of capitalism is an analysis of 'the ghost of dead religious beliefs'. It explains why Giddens (1994, 70) should insist that the 'core of the capitalist spirit was not so much its ethic of denial as its *motivational urgency*, shorn of the traditional frameworks which had connected striving with morality'. Indeed, as Davis (1992) has shown – considering the origins of jokes against the Scots – this compulsiveness did not go unnoticed or unchallenged even from within the thick of modernity. In this respect, there is precious little distance between the *aura sacra fames* in the productive sphere and the endless pursuit for satisfaction that has since colonized the sphere of consumption.

> The capitalist ... was primed to repetition without – once the traditional religious ethic had been discarded – having much sense of why he, or others, had to run this endless treadmill. This was a positive motivation, however; success brought pleasure rather than pain. Hedonism differs from pleasure enjoyed in much the same way as the striving of the entrepreneur differs from economic traditionalism. ... [T]his is why it is much more closely related to the traits upon which Weber concentrated than may seem the case at first blush.
>
> (Giddens 1994, 70)

There is, then, a direct line between the puritan mentality and consumption, which parallels and extends Weber's account, rather than contradicting it. Bauman (2001c, 18) observes that 'under certain conditions irrational behaviour may carry many a trapping of rational strategy and even offer the most immediately obvious rational action among those available'. Both the work ethic and the subsequent systemic role assumed by consumption rest on this kind of 'rationality'; both are particular instances of that general rule. The basic irrationality of endlessly chasing after satisfaction in the consumer market has been rendered 'rational' in precisely the same way that the irrationality of intensified production had been rendered 'rational' at an earlier stage of capitalist history, thus meshing seamlessly with capitalism's overall systemic requirements.[15] This makes the sense in which we are dealing with an 'internal substitution' more readily appreciable. In this light, the crucial difference between the system of industrial capitalism and contemporary consumer capitalism can be seen to be merely the *intensity* of the consumption process.

'Through consumption', writes Appleby (1993, 172), 'people indulge themselves, seeking gratification immediately'. This would seem a simple and apt description of contemporary consumerism. Whilst there has always been a wide variety of strategies of consumption, linked to the stability or otherwise of employment conditions (such that what was rational for one was not necessarily rational for another), the systemic requirements of capitalism have necessarily been differentially served by these varying strategies. Where, previously, 'the ethical principle of delayed gratification secured the durability of the work effort' (Bauman 1999c, 5), capitalism has since moved into a position where 'instant gratification' best serves its purposes. At the same time, although hardly for 'functional' reasons, instant gratification has become the most rational strategy for the majority of consumers to pursue (to put it another way, the number of consumers for whom this is the most rational strategy has dramatically increased). Thus, if the phase of industrial capitalism saw production as the main sphere in which the endless chase after riches was pursued – necessarily complemented by restraint and thrift in other areas of life – this endless chase has now been extended to the sphere of consumption, surmounting previous contradictions with a new, heightened form of contradiction. In a fully fledged consumer-capitalist system, therefore, not one but two 'irrationalities meet, cooperate and jointly self-reproduce through the rationality of sellers' calculations and the rationality of buyers' life-strategies' (Bauman 2001c, 24). Although the operation of the market is largely responsible for ensuring that the 'rationality' of irrational consumerism has taken hold, this should not be regarded as a simple functionalist argument – as if the systemic requirements of capitalism could magically conjure up the kind of behaviour it requires. It is more a case of the market opportunistically expanding to fill the void left by the decline in alternative means of ensuring the continuity of life-experience. It is as much a consequence as a cause of this pattern of change (although, undoubtedly, the market is simultaneously responsible for dismantling the other ways and means of seeking, and prospects of achieving, such continuity of experience).

It is, then, an overall change in social stability that underlies the shift in 'rationality' from deferred to instant gratification. Durkheim (1972, 115) wrote that 'societies are infinitely more long-lived than individuals', suggesting that the transience of an individual's life could draw succour from the permanence of society as a whole. Today, as Bauman (2000) suggests, the relative permanencies have changed places; the lifespan of the individual far outlasts society's once durable institutions. In the absence of a solid social framework, it is increasingly down to the individual to 'compose ... the continuity which society can no longer assure or even promise' (Bauman 2001c, 24). Where it once made sense to resist temptation, in order to be better prepared for the proverbial rainy day, deferring gratification appears increasingly likely to invite regret over missed opportunities. Where society no longer promises permanence, where everything carries an 'until further notice' clause, where assets are likely to turn into liabilities overnight, long-term commitments are best avoided. It is increasingly sensible, under such conditions, to keep one's options open and flexible. 'Get it whilst you can' becomes the most appropriate motto for life in a consumer society (Clarke 1998; 2000). This is precisely where the market comes into its own. The task of constructing some semblance of continuity is hardly delegated to the individual unassisted. The consumer market is dedicated to supplying the necessary means for its execution – though rarely for its accomplishment. The market offers

> the reassuring certainty of the present without the frightening prospect of mortgaging the future. It supplies durability through the transience of its offerings – a durability which no longer needs to be built piece by piece through perpetual effort and occasional self-sacrifice.
>
> (Bauman 2001c, 24)

It offers, in short, the kind of 'continuous discontinuity' which today appears to be 'the only form continuity may take' (*ibid.*, 23). The very form of the market has, therefore, afforded its success; to the extent, as Baudrillard (1998a, 192) suggests, that *'the ludic dimension of consumption has gradually supplanted the tragic dimension of identity'*. The tragic quality of identity lay in the fact that the arrival of the long-awaited maturity date of one's life's-work was always likely to prove, in retrospect, a disappointment (revealing that one's life-journey had, all along, been proceeding towards a mirage rather than the anticipated oasis). The ludic quality of consumption relates to the situation where any move an individual might make is unavoidably unpredictable yet, nonetheless, hardly devoid of consequence. In such circumstances, the market offers a ready-made means for the individual to adopt one particular course of action, try it out for size, gauge its ability to meet requirements, but also to discard it if (and when) it fails to live up to expectations. The market's capacity to offer an endless stream of new commodities, suitable for donning a 'makeshift identity' – fit for the moment, yet easy enough to cast off when the moment is passed – ensures that the 'never-ending process of identification can go on, undisturbed by the vexing thought that identity is one thing it is unable to purvey' (*ibid.*, 25).[16]

There is, in all of this, a clear reflection of the origins of consumerism. Despite its ideological characterization as a form of freedom, consumption necessarily assumes the status of a *task*: something one is *obliged* to do something about. Consumerism inaugurates a 'subtle dialectical game of dependence and freedom' (Bauman 1992c, 24). Specifically, the consumer is forced 'to adapt to the principle of need ... or, in other words, to the ever full and *positive* correlation between a product of some kind (object, good

or service) and a satisfaction through the one being indexed to the other'
(Baudrillard 1998a, 177). As a result,

> *the whole of the negativity of desire,* the other side of *ambivalence,* and
> hence all the things that do not fit into this positive vision *are rejected,*
> *censored by satisfaction itself* ... and, no longer finding any possible
> outlet, crystallizes into a gigantic fund of anxiety.
>
>   (Baudrillard 1998a, 177; cf. Lasch 1980; 1984; Pahl 1995)

This is not, however, the kind of disabling angst that would bring the system
to a grinding halt. On the contrary, the fund of anxiety re-enters and circulates within an expanded system that feeds off and converts its negative
consequences into the means of its own expanded reproduction. The same
goes for waste, crime, pollution, and so on. As Baudrillard (1998a, 42) wryly
observes, we 'have to accept the hypothesis that all these nuisances some-
where enter into the equation as positive factors, as continual factors of
growth, as boosters of production and consumption'.[17] Even so, the
ideology of 'freedom of choice' should not be taken lightly. Ideology rarely
amounts to mystification pure and simple, and freedom of choice is not in
any simple sense an illusory freedom (though this is far from suggesting that
it amounts to a 'real' freedom). Freedom of choice, like all other freedoms,
possesses a double-edged character: freedom entails responsibility. It is,
however, the *individualization* or *social atomization* implied by freedom of
choice that is particularly significant.

'What we have not chosen', writes Kundera (1984b, 85), 'we cannot
consider either our merit or failure'. However, 'whatever is chosen is a
matter of individually made decision and will remain a feather in the individual's cap or a burden on that individual's conscience' (Bauman 1999b,
138–139). The crucial distinction, first established at the dawn of modernity,
'between chosen strategy and unsolicited fate' (*ibid.*, 136) has, finally, been
placed squarely on the shoulders of the individual.[18] The upshot is a society
saturated by an unprecedented degree of uncertainty, which is nonetheless
met by the spontaneous production of various forms of specialist knowledge aimed at managing uncertainty. Whilst unavoidably arousing anxiety,
therefore, the consumer society simultaneously works 'to diminish this
anxiety by the proliferation of caring agencies: innumerable collective
services, roles and functions are created; soothing, guilt-dispelling balms and
smiles are injected into the system' (Baudrillard 1998a, 177). Because the
'choices that are constitutive of lifestyle options are very often bounded by
factors out of the hands of the individual or individuals they affect'
(Giddens 1994, 75), the consumer is typically invited to place his or her *trust*
in some kind of mechanism purposely designed to offer reassurance. Indeed,
as a means of coping with the possibility of frustrated expectations, one
might say that *trust itself* amounts to such a mechanism. This is a fundamentally different arrangement to that found in traditional society (Boyer

1990). Whilst tradition might appear to embrace a 'repetitive' pattern, only to modern eyes does its ostensible alignment with the past foreclose on the future. In fact, tradition accords to a *reversible* time, where neither 'past' nor 'future' 'is a discrete phenomenon, separated from the "continuous present", as in the case of the modern outlook' (Giddens 1990, 105). Tradition

> denies in practice what it avers in theory. It prompts us to believe that the past *binds* our present; it augurs, however (and triggers), our present and future efforts to *construe* a 'past' by which we need or wish to be bound.
>
> (Bauman 1999b, 132)

In other words, tradition assumes a symbolic relation to the world, which ameliorates contingency by placing it beyond human will. As Baudrillard (1993b, 62) puts it, '[s]ymbolic culture has always been lived as a denegation of the real, something like a radical distrust: the idea that the essential happens elsewhere than in the real'. Modernity, however, amounts to precisely the opposite situation: 'all that is required now is to operate in the real' (*ibid.*) – which explains the proliferation of relations of trust.[19]

Luhmann (1979) proposes that 'trust' is a means of overcoming the unbearable level of contingency that would otherwise arise as society grows in scale and complexity, and adopts a 'future-orientated' character. Hence Giddens' (1990, 84) suggestion that modern society is distinctive in the extent to which it is 'structured by trust in abstract systems'. The proliferation of trust therefore indicates a *generalized absence* of confidence, for only when confidence is lacking is trust called upon. Whilst both confidence and trust presuppose a situation of risk, trust additionally 'expresses a commitment to something rather than just a cognitive understanding' (Giddens 1990, 27). Thus, as Simmel (1978) definitively demonstrated for the case of money, trust necessarily involves an element of *faith*.[20] As Luhmann (1988, 97–98) therefore notes, whereas in 'the case of confidence you will react to disappointment by external attribution', in 'the case of trust you will have to consider an internal attribution and eventually regret your trusting choice' (that is, your *misplaced* trust). Considered in this light, the operation of the market would seem to be systematically geared towards ensuring the establishment of trust. The availability of a 'market solution' to each and every problem faced by the individual suggests that the market sustains itself by holding out a promise of *generalized reliability*, replacing numerous individual solutions with a single, general solution. Insofar as this is likely to generate an ever-increasing level of market dependency, consumerism contains an inbuilt, self-perpetuating tendency. But this does not compromise the autonomy entailed by freedom of choice. The market thus accomplishes a remarkably fine balancing act between providing freedom of choice and managing uncertainty, assuring its own reproduction in the process:

> The consumer market is ... a place where freedom and certainty are offered and obtained together: freedom comes free of pain, while certainty can be enjoyed without detracting from the conviction of subjective autonomy. This is no mean achievement of the consumer market; no other institution has gone this far towards the resolution of the most malignant of the antinomies of freedom.
>
> (Bauman 1988b, 66)

Insofar as this situation once more reveals the refracted pattern of the unresolved conflict over productive autonomy, it is destined to gestate further contradictions. Whatever the consumer market may promise, what it finally delivers is an interminable playing out of the compulsiveness unleashed by modernity. The most immediately obvious case in point is the problem of *addiction*.

Campbell (1996, 149) suggests that

> the decline of tradition can just as easily lead to an increase in the extent to which individuals abandon themselves to whim and impulse, or indeed succumb to the appeal of addictions, as it can to an extension of deliberation and hence rational, informed choice.

In fact, the distinction between the two is by no means as clear-cut as this would imply. As is revealed by the sense in which choice has become obligatory, the obsessional mindset of the puritan is equally apparent in choice *and* its obverse. Insofar as the 'only true compulsion left in the society of choosers, the only form of repetitive behaviour oblivious of or blind to all other choices, is the compulsion to choose', all other 'quasi-compulsions are, by Giddens's suggestion, better described as *addictions*' (Bauman 1999b, 135). Despite the significance of the distinction, in other words, addiction and choice lie in almost absolute proximity. It is only the matter of social recognition that distinguishes between the two forms of compulsion. This asymmetrical self-consciousness on the part of society is, however, as crucial as the distinction it underlies. It arises from the way in which modernity as a whole has sought to disguise its own endemic compulsiveness, primarily by drawing a false distinction between its own supposed rationality and the alleged compulsiveness of tradition.[21] Modernity's self-declared commitment to overcoming the limitations of tradition effectively worked by revoking the authority of the past – by redefining the past in terms of the obsolete. By so doing, however, modernity's future-orientated character unwittingly opened the floodgates to the kind of behaviour that can remain locked in the past. Addictive behaviour is the inevitable, if unintended, consequence of a system that demands the kind of constant renewal implicit in the notion of choice. This demand permits the possibility that the active engagement of trust will gel into '*frozen trust*, commitment which has no object but is self-perpetuating' (Giddens 1994, 90):

A world of abstract systems, and potentially open lifestyle choices ... demands active engagement. Trust, that is to say, is invested in the light of selection of alternatives. When such alternatives become filtered out by unexplicated commitments – compulsions – trust devolves into simple repetitive urgency. Frozen trust blocks re-engagement with the abstract systems that have come to dominate the content of day-to-day life.

(*ibid.*, 91)

Addictive behaviour is typically judged irrational on the basis that it negates the opportunities afforded by freedom of choice. But addictions, like choices, are simply 'modes of coping with the multiplicity of possibilities which almost every aspect of daily life ... offers' (*ibid.*). A world where one can choose anything (TV dinners, designer handbags, pet spiders, weekends in Rome) is simultaneously 'a world where one can get addicted to anything (drugs, alcohol, coffee, but also work, exercise, sport, cinema-going, sex or love)' (*ibid.*, 71). For Giddens, therefore, the 'progress of addiction is a substantively significant feature of the postmodern social universe, ... a "negative index" of the very process of the detraditionalizing of society' (*ibid.*).[22] One might add, however, that 'choice' fits the picture in precisely the same way.

To those who set about the task of constructing the kind of reason-guided world in which the puritan would feel at home, the taming of fickle human desires was of the utmost imperative. Consequently, they 'trimmed the "pleasure principle" down to the size dictated by the "reality principle"' (Bauman 2001c, 14). The very desires modernity set about repressing, however, have finally come full circle, to serve the selfsame ends their attempted eradication once hoped to achieve. For the architects of modernity, the 'reality principle' and the 'pleasure principle' were fundamentally at odds. It 'did not occur to either the managers of capitalist factories or the preachers of modern reason that the two enemies could strike a deal and become allies: that pleasure could be miraculously transmogrified into the mainstay of reality' (Bauman 2001c, 16). But this is precisely what has happened. With the emergence of a fully fledged consumer society, the two formerly opposed principles have been reconciled to the extent that they directly feed off one another. 'This', says Bauman (*ibid.*), 'is precisely what the consumer society is about':

enlisting the 'pleasure principle' in the service of the 'reality principle', harnessing the volatile, fastidious and squeamish desires to the chariot of social order, using the friable stuff of spontaneity as the building material for the lasting and solid, tremor-proof foundations of the routine. Consumer society has achieved a previously unimaginable feat. ... Instead of fighting vexing and recalcitrant but presumably invincible irrational human wishes, it made them into faithful and reliable (hired) guards of rational order.

(*ibid.*, 16)

We have, in other words, arrived at an entirely paradoxical state of affairs, by way of a wholly unintended route, from which any kind of imminent departure seems increasingly difficult to imagine. As the haunting ground of the puritan's spectre, where formerly opposing principles meld, dream-like, one into the other, where paradox resolves itself in further paradox, the consumer society can only ceaselessly reiterate the hollow promise that satisfaction is guaranteed. The fact that the market can never deliver on its promise might prompt one to believe that such an arrangement is inherently unstable. Yet numerous countervailing tendencies, and the bulk of the evidence, seem to suggest the opposite. Let us therefore familiarize ourselves with the ways in which the market manages to sustain itself against all the odds, and consider the way in which discontent might manifest itself in a society dedicated to its eradication. This will, at last, lead us to a position where the meaning of 'lifestyle' discloses itself.

Satisfaction and its discontents

> In the consumer society, consumption is its own purpose and so is self-propelling.
>
> (Bauman 2001c, 13)

> Our society thinks itself and speaks itself as a consumer society. As much as it consumes anything, it consumes *itself* as consumer society.
>
> (Baudrillard 1998a, 193)

'Satisfaction at any cost' presents us with an intensely paradoxical situation:

> this society which claims to be – which regards itself as being – in constant progress towards the abolition of effort, the resolution of tension, greater ease of living and automation, is in fact, a society of stress, tension and drug-use, in which the overall balance sheet of satisfaction is increasingly in deficit, in which the individual and collective equilibrium is being progressively compromised even as the technical conditions for its realization are being fulfilled.
>
> (Baudrillard 1998a, 182)

The market operates, as we have previously suggested, as an overarching system that simultaneously arouses and curbs anxiety, holding opposing properties in a kind of permanent dynamic tension. For the most part, this *modus operandi* has served capitalism well, providing it with the kind of self-propelling mechanism that ensures its perpetuation from one moment to the next. It is not, however, as we have seen with reference to its underlying compulsiveness, without its problems. Although its ability to feed off its own negativity produces its self-sustaining potential, consumerism's dedication to ensuring the conformity of the world to a principle of unrelenting positivity ultimately rests on the *denial of ambivalence*; in Baudrillard's (1998a, 184) terms, the denial of symbolic exchange:

> That ambivalence, totalized in *jouissance* and the symbolic function, is split apart, but in going off in two different directions, it obeys a single logic: all the positivity of desire passes into a series of needs and satisfactions, where it resolves itself in terms of managed aims; all the negativity of desire, however, passes into uncontrolled somatization or into the acting out of violence. This explains the profound unity of the whole process: no other hypothesis can account for the multiplicity of disparate phenomena (affluence, violence, euphoria, depression) which, taken together, characterize the 'consumer society'.
>
> (*ibid.*)

Whatever the extent to which the negative symptomatic responses consumerism gestates might be channelled back into the system, they are necessarily insoluble contradictions. If this is a situation the consumer society has made into a strength, it is also likely to be its point of weakness. Let us therefore proceed to uncover the way in which this manifests itself at the level of everyday life, both in terms of the operation of the market and in terms of the form taken by the discontent that issues from this situation. Perhaps the single most important point to stress in advance is that the proclivity consumerism shows for feeding off its own discontent tends to neutralize whatever impact it might otherwise have had. The possibility of *resistance*, in other words, is spontaneously frustrated by the very nature of consumerism. In consequence, discontent is hardly likely to manifest itself in any straightforward form.

As we have already seen, the way in which consumerism relies on simultaneously curbing and arousing anxiety is revealed in the nature of the 'solutions' the market offers in response to the problems afflicting everyday life. Typically, symptom and cure are twin-born. Halitosis, for example, was the problem named by the product promising to eradicate it, whilst tobacco companies are currently striking deals with biotech companies for exclusive rights to any future lung-cancer vaccines (Boseley 2001, 1). The majority of advertising works in precisely this way (Falk 1997b).[23] Typically, advertisements promise to compensate for a dull and uninteresting life by portraying a life that can only be dreamed of (Leiss *et al.* 1986). The raising of expectations this promotes is never in any real danger of outrunning the opportunities for fulfilment spontaneously offered by the market, however. The latest wisdom of the marketing gurus, as Bauman (2001c, 13) records, holds that 'consumers should not ever be allowed to "awake" from their "dreams"'. The ability of advertising to constantly refresh the dreamworld it creates ensures as much. There is little need to look beyond the market when the solution constantly arrives together with the problem – when the answer one is looking for is always – allegedly – just within reach. The point is beautifully illustrated in the exemplary advertising slogan reproduced by Baudrillard (1989a, 194): '*the body you dream of is your own*'.

The 'solutions' to life's problems announced by the advertisers generally involve little more than ensuring possession of whatever 'purchasable tools of bodily training, bodily adornments or other goods defined first and foremost as extensions of, or adjuncts to, the body' best fit the bill (Bauman 1983, 41). Today's advertisements offer both the commodities that promise some form of self-improvement – 'smoother, silkier looking hair', 'skin so soft you wouldn't believe it came out of a bottle', 'all the taste, only half the calories' – and the commodities that grant the kind of momentary release that will, ultimately, call forth such products at a later date – 'naughty but nice', 'spoil yourself', 'deliciously wicked'. Whether savouring the moment by indulging in exquisite pleasures with 'a hint of decadence', or undoing the damage with a product that 'takes off the years' or 'removes the visible signs of ageing', it is generally a question of making the individual feel special, comfortable, more secure, attractive or alluring. Or again, advertisements announce the ability of commodities to deputize for other social skills, offering to take care of family responsibilities ('treat *your* family to … ', 'because they deserve only the best'); ensure friendship and popularity (a successful dinner party is assured by serving a particular brand of after-dinner mints; this brand of snack will ensure a convivial party atmosphere); and generally offering reassurance that all that is needed is the correct purchasing decision ('you can rely on … ', 'trust … ').

Somewhat counterintuitively, advertising often works by providing a kind of retroflex social sanctioning (such that, as Erlich *et al.* (1957) discovered long ago, car-owners tend to take most note of adverts for the brands they already owned, as a means of confirming the wisdom and foresight of their purchasing decisions). Even in its most forthrightly projective promises, advertising works to nurture trust by presenting a particular brand as a 'tried-and-tested' old favourite ('welcome back to butter'), or simply the most popular of its kind (the implication being that the 'nine out of ten owners who said their cats preferred it' can't all be wrong, or that consumers can rest assured that they, too, will find the brand voted 'best cosmetic product on the market' by 'readers of a leading women's magazine' singularly appealing). Little wonder, given the overwhelming promise of the market, that Lieberman (1993, 249) should remark that, '[s]trictly speaking, shopping disorders do not exist'. Problems are precisely what the consumer market offers to overcome, even as it invents them. The ostensibly ironic use of the term 'retail therapy', which implies a certain knowing attitude, contains more than an element of truth. It signals the enormity of the hope placed in the market. The market presents itself as a sphere of limitless opportunities and is driven by this self-belief.

The market does not, of course, exist for benevolent reasons. In pulling off the balancing act it does, the market ensures that the system reproduces itself over time, and generally at an expanding rate. Although the exalted claims of individual products are unlikely ever to deliver the level of satisfaction they promise, disappointment over the 'lack of correspondence between the ostensible and genuine use-value of each individual product' (Bauman 1987, 165) rarely leads to any waning of faith in the market. A kind of general immunity is afforded by the range of alternatives on offer, allowing hope to spring eternal. The conviction that the market can provide whatever solution is needed is thus allowed to emerge unscathed from past disappointments. The constant barrage of 'new' and 'improved' formulations continuously serves to update the products already tried, so that 'the memory of their unfulfilled promises' is permitted to fade along with their demise (*ibid.*). The temporality of fashion accords to precisely this logic. 'Fashion' says Baudrillard (1981, 51) 'embodies a compromise between the need to innovate and the other need to change nothing in the fundamental order'. Or, in Bauman's (1987, 165) reformulation, 'fashion seems to be the mechanism through which the "fundamental order" (market dependency) is maintained by a never ending chain of innovations; the very perpetuity of innovations renders their individual (and inevitable) failures irrelevant and harmless to the [overall] order'. As Barthes (1990, 296) suggests, fashion tends to follow a rhythm that 'remains outside history; it changes, but its changes are alternative, purely endogenous'.[24]

In virtually all cases, the market reveals a remarkably consistent pattern. It exhibits a tendency to work simultaneously in counterposed directions, feeding off its own inherent contradictions in a manner that preserves them, continuously reproducing its own form. There is no hint in this of any dialectical progression, merely the perpetual replication of a system locked into its own circularity: 'modernity is paradoxical, rather than dialectical' (Baudrillard 1987b, 70). By feeding off its own contradictions, consumerism works to leave nothing outside the system it establishes and represents. It has pretentions to becoming a *total system*, which would make the world over in its own image. The most important consequence this carries lies in the way in which any form of resistance to consumerism also tends to be recouped into its structure. As unpalatable as the fact may be – and Baudrillard has endured the wrath of many critics on this point – all apparent alternatives exhibit a pronounced tendency to slot neatly into the existing system. Such would-be anti-consumerist movements as 'downshifting' (Schor 1998) and 'voluntary-simplicity lifestyles' (Elgin 1981; Etzoni 1998; Iwata 1997), for instance, leave the general principle of consumption entirely unscathed (effectively pronouncing that 'less is more'). Likewise, organic produce nestles alongside the non-organic food on the supermarket shelves, destined to become just one more sign of distinction in the universe of consumption (which magically annuls the mutual incompatibility of their production systems). In a consumer society, *everything* is consumed, even resistance to consumption itself. Rather than representing 'a real alternative to the processes of growth and consumption', such 'alternatives' are 'merely the inverted and complementary image of those processes' (Baudrillard 1998a,

180). Taken together, 'the discourse of consumption and its critical under-mining' (*ibid.*, 195), form precisely the kind of counterposed elements on which consumerism feeds. 'Consumption', says Baudrillard (*ibid.*, 193),

> is a myth. That is to say, it is *a statement of contemporary society about itself*, the way our society speaks of itself. And, in a sense, the only objective reality of consumption is the *idea* of consumption ... which has acquired the force of common sense.[25]

Needless to say, this presents a significant problem for theory: where does one stand to gain purchase on a system that admits no possibility of transcendence? The 'mythic' status of consumption ensures that all attempts to provide a critique of the system are liable simply to produce a 'counter-discourse, which establishes no *real* distance, [which] is as immanent [to the] consumer society as any of its other aspects' (*ibid.*, 196).[26] The point is not so much that 'resistance is useless' as that 'resistance is useful'; useful to the very system it seeks to oppose.

If the sense in which 'there is nothing outside consumption' is acknowl-edged, the problem for theory is one of identifying the weak point of the system. Accordingly, Butler (1999, 56) proposes that the only possible way 'to challenge the tautology of the consumer society' is 'not by directly contesting it or proposing an other to it', but 'by asking what allows this self-definition or tautology, what is excluded to bring it about'. It is a ques-tion of discerning 'not so much what is outside or other to the system ... as what is excluded by the fact that it has no outside' (*ibid.*, 46). What is excluded by the conformity of the world to a system that excludes nothing? The question entails the sort of 'double strategy' (Baudrillard 1983, 107) that fights fire with fire, pitches tautology against tautology, or adopts the homoeopathic principle of *similia similibus curentur*.[27] Armed with this way of thinking, Baudrillard suggests that certain extreme features of the consumer society provide the best means of understanding the way in which discontent might arise within a system predicated on satisfaction. Certain medicalized 'consumption disorders', such as anorexia and obesity, provide a first example. Such disorders centre on the body and the issue of bodily control. Thus, Warde (1997, 93), summarizing Bordo's (1992) argument, notes that anorexia 'is as much about the conscious exercise of the will to control hunger and the body as it is about physical appearance. Being thin is an achievement resulting from having autonomously exerted personal control'. For Baudrillard (1988, 39), its significance is revealed by the way in which the apparently diametrically opposed condition of obesity works towards precisely the same end:

> The anorexic prefigures this culture in a rather poetic fashion by trying to keep it at bay. He [*sic.*] refuses lack. He says: I lack nothing, therefore I shall not eat. With the overweight person, it is the opposite: he refuses

fullness, repletion. He says: I lack everything, so I will eat anything at all. The anorexic staves off lack by emptiness, the overweight person staves off fullness by excess. Both are homoeopathic final solutions, solutions by extermination.

Subjective estimations of consumption habits reinforce the point. As Fine (1997, 235) reports, 'the overweight tend to deny that they eat too much, often dramatically under-reporting what they do consume. Anorexics exaggerate what they have consumed, especially the calories contained in those foods associated with being fattening'. These diametrically opposed yet eminently substitutable forms provide an indication of the ambivalence occluded by a system that aims at achieving an unrelenting positivity. It is not, then, the one or the other, anorexia or obesity, to which Baudrillard draws attention (or, for that matter, even the conditions in and of themselves).[28] It is, rather, the co-incidence of the one and the other.

Precisely the same situation is revealed by the fact that the consumer society is 'at one and the same time ... a pacified society and a society of violence' (Baudrillard 1998a, 174). Each phenomenon may be perfectly properly understood in isolation, on its own terms, but it is their co-incidence that is significant. Both stem from particular aspects of the consumer society, but it is the fact that seemingly opposed conditions can arise simultaneously that is important. Take, for instance, violence. The fragile unity of the narcissistic ego, stimulated by the forces of consumerism, incubates a basic defensiveness that engenders potential aggressiveness as its 'correlative tension' (Lacan 1977, 22). Whilst random outbreaks of 'senseless' acts of violence highlight the inexplicable deficit of satisfaction in a society geared to nothing but satisfaction, its mechanism is readily apparent. Its fundamental perversity only appears in the light of the fact that such violence is 'structurally linked to affluence' (Baudrillard 1998a, 178) – in its radical difference 'from the violence engendered by poverty, scarcity and exploitation' (*ibid.*, 177). It is not a protest against lack *as such*, but a manifestation of lack of want, 'the *ease* of life, and the fact that *nothing* is missing' (Butler 1999, 54). Consequently, the violence that accompanies abundance is literally meaningless, in stark contrast to 'violence sanctioned by an objective or a cause', 'the good old violence of war, patriotism, passion and, ultimately, rationality' (Baudrillard 1998a, 178). Such random acts of violence manifest 'the emergence, in action, of the negativity of desire which is omitted, occulted, censored by the total positivity of need' (*ibid.*, 177). It is, nonetheless, poised to be reabsorbed within the system, insofar as media fascination soaks up such violence and replays it as yet another occasion for consumption.

Alongside this irruption of senseless violence, an 'endemic, irrepressible fatigue' (*ibid.*, 181) grips the consumer society. It marks, as Baudrillard suggests, an unconscious, *passive*, refusal of the system. This is a refusal of an entirely different order than the active refusal found in the discourse and practice of 'anti-consumption'. This passive form of rejection is *not negative* (it recalls Bartleby's expression: 'I would prefer not to' [Melville 1977]). Or again, one might say that passive refusal amounts to active nihilism – since '[*t*]*rue passivity is to be found in* ... *joyful conformity to the system*' (Baudrillard 1998a, 183). Lethargy, fatigue, boredom and the like are, in essence, the obverse of senseless violence: ' "passive refusal" is in fact a *latent violence*' (*ibid.*, 182), which therefore stands in precisely the same relation to the system of consumption as violence itself. It, too, as with the mediatization of violence, can just as easily find itself absorbed back into the system, this time as 'a cultural trait of distinction', the kind of ' "consumed" fatigue which forms part of the social ritual of exchange or status' (whereby the workaholic wears his fatigue on his sleeve, for instance) (*ibid.*, 185). Nonetheless, violence and fatigue, as with anorexia and obesity, are again eminently substitutable forms: 'Fatigue, depression, neurosis are always convertible into overt violence, and vice-versa' (*ibid.*, 182).

These extreme expressions of dissatisfaction emanating from the heart of a system that proclaims its ability to neutralize dissatisfaction *per se* are salutary in at least two respects. First, they reveal the false promise on which the system is based and, in particular, the consequences of the denial of the symbolic principle. Second, they reveal (and promote) that system's exhaustion, demonstrating that dissatisfaction cannot, finally, be eradicated by the kind of satisfaction that results from a substitute form of freedom – however closely interwoven anxiety and its relief might have become in terms of the day-to-day operation of the market. If, however, such unconscious reactions – in the paired forms of anorexia and obesity, violence and fatigue, and addiction and choice – reveal the fundamental exclusions of the system of consumption that were first established with the withdrawal of productive autonomy, they are hardly the most common form of response. The most common response takes the form of a daily coping mechanism, which, we are finally in a position to appreciate, is precisely what 'lifestyle' is all about.

Conclusions

> Those who do not owe their place in the world to the symbolic ties of tradition must resort to finding their place in the Other's gaze.
>
> (Calligaris 1994, 65)

'If it is true that the grid of "discipline" is everywhere becoming clearer and more extensive', as de Certeau (1984, xiv) puts it, then 'it is all the more urgent to discover how an entire society resists it, what popular procedures ... manipulate the mechanisms of discipline and conform to them only in order to evade them'. For despite the success of consumerism, it is not the case that consumers are unblinkingly conformist. If explicit resistance is, as

Baudrillard suggests, spontaneously recouped into the system, de Certeau suggests that there are nonetheless ' "ways of operating" [that] form the counterpart, on the consumer's (or "dominee's"?) side, of the mute processes that organize the establishment of the socioeconomic order' (*ibid.*). In other words, de Certeau points to a direct counterpart of Foucault's capillary form of disciplinary power in the 'clandestine forms taken by the dispersed, tactical, and makeshift creativity of groups or individuals already caught up in nets of "discipline" ' (*ibid.*, xiv–xv). Even though the colonization of everyday life by consumerism produces a 'technocratically constructed, written, and functionalized space in which consumers move about, their trajectories form unforeseeable sentences, partly unreadable paths across a space' (*ibid.*, xviii).

> Although they are composed with the vocabularies of established languages (those of television, newspapers, supermarkets, or museum sequences) and although they remain subordinate to the prescribed syntactical forms (temporal modes of schedules, paradigmatic orders of spaces, etc.), the trajectories trace out ruses of other interests and desires that are neither determined nor captured by the systems in which they develop.
>
> (*ibid.*)

Recalling this analogy, Bauman (1992c, 24) reflects that the strange 'dialectic of dependence and autonomy' between consumer and consumer society is indeed 'not unlike that of the grammar and vocabulary of language and formed sentences of speech: the latter are in no way "determined" by the former and move freely within the frame it provides'. The practice of consumption amounts, in other words, to an act of enunciation, which takes place in 'a space filled with unattached signifiers … a space awaiting attribution of meaning' (*ibid.*). The consumer is in no way manipulated, nor consumer behaviour 'determined', even though the consumer is formally deprived of autonomy in all forms but the duplicitous form of freedom of choice. The proliferation of relations of trust, the reliance on specialist knowledge and expertise, and the privatization of the task of constructing some form of continuity of life-experience are unavoidable features of life in a consumer society. But consumers are nonetheless in a position to make of their situation something other than what is intended.

The problem, however, is that, as an 'act through which the presence of the individual … can be confirmed and reasserted' (*ibid.*), consumption and the relations of trust it entails provides an opportunistic means for the perpetuation – and dramatic expansion – of capitalism. The act of consumption is destined to repeat itself, since the human subject is irresistibly compelled to seek a unity and coherence it can never know. The subject is ineluctably marked by a *constitutive* lack, deriving from the nature of the subject as a speaking being, aware of its own mortality, and whilst there is no necessity that that this will translate into the desire to consume, it is this possibility to which the forces of capitalism have long since become attuned. Although all acts of enunciation are individual acts, they are never truly individualized acts. They are undertaken under the gaze of the Other, and forged in a way that seeks the validation of the Other (in the form of relations of trust, for instance, or the sanction of an-other as a stand-in for the symbolic Other, the Other *as such*). Rather than acting to ameliorate the human predicament, therefore, which an acknowledgement of the symbolic principle might hope to achieve, consumerism merely exacerbates it, hiding the fact behind the promise that satisfaction is guaranteed. Thus the necessarily expansive nature of capital, and the insatiability of desire, have become ever more closely wedded together in a mutually self-sustaining, self-perpetuating cycle.

It is in this context that the meaning of lifestyle begins to emerge more clearly. As the characteristic mode of sociality of the consumer society, lifestyle accords to the same dialectic of dependence and choice as the consumer's relation to the overarching system of trust represented by the market. This dialectic is captured in Maffesoli's (1996) notion of 'neo-tribalism'. Maffesoli's 'tribes', as we saw at the end of the previous chapter, amount to ephemeral, spontaneous, self-creating 'clusters' of sociality. They are primarily aesthetically defined phenomena; loose collectivities of co-incident style. One would have to accept, however, Bauman's (1992c, 25) revision of Maffesoli's opinion of the 'unambiguously anti-individualistic impact of the neo-tribal phenomenon'. Maffesoli's (1997, 24) formulation adheres rather too literally to a one-sided understanding of aesthetics deriving from 'the etymology of the word, [which means] the ability to feel emotions and sensations collectively. It is to vibrate together in harmony'. For Maffesoli (*ibid.*):

> In contrast to modern individualism, collective sensations or 'vibrations' are the kind of nuclear experiences that give meaning and structure to the postmodern era. The linear evolution that modernity imagined to persist and lead to a total dissipation of the imaginary and the collective, is thereby broken.

Whilst this accurately captures the kind of 'collective sensations' that spontaneously coalesce into distinctive, though temporary, forms of lifestyle, it is difficult to see this process as marking a fundamental *contrast* to modern individualism. It would be more accurate, as Bauman (1992c, 25) proposes, to observe that 'the variegated, chequered and fluctuating tribal scene is engaged with the privatized individual in a subtle dialectical game of dependence and freedom'. In other words, the 'tribe-forming and dismantling sociality is not a symptom of declining individuality, but a most powerful factor in its perpetuation' (*ibid.*). Indeed, one might say that it is the obligatory nature of individual choice that spawns the outbreak of neo-tribal lifestyles in the first place. In a situation where the individual is faced with an overwhelming supply of rival advice and expertise which 'exceeds the absorptive capacity of individual attention' (*ibid.*), neo-tribal allegiance provides a guiding hand to cut the competing options down to size. Identification with one or other 'tribal gathering' renders the daunting task of choice itself more manageable. 'It is in this difficult yet indispensable matter of choice that tribes form a critical function, and they sanction global *lifestyles*, each offering its own structure of relevances' – thus providing the individual with an underlying rationale for making *this* choice rather than *that*.

> One can say that the role of neo-tribes consists in selecting a number of disparate and scattered, partial 'problems of survival', and assembling the samples into comprehensive and relatively cohesive life-models. For the individual, joining a tribe means adopting a [particular] lifestyle; or, rather, the road to a coherent lifestyle leads through the adoption of [a] tribally sanctioned structure of relevances complete with a kit of totemic symbols.
>
> (*ibid.*)

Were it not for the importance it rightly places on the ephemerality of particular tribes (and thus particular lifestyles), this bears a striking similarity to the way in which Douglas (1996) regards consumer choice as providing a mechanism that ensures a basic coherence across different categories of consumption (food, clothes, health care, home furnishings, reading material, etc.). The coherence arises, Douglas (1996, 82) suggests, from the kind of conscious social alignment that knows what it is repelled from more than it knows what it is that it wants or needs:

> 'I wouldn't be seen dead in it', says a shopper, rejecting a garment that someone else would choose for the very reasons that she dislikes it. ... Shopping is reactive ... but at the same time it is positive. It is assertive, it announces allegiances.

Lifestyle, then, we can finally propose, is the hallmark of a society characterized by an immense array of competing opportunities and advice, amidst which individuals must necessarily position themselves, by the choices they make, in order to impose some degree of continuity onto their individual life-experiences. Despite the activities of the marketing industry, lifestyle is not the kind of phenomenon one can attempt to categorize and typologize without losing sight of its underlying dynamism. It is a *form*, the content of which is secondary to – and relatively unimportant in comparison with – its existence in the first place. Lifestyle as a social form is, moreover, intimately connected with the city. The city initially saw an unprecedented growth in consumerism as a mechanism of differentiation.

Lifestyle itself is a response to the conditions to which this, in turn, gave rise; a ready means of coping with the impossible task of weighing all the options vying for attention. The transformation of the experience of city life into a kaleidoscopic array of competing lifestyles faithfully reflects the dialectic of dependency and choice characterizing consumerism, which in turn recalls the conditions of its birth. The form of lifestyle, that is to say, entails that the compensatory reaction to the unfulfilled resistance to disciplinary power has reconfigured social life as a whole in its own image. As such, it is destined to remain riven with contradictions. In accordance with this, there is one final paradox we might note with respect to contemporary city life. Lifestyle allows for a certain promiscuity on the part of the individual. Given the entirely voluntary nature of neo-tribal membership, it is only the psychological difficulty of juggling different 'lifestyle identifications' that restricts the adoption of multiple lifestyles. The postmodern city provides an ideal arena for the donning of different lifestyles, at different times and in different places. Lifestyles themselves are likely to be context-dependent, suggesting that the consumer's individualized task, cut down to size by lifestyle as a social form, reflexively displaces itself over space and time. Continuous discontinuity is in fact, characteristic of postmodern space-time itself.

6 Minimal utopia

From this point on, the problem in hand is not one of changing how life is lived, which was the maximal utopia, but one of survival, which is a kind of minimal utopia.

(Baudrillard 1998b, 6)

Premature conclusion

Obscenity is another world.

(Baudrillard 1993b, 62)

Some time ago, Baudrillard (1987c, 126) declared that there is 'no more system of reference to tell us what happened to the geography of things'. Such pronouncements typically throw geographers into a state of consternation (cf. Philo 2000). Yet geographers need not worry; or, at least, they need only ensure that their concerns are not misplaced. Baudrillard's proclamation is not exactly a claim that the world is no longer differentiated or that its differentiation need no longer concern us, which is what most geographers appear to take it to mean. Rather, it is a claim about the relations between space, time and (post)modernity – which certainly involves a claim about the *reality* of the world, though not as that term is conventionally understood. As the present chapter aims to demonstrate, Baudrillard's declaration expresses a conviction that follows from his understanding of the consumer society, even if he has long since abandoned that particular focus and moved on to consider the world in terms of 'hyperreality', 'simulation', and 'simulacra' (Baudrillard 1994a). The system of consumption we have been addressing carries a number of implications that are aptly summed up by allusion to these terms. The implications are indeed profound for our understanding of geography, perhaps requiring the development of a new kind of geography: a *pornogeography*. For the simplest way of expressing them, in the first instance, is by recourse to Baudrillard's notion of the *obscene*. The understanding this conception encapsulates is, to some extent, already implicit in existing accounts of the postmodern city. It has, though, seldom been made explicit. As a result, most accounts of the postmodern city dramatically underestimate the extent to which it departs from the city in its modern form. The notion of the obscene, in Baudrillard's specialist terminology, relates closely to the kind of situation we have described in the preceding chapter. It relates to the intimate connection that has developed between the arousing and curbing of anxiety; to the scrambling of cause and effect; to the spontaneous deconstruction of formerly distinct processes which have become irredeemably intertwined; to the way in which modernity set out on a path that aimed to increase the *reality* of the world – to eradicate its potentially deceptive *appearance* – but has only succeeded in blurring the firm and rigid boundaries it sought to inscribe. This process is, in Baudrillard's eyes, fundamental. His deployment of the term 'obscenity' signals the way in which modernity has been overrun by its unintended consequences and transmuted into postmodernity. Postmodernity, we might propose, is the obscene form of modernity. And as Baudrillard (1990b, 31) forebodingly suggests: 'Obscenity has an unlimited future'.[1]

Obscenity relates, first and foremost, to 'the loss of scene,' and all that this notion implied (Baudrillard 1993b, 62; cf. Virilio 1991). 'The "scene"', as Baudrillard suggests (1993b, 61), amounts to the 'possibility of creating a space where things can transform themselves, to play in another way, and not at all in their objective determination'. It is 'an enchanted space', a space which 'is arbitrary and that does not make sense from the point of view of conventional space' (*ibid.*). The scene is, in other words, 'a space of freedom from convention and a space that one can take a distance from in order to put oneself outside the realm of ... determinations rather than be overwhelmed, swept over, incapacitated or drowned' (Bauman 2002, 280). It creates room for manoeuvre by creating a 'perfectly capricious division of space' (Baudrillard 1993b, 62), thus allowing for something to take place – though not something that might be predicted and planned in advance (as in the case of production: *pro-ducere*). Rather, the scene is given over to the power of *illusion*, in precisely the same way that play, being unconstrained by reality, is given over to illusion. Play, as we have previously noted (Chapter 3), is never unconstrained as such. On the contrary, it necessarily takes place according to the rules of the game. Yet the rule, unlike the law, is an ineluctably *arbitrary* form: 'pleasure comes when one cuts up very arbitrarily a kind of terrain where one permits play in any possible fashion, in another way, where one will be outside the real, outside the stupid realistic constraints of conventional space' (*ibid.*). The scene, therefore, relates to illusion in a pure, *aesthetic* sense: in terms of the 'art of appearance ... of making things appear. Not producing them, but making them appear' or conjuring them up (*ibid.*, 55). Politics, in its pure form, related to the power of illusion, as did the notion of utopia. The eruption of the obscene, however, implies that there is no longer any room for manoeuvre: it amounts to 'a power of disillusion and of objectivity' (*ibid.*, 60). Being marked by 'the total promiscuity of things', the obscene 'destroys distance' and 'doesn't recognize rules any more' (*ibid.*). For Bauman (2002, 280), obscenity suggests a dense crowd inside which nothing can be seen at a distance, examined and contemplated; no place to breathe freely and take a longer breath, pause and ponder, see what is what and what one could do to make it into something else.

The obscene is thus a condition – a *pathological* condition – that abolishes the critical distance that was established in and by the scene. It is a confusion arising from the absolute proximity of things, which is precisely what the scene managed to avoid. Whilst the scene accords to the 'mode of challenge, of seduction, of play' (Baudrillard 1993b, 57), the obscene permits none of this. It denies the possibility of creating an 'enchanted space', denies critical distance 'and the possibility of playing on that distance' (*ibid.*). All that remains is 'the idea that the world is real and that all that is required now is to operate in the real. There is not even a "scene" of utopia ... utopia has gone into the real, we are in it' (*ibid.*). Politics, in other words, has passed into the obscene, whilst utopia today implies 'the radical deconstruction of all the places of politics' (Baudrillard 2001, 59).

The portentous passage from 'scene' to 'obscene' relates to modernity's overarching commitment to reality: to the *forced realization* of the world. As Baudrillard (1983, 84) suggests, 'the real, which never existed and only came into view at *a certain distance*', underlies the transition. Reality's fateful appearance on the scene was destined to pitch itself against the power of illusion, to which it owed its existence in the first place. Where previous cultures have harnessed the power of illusion, modernity, perhaps uniquely, has harnessed itself to the reality principle, equating the *essential* with the *real*. It has sought to instate the real as the basis of an *irreversible* order, an order of finality, in stark contrast with the reversibility implied by illusion. At its birth, in the figurative, Quattrocento space of the Renaissance, the real was hardly granted the powers it was subsequently to achieve. 'When it invented itself as a representation, its usage ... was quite delirious and not at all representative, economic, as it was later' (Baudrillard 1993b, 62). It was nonetheless to acquire this kind of force by harnessing the power of production (again, in the literal sense of that term). It proceeded by ensuring that reality might measure up to its representation and be *made* to measure up to its representation 'by abolition of the distance between the real and its representation' (Baudrillard 1983, 85). The ultimate, paradoxical consequence is an *excess* of reality, a hyperreality which 'puts an end to the real as referential by exalting it as model' (*ibid.*). It is this situation that marks the eruption of the obscene. The detail of this argument will be spelled out further below. But since this is a premature conclusion – like deaths, all conclusions are premature – let us begin by drawing out three crucial points that arise from the detail of that argument.

Let us stress, first of all, that despite Baudrillard's occasional reticence to acknowledge the fact, the obscene order of hyperreality is directly related to the development of the consumer society. Consumerism offers a prime instance of the way in which the irreversibility that modernity sought to impose has inexorably curved back on itself, being unable to evade the reversibility of symbolic exchange. This much should be evident from our discussion so far. Second, we might single out the implications this holds for geography. The obscene, as the characteristic form of postmodern space, is a direct manifestation of the strange, liminal reversal to which modernity has been subject. Although it may appear banal, the most pertinent conclusion we can draw from this is that the postmodern city can no longer be considered a city in the conventional sense. The same conclusion litters a wealth of books and articles promising to make sense of the notion, but most of those renditions radically underplay, if they do not reject outright, the sense of obscenity that derives from a reality glut. Finally, we might stress the underlying pattern that these two points exhibit. Modernity operated by means of inscribing boundaries, by dividing a totality into two and defining the adequacy of the one side in terms of the inadequacy of the other. It invariably camouflaged its *modus operandi* as an autonomous act of definition, rather than an act of division, exclusion and attempted eradication. But what was excluded, denied a place, or otherwise forced to disappear, has ultimately rebounded in an excessive, obscene presence. This is the upshot of the attempt to impose an irreversible order. What modernity attempted to eradicate has not been erased without trace. Its act of disavowal has merely resulted in a kind of excrescence. The consumer society and the postmodern city are the prime sites of this process. This has been the underlying message of this book. With the basis of this understanding now in place, let us finally outline its full implications for the postmodern city. We begin with a striking example of the situation to which we have been alluding.

City visions

> We live in a *referendum* mode precisely because there is no longer any *referential*.
>
> (Baudrillard 1993a, 62)

In the middle of the nineteenth century, many of northern England's prosperous industrial cities attempted to mark their arrival on the scene by building great civic monuments deemed worthy of their achievements. As powerhouses of the industrial revolution, Manchester, Leeds, Bradford and Rochdale, each in jealous competition with its neighbours, built extravagant town halls, announcing their own importance and symbolizing their confidence in the future they were actively creating.

> The boisterous, self-congratulating town elders who built Leeds Town Hall ... as the monument to their own miraculous rise-in-time ... engraved their moral principles around the walls of its assembly hall. Alongside other commandments there is one most striking by its self-

confident brevity – 'Forward!' Those who designed the town hall had no doubts as to where 'forwards' was.

(Bauman 1997a, 86)

The expression of faith in progress over time was, as Briggs (1968, 176) notes, complemented by the 'words "Europe – Asia – Africa – America" round the vestibule', which 'reminded the inhabitants of Leeds that they were part of an Empire'. As a typical exercise in Victorian civic pride, the erection of Leeds Town Hall was designed to assure the city of its place in the world and to locate it at the forefront of the march of history.[2] Its mottoes perfectly encapsulate, moreover, the properties of modern space-time; particularly the cross-transference of the characteristics of space onto time, and vice-versa. As Bauman (1997a, 86) notes, modernity 'furnished time with certain traits which only space possesses "naturally": modern time had direction, just like any itinerary in space'. Having been given a sense of direction, 'forward' was the only rational course to follow. Likewise, modernity projected its sense of historical progression onto space, representing contemporaneous spatial differences as temporal ones, portraying certain areas as 'backward' and unequivocally placing Europe at the forefront of history (cf. Heffernan 1994). If the modern space-time framework inspired the confidence necessary for plotting out a path of progression, however, the late twentieth-century counterpart of this bold Victorian expression of faith was nothing like so self-assured.

In the closing decade of the twentieth century, Leeds formally declared itself a 'European city' and a '24-hour city', and in 1997, 'Vision for Leeds' – a major public consultation exercise, billed as the largest the city had ever seen – was launched as a means of determining the city's priorities (Leeds Initiative 1999). Vision, it seems, is no longer a matter for visionaries but something to be determined by market research. Unsurprisingly, opinion over 'Vision for Leeds' has been divided. Whilst some have held it to be a noble attempt at participatory planning, others remain suspicious that it merely serves to legitimate a new form of urban governance – a form based on public/private partnership, adopted under the 'Leeds Initiative' launched in 1990.[3] Many of the aims on which the populace were asked to pass comment bear a striking resemblance to those associated with an 'entrepreneurial' mode of governance (Harvey 1989b). But whilst the public's responses cannot be read as unequivocally supporting these aims, they cannot strictly be regarded as oppositional, either (Figure 3). The fact that 69 per cent of respondents believe 'improving the international image of Leeds as a location for business' to be important, whilst only 9 per cent consider this unimportant, seems to provide a remit for a capital-led growth regime to continue along already-established lines. But such a remit is, in any case, unnecessary: the forces of globalization hardly depend on widespread local support to accomplish their aims (which tends to mean, granting the circularity of the argument, that public acquiescence can readily be obtained). Equally, however, some 93 per cent of the respondents to

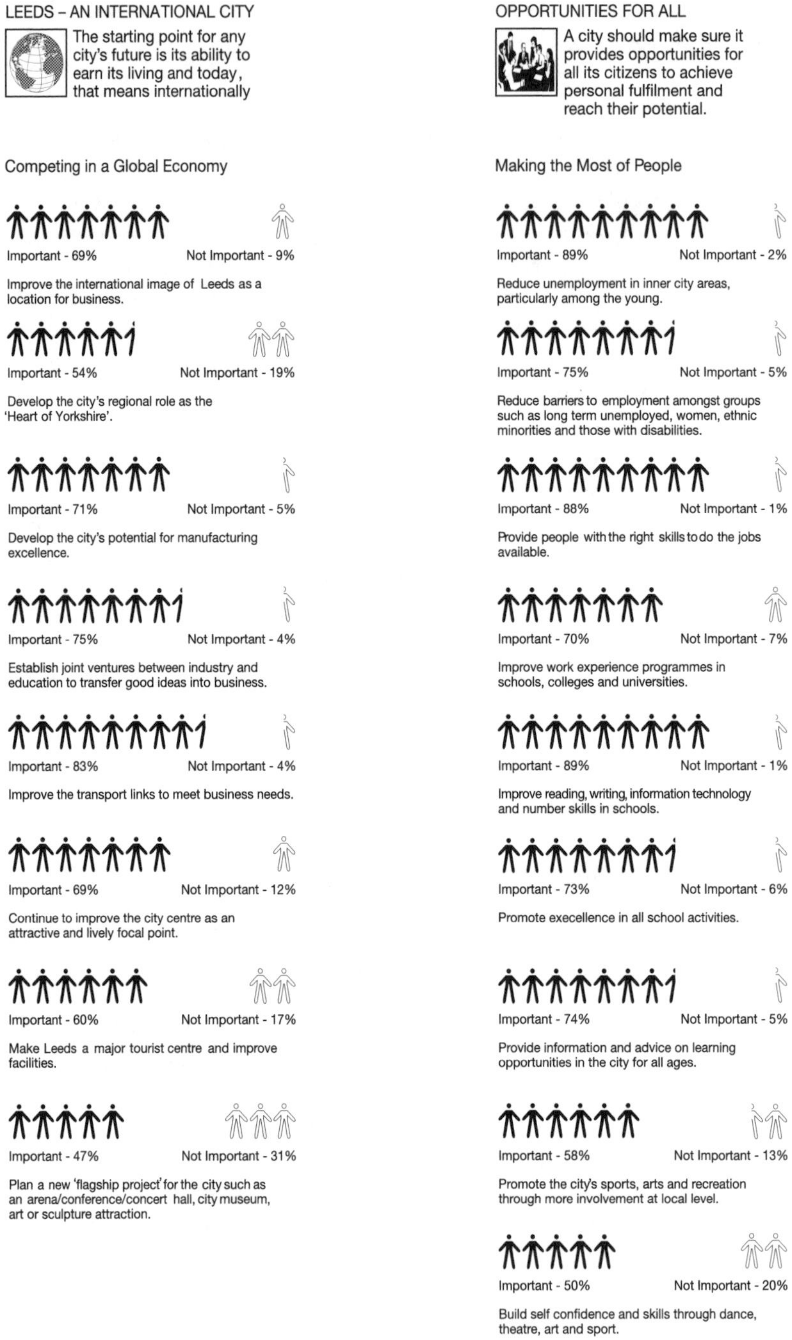

LEEDS – AN INTERNATIONAL CITY

 The starting point for any city's future is its ability to earn its living and today, that means internationally

OPPORTUNITIES FOR ALL

 A city should make sure it provides opportunities for all its citizens to achieve personal fulfilment and reach their potential.

Competing in a Global Economy

Important - 69%   Not Important - 9%

Improve the international image of Leeds as a location for business.

Important - 54%   Not Important - 19%

Develop the city's regional role as the 'Heart of Yorkshire'.

Important - 71%   Not Important - 5%

Develop the city's potential for manufacturing excellence.

Important - 75%   Not Important - 4%

Establish joint ventures between industry and education to transfer good ideas into business.

Important - 83%   Not Important - 4%

Improve the transport links to meet business needs.

Important - 69%   Not Important - 12%

Continue to improve the city centre as an attractive and lively focal point.

Important - 60%   Not Important - 17%

Make Leeds a major tourist centre and improve facilities.

Important - 47%   Not Important - 31%

Plan a new 'flagship project' for the city such as an arena/conference/concert hall, city museum, art or sculpture attraction.

Making the Most of People

Important - 89%   Not Important - 2%

Reduce unemployment in inner city areas, particularly among the young.

Important - 75%   Not Important - 5%

Reduce barriers to employment amongst groups such as long term unemployed, women, ethnic minorities and those with disabilities.

Important - 88%   Not Important - 1%

Provide people with the right skills to do the jobs available.

Important - 70%   Not Important - 7%

Improve work experience programmes in schools, colleges and universities.

Important - 89%   Not Important - 1%

Improve reading, writing, information technology and number skills in schools.

Important - 73%   Not Important - 6%

Promote execellence in all school activities.

Important - 74%   Not Important - 5%

Provide information and advice on learning opportunities in the city for all ages.

Important - 58%   Not Important - 13%

Promote the city's sports, arts and recreation through more involvement at local level.

Important - 50%   Not Important - 20%

Build self confidence and skills through dance, theatre, art and sport.

Figure 3 'Vision for Leeds' – the results

Source: Courtesy of Leeds Initiative

A GOOD PLACE TO LIVE

 People experience much of the city through where they live. They need neighbourhoods where they can achieve their ambitions and enjoy life with those around them.

Better Neighbourhoods and Confident Communities

Important - 78%   Not Important - 3%

Deal with specific housing problems such as damp and poor heating.

Important - 93%   Not Important - 1%

Make neighbourhoods more attractive and safer places to live.

Important - 82%   Not Important - 2%

Improve provision of, and access to, good quality health care for all.

Important - 49%   Not Important - 18%

Launch campaigns to encourage a heathier lifestyle.

Important - 63%   Not Important - 10%

Improve childcare facilities and encourage employers to provide childcare.

Important - 86%   Not Important - 2%

Help people take pride in their neighbourhood and become more involved in dealing with local problems.

Important - 89%   Not Important - 1%

Work together to reduce crime.

Important - 89%   Not Important - 2%

Improve street lighting, open spaces and transport access to make people feel safer.

Important - 79%   Not Important - 3%

Improve local activities (such as youth clubs) to meet the needs of young people.

NB: The above percentages omit figures for people who did not provide an answer to that particular question.

CONCERN FOR THE ENVIRONMENT

 In making our plans for the city we must be aware of the consequences of our actions over a much longer timescale.

Ensuring Sustainable Development

Important - 53%   Not Important - 8%

Involve local communties and businesses in environmental projects in areas such as the Lower Aire Valley

Important - 60%   Not Important - 15%

Improve the appearance of the main routes into the city (by road, rail and water).

Important - 76%   Not Important - 6%

Improve existing and create new green space in inner city areas.

Important - 83%   Not Important - 2%

Explore new ideas to reduce the amount of waste we create and to dispose of it safely

Important - 67%   Not Important - 8%

Mount a publicity campaign to encourage everyone ro reduce waste and provide assistance to reduce waste.

Important - 91%   Not Important - 3%

Develop a transport system which encourages a greater use of public transport - better rail and bus services, park & ride schemes and better links between them.

Important - 78%   Not Important - 3%

Improve access to public transport for people with disabilities.

Important - 59%   Not Important - 18%

Run a publicity campaign to increase use of public transport and reduce car usage.

Important - 52%   Not Important - 19%

Make it possible for more people to live in or near our town and city centres.

Important - 61%   Not Important - 10%

Improve the quality and design of new housing and other buildings.

the questionnaire regarded 'making neighbourhoods more attractive and safer places to live' as important, representing the single highest score on any individual question. Only 1 per cent of respondents saw this as unimportant (presumably reflecting the fact that they already live in such areas, rather than expressing a preference for unattractive and dangerous residential environments). But could anything other than a residual failure on the part of the public to appreciate the global forces shaping the late-twentieth-century city be read into the 23 per cent higher level of public approval for the 'neighbourhood' as opposed to the 'international' question? What, in fact, do such statistics reveal?

'Tests and referenda', according to Baudrillard (1993a, 62), 'are perfect forms of simulation'. The sense of 'simulation' appealed to here does not amount to a simple divergence from reality, a basic mismatch between an underlying reality and its inaccurate representation (or misrepresentation). It implies, rather, the dissolution of referentials in the wake of a different order of reality (hyperreality). Excusing the inelegance of the terms he is forced to adopt, Kundera (1984a, 127–128) beautifully elucidates this sense of 'hyperreality' and 'simulation' in describing the eclipse of reality by 'imagology' (which comes close on the heels of the eclipse of ideology by reality):

> All ideologies have been defeated: in the end their dogmas were unmasked as illusions and people stopped taking them seriously. For example, communists used to believe that in the course of capitalist development the proletariat would gradually grow poorer and poorer, but when it finally became clear that all over Europe workers were driving to work in their own cars, they felt like shouting that reality was deceiving them. Reality was stronger than ideology. And it is in this sense that imagology is stronger than reality, which has anyway long since ceased to be what it was for my grandmother, who lived in a Moravian village and still knew everything through her own experience: how bread is baked, how a house is built, how a pig is slaughtered and the meat smoked, what quilts are made of, what the priest and the schoolteacher think about the world; she met the whole village every day and knew how many murders were committed in the country over the last ten years; she had, so to speak, personal control over reality, and nobody could fool her by maintaining that Moravian agriculture was thriving when people at home had nothing to eat. My Paris neighbour spends his time in an office, where he sits for eight hours facing an office colleague, then he sits in his car and drives home, turns on the TV and when the announcer informs him that in the latest public opinion poll the majority of Frenchmen voted their country the safest in Europe (I recently read such a report), he is overjoyed and opens a bottle of champagne without ever learning that three thefts and two murders were committed on his street that very day.

As Kundera (*ibid.*, 129) suggests, 'the findings of polls have become a kind of higher reality'. Or again: 'Public opinion polls are a parliament in permanent session, whose function is to create truth, the most democratic truth that has ever existed' (*ibid.*). Whatever else 'Vision for Leeds' might reveal, therefore, it reveals most clearly that the 'entire communications system has passed from a complex syntactic structure of language to a binary system of question/answer signals – perpetual testing' (Baudrillard 1993a, 62).[4] The unilateral questions posed by 'Vision for Leeds' amount to the construction of a 'democratic truth'; an exercise in hyperreality.

The unilateral question was, of course, no stranger to modernity. However, where it once amounted to a classic crystallization of coercive force, 'it is precisely not an interrogation any more, but the immediate imposition of a meaning which simultaneously completes the cycle. Every message is a verdict' (*ibid.*). Once delivered, it resists all further interrogation. Thus, whilst proponents and critics alike might endlessly search for meaning in the responses prompted by such exercises – not to mention the silences and aporia they reveal – ultimately, 'they have no meaning because they can have all possible meanings' (Baudrillard 1998b, 7). Exercises such as 'Vision for Leeds' allow 'for all possible interpretations, even the most contradictory – all true, in the sense that their truth is to be exchanged, in the image of the models from which they derive' (Baudrillard 1994a, 17). Meaning, in other words, disappears in the excess of meaning. Approaching a certain limit, parameters become conflated with their opposites, not in the sense of a dialectical resolution (*Aufhebung*), but in terms of a short-circuiting of poles or scrambling of terms: the important coincides with the unimportant; the relevant with the irrelevant; the meaningful with the meaningless. 'There is a kind of reversible fatality for systems', writes Baudrillard (1993b, 91), 'because the more they go towards universality, towards their total limits, there is a kind of reversal that they themselves produce, and that destroys their own objective'. As everything becomes equally important, everything becomes equally unimportant. When everything is privileged, nothing is. This is the epitome of postmodernity. Everything is true and yet nothing is true.

The hypothesis that our current situation exhibits this strange reversibility entails an historically and, indeed, geographically specific argument. Insofar as it sought to accomplish the 'unconditional realization' of the world (Baudrillard 1998b, 6), modernity was destined to become a victim of its own success; to be 'transformed into its own negation' (Bauman 1993a, 42). In its attempt to exhaust the world's possibilities, to reduce the illusion of the world in the name of solid, durable reality, modernity could not but result in the kind of excess of reality that undermines the certainty and trustworthiness of reality itself. Paradoxically, therefore, the 'liminal state of [modernity's] anti-decomposition drive' (*ibid.*, 39) is one of *decomposition* – but not 'the decomposition we normally think of when we hear the word' (*ibid.*, 38). Its pathological status is not the result of a deficiency or lack, which was the diagnosis that fuelled modernity in the first place. It is the result of precisely the opposite tendency: a kind of monstrous process of hypertrophy and metastatic proliferation (*Steigerung*). Modernity's teleological pursuit has given rise to the kind of 'malignant growth that displaces "normal" tissue through an uncontrolled and unmotivated division' or 'excessive reduplication' (Doel 1999, 60, 61). If modernity was characterized by growth, its liminal, postmodern condition takes the form of an outgrowth (*excroissance*). It represents an hypertelic situation. It is in this sense that it marks the 'end of history'. Having 'no destiny, no human purpose beyond endless replication ... [it] has no work to finish and no reason to die' (Burroughs 1988, 60). Hence Kundera's (1984a, 129) suggestion that 'the reign of imagology begins where history ends'. For where the ideological machinery of modernity sought to bring about the kind of revolution that would steer the course of history in one direction or another, the 'wheels of imagology turn without any effect on history' (*ibid.*). Or, as Baudrillard (1998b, 7) sardonically proposes, events 'no longer have any resolution except in the media (in the sense in which we speak of the resolution of an image); they have no political resolution'.

If it is true that 'the myth of progress seems to have been largely exhausted' (Calinescu 1987, 247), this is not because of some particular event leading to widespread disaffection and loss of faith (though it is tempting to think in such terms), but as part and parcel of the untrammelled realization of the world, which progressively undermines its own sense and purpose. One suspects, however, that it is the sense of *delayed arrival* that casts growing doubt on the whereabouts, or even the existence, of the destination. The 'crucial moment is that brutal instant which reveals that the journey has no end, that there is no longer any reason for it to come to an end' (Baudrillard 1988, 10), whence the overwhelming sense of exhaustion derives. As a search for clues to the destination of this mystery tour, therefore, 'Vision for Leeds' provides a perfect allegory of the end of history. It reaches beyond superficial versions of that thesis, revealing that 'in a non-linear, non-Euclidean space of history the end cannot be located' (Baudrillard 1994b, 110). The end *as such* was always already a function of linear time. If linear time curves back on itself, the very possibility of the end is disrupted. This entails, of course, the end of the end itself:

> when we speak of the 'end of history', the 'end of the political', the 'end of the social', the 'end of ideologies', none of this is true. The worst of it all is precisely that there will be no end to anything, and all these things will continue to unfold slowly, tediously, recurrently, in that hysteresis of everything which, like nails and hair, continues to grow after death.
>
> (*ibid.*, 116)

This sense of hysteresis (the continuation of the effect in the absence of the cause) perfectly encapsulates the situation of the postmodern city. As

Benjamin (1985, 50) once wrote: 'That things "just go on" *is* the catas-
trophe'. Perhaps each society gets the utopia it deserves.

The meltdown of modernity, or history's vanishing act

> Everywhere, in every domain, a single form predominates: reversibility,
> cyclical reversal and annulment put an end to the linearity of time.
> (Baudrillard 1993a, 2)

> Everything becomes immortal, and nothing is. Only transience is durable.
> (Bauman 1993a, 31)

Not recognizing something for what it is is both comforting and dangerous,
in equal measure. It absolves one of the arduous task of straining to detect
the unfamiliar amongst the outlines of the familiar. It thereby grants,
however, a decidedly false sense of security. This is precisely the difficulty of
Harvey's (1989a) take on the postmodern city, which spurns all suggestion of
hyperreality in favour of reaffirming a firm distinction between solid reality
and insubstantial appearance.[5] For Harvey (1987, 279), the postmodern is
'nothing more than the cultural clothing of late capitalism', a superstruc-
tural manifestation of underlying political-economic change: a new regime
of 'flexible accumulation' and its attendant 'mode of regulation' (Boyer
1990; Lipietz 1986). As such, the postmodern needs only to be brought to
account before the established precedent of historical-materialist analysis to
be revealed for what it is. All fashionable pronouncements of 'incredulity
towards metanarratives' (Lyotard 1984, xxiv) dissolve into the comfortingly
familiar form of old-fashioned ideological rhetoric once the Marxian metanarrative is revealed as still having teeth. The legitimacy of this argument
lies in the cogency of its demonstrable truth, and Harvey's (1989a) is indeed
the most cogent demonstration one could hope for. The problem, however, is
that the 'incredulity towards metanarratives' proceeds from the opposite
direction: from the impossibility of finding anything solid enough into
which to bite. The question of legitimacy is, moreover, simply an *unnecessary* defence in the face of incredulity – incredulity being what it is. The only
real response to such postmodern objections is outright dismissal, which
Harvey (1989a) manages with considerable aplomb. The danger, however, is
that this can easily amount to the theoretical equivalent of whistling through
the graveyard. With an eye towards the dangers of remaining in denial, it is
worth taking the notion of hyperreality seriously, if only for the sake of
argument. For hyperreality scrambles the very logic that is called upon to
usher it away – out of sight, and out of mind. Harvey's (1987; 1989a; 1989b;
1992) repeated declaration of 'business as usual' for urban theory amounts
to a blunt refusal to accept the novelty of the postmodern situation. Yet its
novelty cannot be so easily refuted. To determine the seminal difference that

history's disappearing act makes requires a detailed exploration of modern space-time, as well as its postmodern decomposition. Let us begin, therefore, with a re-examination of modernity.

The chief legitimizing myth of modernity was not an *aetiological* myth, a foundational 'myth of origin' (though there are many such myths from which to choose). It was, rather, a 'peculiar (and in historical terms idiosyncratic)' *teleological* myth, driven by *potential future accomplishment*: modernity was premised on 'an idea (of freedom, of wisdom, of justice, of equality, or whatever) which is universal but whose universality lies in the future' (Bauman 1997a, 40, 41). In other words, modernity necessarily cast the present as inadequate with respect to what yet might be. This historically exceptional, future-orientated character granted knowledge a uniquely powerful role. If the present is cast as but a pale shadow of future glories, then the ability to set the course towards that pre-ordained destiny, and the knowledge to ensure it stays on course, is afforded an unassailable centrality. If, however, that legitimizing myth loses its legitimacy, its centrality is effectively – and unceremoniously – disarmed.

The conditions where the legitimacy of knowledge loses its authority are those brought about by the reversibility to which modernity has become subject. If doubts linger with respect to the veracity of the 'end of history', this is largely because – beyond the putative end – nothing much appears to have changed. The immediate perception is not, despite frequent claims to the contrary, one of radical discontinuity, bearing no semblance of continuity whatsoever. This, however, is precisely the point. The end of history, in the only credible form of that thesis, involves its *disappearance*. History has not, and cannot, come to a *mortal* end, since the dissolution of linear time itself undoes the particular sense of mortality, releasing it from the holding formation that fashioned its erstwhile meaning. Linear time signals the very idea of mortality and defines it as *irreversible*: linear time is the time of 'no return'. Consequently, the dissolution of linear time also puts an end to the end as such; it entails only its disappearance. In stark contradistinction to death, disappearance entails no certainty of irreversibility: 'What has disappeared has every chance of reappearing … [whereas] what dies is annihilated in linear time' (Baudrillard 1990d, 92). This distinction is a crucial one: 'Death and disappearance are two sharply distinct modes of "ceasing to be". As sharply distinct are the two worlds in which one or the other gains prevalence' (Bauman 1993a, 32).

Modernity was the name for a world overwhelmed by the foreboding presence of mortality. All the traits of modernity announce as much. If this is not always immediately apparent, it is only because modernity has, since its very inception, mounted a sustained attempt to dissimulate this unpalatable truth. Mortality has always been a profoundly uncomfortable truth for modernity: the threat of permanent cessation of being dealt a fatal blow to modern ambitions, humiliating its self-proclaimed omnipotence. Little wonder, therefore, that modernity should have constantly censured death –

deflecting our gaze from it; keeping it 'hidden [away] beneath the difficulty of living' (Baudrillard 1990d, 67). Mortality, as Bauman (1993a, 29) says, was deconstructed on a daily basis: the 'characteristically modern way of defusing the perennial horror of mortality' was 'to dissemble the intractable irreversibility of death into the infinite chain of all-too-human, practical, mundane tasks and worries'; to break down an irrevocable, unspeakable certainty into a series of manageable tasks designed to detract attention from the inevitability that ultimately lay in wait. Under modern conditions, therefore, the inevitability of transience – of mortality – was constantly exorcized from the lifeworld. 'But what if, for a change, transience itself becomes the acknowledged norm of life? If not mortality, but *immortality* is deconstructed?' (*ibid.*, 30). If the end is merely a vanishing point, rather than an instance of fatality pure and simple, this changes everything.

Births and deaths, origins and finalities, lose their meaning if their defining features are stripped away; which is to say, if linear time should happen to curve back on itself. Birth, according to the dictates of linear time, guarantees only an impermanent existence. We are born to die. It is our destiny to be 'here today, gone tomorrow'. Births or origins are profoundly revocable. Deaths or endings, however, are fundamentally irrevocable: they are final and permanent, once and for all. If time were to lose its irreversible linearity, however, the meanings of origins and finalities would alter radically. If time assumes a cyclical, reversible form, births assume an unheard-of permanence and irrevocability. Once they arrive, things are here to stay for good. Their absence may occasionally be noted, but there would be little reason, if any, to assume that this is anything other than a temporary disappearance, holding all the likelihood of imminent, or eventual, return. What seem like *endings*, in other words, are no longer final, irrevocable, and marked by the weight of certainty – just as what seem like births are no longer revocable, marked only by the certainty of their transitoriness and impermanence. The putative transformation of irreversible, linear time into reversible, cyclical time would entail, moreover, a concomitant transformation of space. Space, acting in unison with linear time, is inviolably marked by *scarcity*: space-time acts in something like the manner of a conveyor belt, permitting the process of succession to proceed by tipping off one end whatever is needed to make room for the new arrivals at the other. A curvature in time, however, results in 'the saturation of a limitless space' proper to a process of proliferation (Baudrillard 1998b, 5).[6] A curvature in time and a distension of space scramble the meanings of once set and certain terms. Life, strung out eternally, assumes the pallor of death; whilst death no longer augurs the end, since everything is, from the start, assured of a posthumous life. However, where 'immortality is, so to speak, a birth right ... it has none of the attractions' that formerly surrounded it: 'It is not a challenge to be taken up, a task to be performed, a reward to be earned. ... [I]mmortality dissolves in the melancholy of presence, in the monotony of endless repetition' (Bauman 1993a, 33). Life and death, in other words, lose the qualities that formerly set them apart. What was once distinct dissolves into an indistinct, eternal limbo – as the transitoriness of being and the permanence of cessation of being meld into one another and lose their formerly distinct characteristics.

It is a commonplace to assert that modernity planted *reason* at the cornerstone of its elaborate edifice. It is less common to reflect, however, that the centrality afforded to reason was conferred directly by the properties of linear time. If time exhibits a subtle curvature, the centrality of reason shifts position. The attention the voice of reason commands reaches its zenith in the face of irreversible events: events which, once they have taken place, cannot be undone. If, however, nothing is irreversible, if immortality itself assumes a transitory character, the vigilance required of reason becomes a far less pressing matter. The capacity of reason to anticipate and militate against the most disastrous consequences loses its impetus if nothing is assured of a solid, irreversible existence in the first place. (Or, more precisely, it looses its impetus where everything is assured of such a status, since its status cannot but be undermined by its universality: if everything is permanent, nothing is privileged in this regard; everything is equally (im)permanent.) Likewise, if objects lose their solidity – as their temporary permanence dissolves into a state of permanent transience – the threat of being eliminated once and for all is also vanquished. With it, reason, as the principal safeguard against permanent cessation of being, loses both its urgency and authority. The power of reason to calculate, plan, and control is vital in the face of a world dominated by irreversible events and finite objects. Reason is effectively disarmed in a world where events no longer possess an irreversible nature and where the very solidity of reality crumbles underfoot. 'Only finite objects, only irreversible events have the obstreperous solidity we call "reality"' (*ibid.*, 33). Only in a world where 'solid reality' holds sway does reason assume a privileged position.

Reason thrived on the need to separate fickle, insubstantial appearances and tough, durable reality, and to constantly vet that separation. It unceasingly sought to uphold the distinction between the *real* and the merely *apparent*; to delve beneath the (deceptive) appearance of things and uncover their (hidden) truth. Where unanticipated or uncontrolled occurrences can have the direst of consequences, such a faculty is of the strictest necessity. First impressions are never to be trusted, for fear they are merely duplicitous (mis)representations. Things 'cannot be taken at face value, and not taking them at face value is, literally, a matter of life and death' (*ibid.*, 33). Where events are not charged with the properties ascribed by the irreversibility of time, however, reason is rendered superfluous. The curvature of time erases the very meaningfulness of the demarcation between appearance and reality. In such a situation, appearances are no longer *mere* appearances, capable of disguising what is *really* the case. Representations are no longer mere representations. They are not, in fact, representations at all (a representation being a derivative image of whatever it represents). Where everything is equally insubstantial, everything is equally substantial. It makes no sense to regard some things as more trustworthy or durable than others in a world where everything is, from the start, no more real (and no more false) than anything else. Where there is no durable, trustworthy kernel, capable of retaining its constancy against flighty, untrustworthy appearances, the distinction between reality and representation breaks down. First impressions no longer have anything about them that is likely to prove any less reliable than second or third impressions. Indeed, lasting impressions are unlikely to be forthcoming at all. The erasure of the distinction between appearance and reality is not, however, a simple reversion to a situation where the *appearance* of the world resumes prominence. The category of the real can hardly be done away with and forgotten. On the contrary, the distinction dissolves in the opposite direction, in the direction of hyperreality. The sense of 'simulation' proper to hyperreality does not, therefore, imply a degree of dissimulation, falsity or deception, which would be to obliquely construe the presence of a solid and durable reality elsewhere. The situation is one of

> simulation, in the sense that, from now on, signs are exchanged against each other rather than against the real (it is not that they just happen to be exchanged against each other, they do so *on condition* that they are no longer exchanged against the real).
>
> (Baudrillard 1993a, 7)

Simulation implies, in other words, a world where there are no longer any grounds for taking the so-called 'representation' as inferior to or derivative of the thing it (purportedly) 're-presents'; a world where reality and that which feigns it lose their distinction. The erstwhile representation becomes 'its own pure simulacrum' (Baudrillard 1994a, 6) – a simulacrum being a 'copy' for which there is no 'original', or where the supposed 'copy' *is* an 'original' in its own right. In such a world,

> all things stand ultimately for nothing but themselves – there is no division between things that mean and things that are meant. More exactly, each such division is but momentary, protean, and ultimately reversible. It is just by linguistic inertia that we still talk of signifiers, bereaved of signifieds, as signifiers; of signs which stand but for themselves, as 'appearances'.
>
> (Bauman 1993a, 36)

A world lent consistency by the play of simulacra is not, therefore, one in which the responsibilities of thought are wilfully abrogated. The postmodern world 'is not actively, programmatically, wilfully, consciously anti-critical; it is not conservative in the sense of ideological commitment to preservation or restoration' (*ibid.*, 40). Rather, it is a world in which the very possibility of critique is rendered impotent by the liquidation of the situation in which the legitimizing myth of modernity could sustain itself. It is not as a result of smug self-satisfaction with the present state of affairs that the occasion for critical thought evaporates. It is a result of the abnegation of very conditions in which critique could gain a toehold. Thus, however frequently and convincingly the arguments for the *legitimacy* of critique might be made (Harvey 1989a; 2000; Habermas 1983; 1985; 1989), it is the very *possibility* of critique that is at issue. The short-circuiting of the distinction between appearance and reality disarms the ability of reason to uncover a solid, durable, intransient reality – the counterpart of which, in the postmodern world, is simply *there*: empty, unintelligible, obscene. Hyperreality thus signals an entirely different logic to the logic of the modern world. It is sustained through seduction, and the 'power of seduction derives from its immunity to the sort of interrogation which one would address to an appearance' (Bauman 1993a, 36).

There is evidently, in all of this, a strange resurgence of everything that modernity tried to disavow. The art of seduction, the mode of illusion and challenge, of symbolic exchange, is precisely what the reality principle opposed itself to. Yet the distinction between the reality principle and the principle of seduction has undone itself, preserving both in a mutant, adialectical form. The result is an aleatory world in which the power of illusion reappears, but no longer in its pure form, uncontaminated by the reality principle. In such a world, life once more assumes the character of a game, but in the paradoxical sense of a *managed* game. This is precisely the situation presented by the consumer society, as a society in which choice itself has become obligatory. To suggest that the postmodern world is recast in the image of the game is not, then, a suggestion that has any bearing on the so-called seriousness of the world: games 'do not belong to the realm of fantasy' (Baudrillard 1990b, 148). The situation accords, rather, to the principle of indeterminacy aptly captured in Dostoevsky's (1991, 266) formula: 'one turn of the wheel and – everything changes'. It is vital to appreciate this, if we are to understand its implications in relation to consumerism: the situation is one in which modernist thought and practice can gain no foothold. Let us first stress that the game neutralizes both determinacy *and* contingency. If the reality principle is undone by the short-circuiting of appearance and reality, this does not augur a realm of pure contingency and randomness. Contingency itself was always 'the mirror image of the principle of causality' (Baudrillard 1990b, 146). It amounts to the by-product of a law-abiding determinacy, the residual element of the attempt to render the world a series of orderly, determined sequences of cause and effect. The ludic world 'is built of networks of symbolic relations – not contingent connections, but webs of obligation, webs of seduction' (*ibid.*, 144). Because the game involves neither determinacy nor chance, the 'player cannot determine the outcome', but 'nor are the player's moves devoid of consequence. … This world offers no certainty – but no despair either; only the joy of a right move and the grief of a failed one' (Bauman 1993a, 38). The ludic world thus solicits a response from the player, but offers no assured system for attaining the desired results. Little wonder, therefore, that in 'the stage of the aleatory processes of control, even revolution becomes meaningless' (Baudrillard 1993a, 3).

One 'does not blame the rules for losing a game. The remedy for a defeat in the game lost yesterday is to win the game played today or tomorrow' (Bauman 1993a, 44). Whereas modernity actively gestated its own discontent, the postmodern world effectively neutralizes disaffection with the promise that '[o]ne only has to play one's hand right' (Baudrillard 1990b, 144). Whereas the law actively inscribed a margin that opened up the possibility of transgression, the ludic world entails that the one thing 'the player cannot do is opt out from playing altogether' (Bauman 1993a, 37). One can, perhaps, switch games, just as one can switch lifestyles. But each game is in absolute proximity to another, and 'together the games leave nothing outside' (*ibid.*). Each game, moreover, consists of 'an indefinitely reversible cycle' (Baudrillard 1990b, 147). It accords to the time of 'eternal return'. When it is over, it is ready to begin again: 'games are characterized ... by their capacity to reproduce a given arbitrary constellation in the same terms an indefinite number of times' (*ibid.*, 146). Just as there is no assured strategy for winning, therefore, there is no possibility of escape. The fact that the game starts again from square one discharges the build-up of the disaffection that might lead to calls for the rules to be overhauled. The promise of personal success thus militates against the kind of emancipatory drives that modernity spawned as a result of its inbuilt tendency to gestate and accumulate discontent. The fact that the game is of a zero-sum character – that there are winners and losers, the winners at the expense of the losers – ensures that the 'condition of survival postmodern-style is refusal of solidarity' (Bauman 1992c, 31). Not only does the privatized promise render discontent 'immune to collectivization', it is reinforced by the fear of losing. As a result, postmodernity 'enlists its own discontents as its most dedicated storm-troopers' (Bauman 1993a, 44).

Consumerism has proved itself one of the most effective means yet discovered for undermining the build-up of solidarity.[7] In the obscene world of simulation, 'the most insidious of simulacra, a true meta-simulacrum, is ... constraint dressed as free choice' (Bauman 1993a, 43). The disorienting freedom of consumerism marks 'the beginnings of a "human liberation" that is to be achieved instead of, and in spite of, the failure of social and political liberation' (Baudrillard 1998a, 85). For freedom of choice represents a freedom that 'defines away' all other questions of freedom – not least the freedom to influence or determine the nature of the system in the first place. It is a situation that encourages the perception that the ' "system" has done all it possibly can ... [and that] the rest is up to those who "play" it' (Bauman 1996a, 34). Of course, in a 'world which makes freedom into necessity (having first made necessity feel like freedom) disaffection is aplenty' (Bauman 1993a, 43). But whereas 'in the oppressive modern world of naked necessities ... agony and horror blended into projects of collective emancipation, in the seductive postmodern world of ostentatious liberty they stay apart in a loose heap of non-additive personal tragedies' (*ibid.*, 43–44); 'merely a hopeless imbroglio', as Baudrillard (1994b, 32) once put it. Such is our 'minimal utopia', ruled by freedom of choice and driven by the seductive power of consumerism.

From universalization to globalization

> the project of modernity (the realisation of universality) has not been forsaken or forgotten, but destroyed, 'liquidated'.
>
> (Lyotard 1992, 30)

> By crossing into a space whose curvature is no longer that of the real ... the era of simulation is inaugurated by a liquidation of all referentials.
>
> (Baudrillard 1994a, 2)

In an attempt to allude to the nature of hyperreality, Baudrillard (1994a, 1) once proposed that the 'territory no longer precedes the map, nor does it survive it' – that today 'the map ... precedes the territory' and indeed 'engenders the territory'. This is, once again, the kind of image that arouses consternation on the part of geographers. But Baudrillard himself is hardly satisfied with this metaphor: 'it is no longer a question of either maps or territories. Something has disappeared: the sovereign difference, between one and the other' (*ibid.*, 2). The disappearance of the real does not, in other words, imply that its position has simply been *usurped* by the representation, which is what Philo (2000) appears to take it to mean. The notion of hyperreality, as we have repeatedly stressed, implies the *short-circuiting* of two formerly distinct, if correlated, terms. Strictly speaking, then, the growing distance between the modern city and its postmodern form implies the diminishing distance between the real and its representation: the simultaneous disappearance of the real into the representation, and of the representation as an ideal against which the real might be measured. All too frequently, however, there has been a tendency to grasp only the first aspect of this process. This has typically been rendered as something like the hegemony of representation, which translates into an erasure of local difference via some kind of homogenizing tendency ('McDonaldization', 'Coca-colonization', and the like). It is against this image that most geographers insist on the persistence of 'real geography', emphasizing the continuity of spatial differentiation and the uniqueness of place even whilst conceding the power of homogenization (Cook and Crang 1996; Relph 1976; Sack 1988; 1993). Such a rendition is far too simple, however, and typically invites solutions that are destined to become caught up in the system they seek to oppose (to what extent, for example, is the 'slow cities' movement, emphasizing vernacular architecture, local produce, and so on, simply likely to result in an increase in tourism?). We are afforded a better appreciation of the situation if we focus attention on the second aspect of the process; the disappearance of the representation into the real. This is, in fact, where most recent work on the political economy of urban change has managed to position itself, almost by default. If we take this work as a point of departure, it is possible to connect it to the account of the postmodern outlined above, providing for a better understanding of the way in which the latest urban transition has been affected by hyperreality. This will allow us to appreciate that spatial differentiation and place have not been effaced by homogenizing tendencies, but have themselves been caught up in the thrall of hyperrealization.

A broad consensus of opinion has characterized the latest urban transition in terms of the emergence of the entrepreneurial city (Mellenkopf 1983; Boyle 1995; Feagin 1988; Harvey 1989b; Hall and Hubbard 1996; 1998; Jessop 1997; 1998; Leitner 1990; Roberts and Schein 1993). One of its principal defining features is a new mode of urban governance, which typically involves some kind of 'public–private partnership focusing on investment and economic development, with the speculative construction of place rather than amelioration of conditions within a particular territory as its immediate (though by no means exclusive) political and economic goal' (Harvey 1989b, 8). It involves, in short, an overriding commitment to economic growth. Critics have pointed to the wholesale depoliticization of local economic development and erosion of local democracy this has frequently involved, as well as an inbuilt tendency to present particular private interests, particularly those of property developers and business leaders, as the public interest (Barnekov *et al.* 1989; Clavel 1986; Cox and Mair 1988; Deakin and Edwards 1993; Eisinger 1981; Goetz and Sydney 1994; Harding 1992; Judd and Ready 1986; Logan and Molotch 1987; Molotch 1976; 1988; Molotch and Logan 1985; Nickel 1995; Parkinson and Judd 1990; Peck 1995; Peterson 1981; Stone 1989; Turock 1992; Ward 1997). The context of this transition is widely held to be the increasingly mobile and extraterritorial nature of capital, ever more commonly referred to by the shorthand term, 'globalization'. The adequacy of that term, however, has been much debated, not least because of the way in which it is at the local level that the implications of processes operative at a higher level of resolution are felt and, increasingly, managed (Beauregard 1995; Goetz and Clarke 1993; Jonas 1994; Lake 1990; Lovering 1997; Marcuse and van Kempen 2000; Smith 1992; Smith and Feagin 1987; Swyngedouw 1997; Wood 1998). The somewhat inelegant and incongruous-sounding neologism, 'glocalization' – originally derived from the Japanese term *dochakuka*, roughly meaning 'global localization' (Robertson 1992, 174) – has often been drafted in, in order to highlight the simultaneous process of globalization and localization (Robertson 1995). It is important to underscore the central point this terminology embodies: 'globalization' and 'localization' each derive their sense from the unholy alliance that, together, they form. They attain their meaning in circumstances that implicate both in contradictory, asymmetrical, yet at the same time mutually reinforcing ways.

Globalization itself is commonly, though not entirely helpfully, framed as a primarily economic process. Given that much of the globe has long been economically integrated into a world-system (Wallerstein 1974; 1980), it is also often objected that there is nothing new to globalization, beyond a merely semantic shift. However, as Harvey (2000, 68) proposes, the process gains its specificity from certain *emergent* structural properties, which arise from the intensification of a constant underlying process; it amounts to a 'qualitative translation wrought on the basis of ... quantitative shifts'. Whatever the veracity of this formulation, which is not without its merits, a sharper sense of the meaning of the term emerges from the striking contrast between the 'global' and the 'universal'. 'Globalization and universality don't go together', writes Baudrillard (1998c, 11): 'Globalization is the globalization of technologies, the market, tourism, information. Universality is the universality of values, human rights, freedoms, culture, democracy. Globalization seems irreversible, the universal might be said, rather, to be on the way out'. In other words, the rise of the 'globalization' discourse is, as Bauman (2001c, 299) suggests, intimately related to the demise of the modern ambitions encapsulated in the notion of 'universalization':

> Together with such concepts as 'civilization', 'development', 'convergence', 'consensus' and many other terms of early- and classic-modern debate, universalization conveyed the hope, the intention and the determination of order-making. Those concepts were coined on the rising tide of modern powers and the modern intellectual's ambitions. They announced the will to make the world different from what it was and better than it was, and to expand the change and the improvement to global, species-wide dimensions. It also declared the intention to make the life conditions of everyone everywhere, and so everybody's life chances, equal. Nothing of that has been left in the meaning of globalization, as shaped by the present discourse.

Globalization refers, above all, to *effects* rather than *intentions*. It 'is not about what we all, or at least the most resourceful or enterprising among us, wish or hope *to do*. It is about what is *happening to us all*' (Bauman 1998b, 60). Thus globalization takes shape, as an untamed and untameable force, in the wake of the growing impotence of modernity's once-vaunted ordering capacity. With diminishing prospects of universalization, global disorder looms large (Jowitt 1992). It is against this backdrop that the emergence of the entrepreneurial city takes shape.

The impact of globalization on the city is not, however, as direct as is often envisaged. The image of hypermobile capital coming down to earth in an uneven and unpredictable way, leaving individual places increasingly stranded in space and at the whim of global forces, is not so much incorrect as irredeemably partial. What this image misses out is the significance of the transformation of political space that is both a primary consequence and a principal cause of globalization. The predicament of the nation-state is particularly vital to understanding the impact of globalization on the city. Of all modern institutions, the nation-state embodied the greatest ambitions to stand firm against the rising tide of chaos. Its primacy lay in its claims to territorial sovereignty. As Bauman (2001d, 300) notes:

> the very meaning of 'state' has been precisely that of an agency claiming the legitimate right and the resources to set up and enforce the rules and the norms binding the run of affairs over a certain territory; the rules and the norms hoped and expected to turn contingency into determination, ambivalence into *Einduetigkeit*, randomness into predictability – in short, chaos into order.

The state held a monopoly on the means of ordering its sovereign territory, uniting economic management, political authority and cultural hegemony within a single overarching framework. The state's calls upon resources were, as a consequence, formidable. This factor was responsible for strictly limiting the sum total of nation-states that might feasibly be sustained. Not all collectivities were equally viable propositions, and in consequence state formation typically involved 'the suppression of state-formative ambitions of many lesser collectivities' (Bauman 2001d, 300; cf. Smith 1986). Alongside active suppression, disputes over nationhood frequently relied on the *invention* of national traditions, designed to unite lesser collectivities into a more-or-less coherent, if imagined, national community (Anderson 1991; Hobsbawm and Ranger 1983). Although this process proceeded along conflictual lines, and periodically involved explosive contestations of sovereignty, it nonetheless established an overarching global order in relation to which no territory or collectivity could reasonably remain ambivalent. Indeed, the 'meaning of "global order" … boiled down to the sum-total of a number of local orders, each effectively maintained and efficiently policed by one, and one only, territorial state' (Bauman 2001d, 301). Although such a system made border skirmishes inevitable, it contrasts starkly with the 'new world disorder' brought about by globalization (Jowitt 1992).

Harvey (2000, 68) claims that 'globalization is undoubtedly the outcome of a geopolitical crusade waged largely by the United States'. Whilst there can be little doubt that, as the single remaining superpower, the United States' position is of fundamental importance, Harvey's formulation inadequately captures the essential indeterminacy ushered in by the collapse of the Eastern Bloc. Its position is as much determined by, as it is a determinant of, the new world disorder, its greater capacity for fending off indeterminacy notwithstanding. As Baudrillard (1993a, 69) once put it: 'Two superpowers are necessary in order to keep the universe under control: a single empire would crumble by itself'. In other words, any totality relies on a form of competitive opposition to hold it together; a dual structure or '*binary regulation*' is the apex of any 'unitary system' (*ibid.*). The disintegration characterizing globalization is a direct consequence of the absence of this structure. The strategic balance of power between the two superpowers represented the continuation by other means (more accurately, at another scale) of precisely the same process that initially saw the formation of an interlocking system of territorial nation-states.

> Non-alignment, refusal to join one or another of the super-blocks, sticking to the old-fashioned and increasingly obsolete principle of supreme sovereignty vested with the state, was the equivalent of that 'no man's land' ambivalence which was fought off tooth and nail, competitively yet in unison, by modern states at their formative stage.
>
> (Bauman 2001d, 301)

The politically divided world conjured up an 'image of totality' (Bauman 1998b, 58), giving every part of the globe significance in terms of the overall balance of power. No part could hope to remain a matter of indifference to the major superpowers. However, the sense of 'absent totality' brought about by the collapse of the strategic power-balance reconfigures the world as an indeterminate 'field of scattered and disparate forces, sedimenting in places difficult to predict and gathering momentum impossible to arrest' (*ibid.*).

For a time, the image of totality detracted from the changing situation of the nation-state itself. The collapse of the Eastern Bloc revealed what had hithertofore been hidden; namely, a fundamental change in the viability of the nation-state in terms of its former territorial principle. The conditions under which the pattern of nation-states had first established itself could hardly have been expected to remain stable or static over time. As the emancipation of capital from space gathered pace over the course of the twentieth century, the distance between the initial conditions under which the world order was set and the current conditions reached a critical point. The consequence, as Veltz (2000, 34) puts it, was that it became increasingly infeasible for 'nation-states to define and implement coherent and efficacious economic policies and, more generally, maintain their role as first-rank social and cultural reference points'. As the boundaries of the nation-state were rendered increasingly permeable by the hypermobility of capital, aided and abetted by the 'information revolution', economic activity assumed an increasingly footloose character. It is in this context that the first pronouncements of the 'end of geography' from those not taking their cue from the likes of Baudrillard were issued (O'Brien 1992). Again, geographers were amongst the first to demur (Corbridge *et al.* 1994; Leyshon and Thrift 1997; Martin 1999; Tickell 2003), and it is easy to see why this might be regarded as a misleading diagnosis. The *deterritorializing* forces of global capital can be seen to be predicated on, and positioned to give rise to, a concomitant *reterritorializing* tendency, as the holding-formation established by the system of modern nation-states undergoes fragmentation at lower spatial scales (cf. Luke 1998; Ohmae 1995). This spatial restructuring is evident in the growth of regionalism and devolution (Bogdanor 1999; Deas and Ward 2000; Evans and Harding 1997; John and Whitehead 1997; Jones and MacLeod 1999; Keating 1997; MacLeod 1999; Tewdwr-Jones and McNeill 2000), as well as in relation to the city itself (Mayer 1992; 1995). It is most evident, however, in the clamour from those whose collective identities had formerly been subjugated to 'invented national traditions' for states of their own. Far from having been abandoned, therefore, the territorial principle has become increasingly popular with the demise of state sovereignty. New nations, once 'much too small and inept to pass any of the traditional tests of sovereignty, but now demanding ... the right to legislate and police order on their own territory' (Bauman 2001d, 302), have been given a new lease of life by the increasingly extraterritorial character of economic activity. The diminishing economic and military requirements for statehood have prompted a scramble for sovereignty by would-be elites, eager to accentuate ethnic distinctions that had not previously been emphasized. Hence Baudrillard's (1994b, 32) caustic remark that, at 'the rate we are going, we shall soon be back at the Holy Roman Empire'.

Despite appearances, the wave of reterritorialization is hardly at odds with the deterritorialization ('globalization') of economic activity. The two are 'not opposed and hence mutually conflicting and incompatible trends; they are, rather, coeval factors in the ongoing rearrangement of various aspects of systemic integration' (Bauman 1993b, 232). There is 'an intimate kinship, mutual conditioning and reciprocal reinforcement between "global-ization" and the renewed emphasis on the "territorial principle"' (*ibid.*, 303). The 'globalization' of the economy requires little more than the dismantling of everything that might restrict the freedom of movement of capital, and the tendency towards reterritorialization at a different scale is entirely compatible with this end.[8] Weak states are far more amenable to being 'reduced to the (useful) role of local police precincts, securing a modicum of order required for the conduct of business, but need not be feared as effective breaks on the companies' freedom' (Bauman 2001d, 303). Explicating the situation, Bauman (1998b, 33) draws on Crozier's (1964) demonstration of the 'intimate connection between the certainty/uncertainty scale and the hierarchy of power'. As a general rule, holding on to the reins of power involves the manipulation of uncertainty: maximizing certainty for oneself and maximizing uncertainty for everyone else; giving oneself the maximum degree of freedom to act whilst minimizing the leeway of the others. Where the dominant position was once held by the state, 'it is now world capital and money that are the focus and the source of uncertainty' (*ibid.*). The present predicament of the state, in other words, is underpinned by the fact that the certainty on which it thrived is increasingly withheld by other interests. Consequently, the state is increasingly characterized by political authority alone. In the absence of national economic management and cultural hegemony, it is unsurprising that a new series of economic imperatives, together with a search for some kind of cultural identity, should have been transmitted downwards from the nation-state to the city.

The rise of the entrepreneurial city and the increasing focus on the culture of cities associated with its 'postmodernization' are, therefore, dual aspects of the same underlying situation. It is not that one process is a manifestation or reflection of the other. They are the twin effects of a reconfiguration of the balance of power between economy and state, reducing the latter to a focus, and a source, of political authority alone – devoid of its former economic powers and increasingly stripped of cultural hegemony.[9] The upshot of the globalization of economic activity is that cities are increasingly pitched as *rivals*, in direct competition with one another for the same investment, jobs, grants, and so forth. Whilst the situation might generate individual success stories (and media attention), on aggregate it does little more than embroil cities in a zero-sum game. Indeed, the widespread perception that the situation does amount to a zero-sum game encourages the adoption of an entrepreneurial stance in the first place. It is not so much growth *per se*, therefore, but a kind of *simulated* growth that fuels the entrepreneurial city. The situation fosters the all-too-familiar attitude, this time at the level of urban governance, that one only needs to play one's hand right. It places the matter of whether one deserves to have been dealt a better hand in the first place beyond question, rendering it purely academic. The situation that manifests itself in consumption, in other words, is also increasingly evident for the city. Both are manifestations of the situation that remains in the wake of the liminal reversibility finally resulting from modernity's original ordering drive.

The glocalization of urban culture

> Glocalization ... polarizes mobility – that ability to use time to annul the limitation of space.
>
> (Bauman 2001d, 307)

If place has, paradoxically, been rendered increasingly significant by the impress of globalization, the situation has spontaneously spawned a proliferation of attempts by particular places to assert some kind of distinctive cultural identity – typically highlighting local landmarks, historical events or connections, specially constructed 'attractions' and 'facilities', and the like. All this attention to uniqueness has, of course, tended to ensure that erstwhile, unacknowledged local differences are increasingly encountered merely as variations on a theme. The situation has, nonetheless, fuelled a preoccupation with the *culture* of cities, not only in academic circles but increasingly in urban policy and governance as well. At the same time, the forces of globalization have introduced a new form of social differentiation, ensuring that any given city is increasingly heterotopic: the cultural gloss designed to make cities attractive propositions for inward investment increasingly sits uneasily alongside the presence of a residualized 'underclass'. Spatial differentiation, then, is far from having been wiped out by the homogenizing forces of a mass consumer culture, as was once feared (Peet 1986). Even so, as Appadurai (1990) implies, its persistence hardly suggests continuity with the past. In fact, it is a particularly weak conception that identifies 'culture' with the 'local', and 'differentiation' with 'authenticity'. The apparently increasingly diverse mosaic of regional and local cultures owes its very existence to the globalization of economic activity, which operates in space in much the same way as it operates for the individual consumer. It presents a global 'matrix of possibilities', from which 'highly varied selections and combinations can be, and are, made' (Bauman 2001d, 304). A common combinatorial source automatically ensures a variegated cultural topography. Despite the 'ostensibly world-wide availability of cultural tokens', the selection of possibilities tends to be made *locally*, the global matrix providing a plethora of 'new symbolic markers for the extinct and resurrected, freshly invented or as yet [only] postulated ... identities' (*ibid.*) that might serve to put a place 'on the map' (Ashworth and Voogd 1990; Gold and Ward 1994; Jessop 1998; Kearns and Philo 1993; Paddison 1993). Precisely the same situation that characterizes lifestyle, in other words, is increasingly evident for the city itself; the overall result being the strangely familiar pattern of 'major' convention centres, 'unrivalled' shopping and entertainment complexes, and 'unparalleled' visitor attractions that is claimed as 'unique' by each city in turn. The local is, finally, a truly global phenomenon; it is to be found everywhere.

The recommodification of local culture has, of course, attracted considerable criticism. The list of reckoning includes the selective reanimation of the past accomplished by the heritage industry (Hewison 1987; Lowenthal 1996; Graham *et al.* 2000; Baker 1988; Bagnall 1996; Hudson 1983; Hall 1995; Watson 1991); the pursuit of so-called cultural strategies geared towards urban regeneration (Bassett 1993; Bianchini and Parkinson 1993; Griffiths 1995; Landry and Bianchini 1995; Lovering 1995; Montgomery 1995);[10] speculative financial commitment to flagship projects (Bianchini *et al.* 1992) that are as likely to be loss-making as profit-generating; a plethora of waterfront development schemes (Goss 1998); and so on. Much of the critique directed at such schemes, however pertinent it may be, involves an unfortunate implicit appeal to a sense of *Gemeinshaft* that was, from its very inception, bathed in the glow of nostalgia. What glocalization implies is not simply an uneven contest between the global and the local (however true this might be). Its conceptualization in such terms merely invites a critique of the dominance of the representation over the real, together with an untenable assertion of an erstwhile authenticity. What such critiques tend to miss is that the increasingly vociferous claims and frantic activity directed towards promoting the 'uniqueness' of urban places is merely a side effect of a fundamental redistribution of privileges and deprivations, itself defined in spatial terms – that is, in terms of *differential mobility*. It is this situation that underlies the fate facing individual places in the first place. It implies that some (individuals, groups, organizations) are increasingly footloose, whilst others are left stranded in space; that the concerns of increasingly polarized sections of society have diverged radically. Whilst one would therefore be 'right to perceive glocalization as the concentration of capital, finance and all other resources of choice and effective action', glocalization also involves 'a process of world-wide *restratification*' and a '*concentration of freedom* to act' (Bauman 2001d, 304–305). Glocalization means, above all: globalization for some, localization for others.

Freedom to act is dependent on the resources to back up that freedom. The concentration of freedom to act is, therefore, a direct function of the concentration of wealth. At the same time, however, that freedom is portrayed as being the appropriate means for solving the problems to which it in fact gives rise. Freedom to act is represented as the key to wealth creation, 'the hothouse in which wealth would grow faster than ever before' (*ibid.*, 306) (the implication being, as ever, that the increasing size of the cake will guarantee a larger slice for everyone). The neo-liberal fairy-tale is credible, however, only for those whose interests it serves: for those who no longer have any long-term concern with the integrity of place, but only with their own ability to transcend place. Consequently, what 'is free choice for some is cruel fate for others' (*ibid.*, 305). As the growing number of references to the 'divided' or 'dual' city make clear (Fainstein *et al.* 1993; Mellenkopf and Castells 1991), the experience of glocalization is sharply divided, and shows precious little prospect of abating (Harvey 1992; Laws 1995; Merrifield and Swyngedouw 1997).[11] This is evident from Harvey's (2000, 68) characterization of globalization as a dangerous admixture, an essentially retrogressive phenomenon coupled to a hitherto unknown intensity:

> if there is any real qualitative trend it is towards the reassertion of early nineteenth-century capitalist values coupled with a twenty-first century penchant for pulling everyone (and everything that can be exchanged) into the orbit of capital whilst rendering large sections of the world's population redundant in relation to the basic dynamics of capital accumulation.

This touches on the nub of the issue. The emancipation of capital from labour decisively breaks the link between rich and poor (Bauman 1998a; Wilson 1997). Where the 'rich [once] needed the poor to make and keep them rich', this is simply no longer the case (Bauman 1998a, 306). The novelty of the situation facing the 'new poor' derives from their structurally redundant position. In an unprecedented and unequivocal way, today's poor are marked out as essentially useless. The only practical purpose their sorry

fate can serve is as an object lesson for those lucky enough to find themselves – for the present, at least – in the 'two-thirds society' of consumers.[12]
The undeserving poor, stigmatized by their fate, are overwhelmingly
portrayed not only as a drain on resources but as a threat to the rest of
society (Gans 1995). The protean nature of the supposed threat is one of its
principal qualities: whatever specific threats may be singled out from the
heterogeneous underclass, they are invariably seen as threatening the freedoms enjoyed by fully paid-up members of the consumer society. The
intimate connection between the freedom of one section of society, and the
lack of freedom available to the rest, is rarely recognized for what it is.
Indeed, although the 'globalizing and localizing trends are mutually reinforcing and inseparable', 'their respective products are increasingly set apart
and the distance between them keeps growing, while reciprocal communication comes to a standstill' (Bauman 2001d, 306). This is particularly evident
in the way in which the freedom to act translates into mobility, and hence
into a lack of any real commitment to place.

Freedom to act amounts to the capacity *not* to remain tied in place, the
ability to traverse space at will. Lack of freedom involves the inability to
overcome the constraints of space; to confinement without the necessity of
prison walls.[13] The 'globals' and the 'locals' inhabit increasingly separate
worlds, characterized by their differential spatial and temporal opportunities
(Bauman 1998b; cf. Kanter 1995; Hannertz 1990):

> If for the first world, the world of the rich and the affluent, the space
> has lost its constraining quality and is easily traversed in both its 'real'
> and 'virtual' renditions, for the second world, the world of the poor, the
> 'structurally redundant', real space is fast closing up – the deprivation
> made yet more painful by the obtrusive media display of space conquest
> and the '*virtual* accessibility' of distances untouchable in the non-virtual
> reality.
>
> (Bauman 2001d, 306)

As the world is increasingly made to the measure of those possessing the
freedom of mobility, the ability to engage with the world at all is, for those
lacking that freedom, dramatically diminished. Indeed, the growing
distance between the globals and the locals is not measured primarily in
terms of physical space, nor in terms of social space, but increasingly in
terms of virtual space (though not necessarily in the narrow, technological
sense of that term: Doel and Clarke 1999b). The familiar modern pattern
of urban residential differentiation has hardly been done away with, of
course. The social-and-physical space of the postmodern city runs the full
gamut, from 'no-go' areas to exclusive enclaves (Dear 2000; Dear and
Flusty 1998; Ellin 1996; 1997). Indeed, the pattern is heightened all the
more by the polarization of mobility, which increasingly stratifies society in
terms of the differentiation of space with respect to time. At one end of the

spectrum, the globals increasingly 'live in *time*', as Bauman (2001d, 307) suggests: 'space does not matter for them, since spanning every distance is instantaneous'. Hence the sense in which time becomes experienced, as Jameson (1983) put it, as a 'perpetual present': an *episodic* time, divorced from the past and the future, and always in short supply, 'since each moment of time is non-extensive – an experience identical to that of ... time "full to the brim" ' (Bauman 2001d, 306). At the other end of the spectrum, the locals 'live in *space* – heavy, resilient, untouchable – which ties down time and keeps it beyond [their] control' (*ibid.*, 307). Their time is empty and monotonous, both abundant and redundant. They are unable to control it, but 'neither are they controlled by it, unlike their clocking-in, clocking-out ancestors, subject to the faceless rhythm of factory time. They can only kill time, as they are slowly killed by it' (*ibid.*). The qualities of space experienced by those subject to the decomposition of time are the precise opposite of those who inhabit an increasingly globalized hyperspace. Hyperspace carries all the ramifications of hyperreality, 'where the virtual and the real are no longer separable, since both share and miss in the same measure that "objectivity", "externality" and "punishing power" which Emile Durkheim listed as the symptoms of "reality" ' (*ibid.*). For those for whom hyperspace remains inaccessible, however, the harsh reality of space, with all the implications of constraint it carries, remains all too real – more real than ever, in fact, in the face of the freedom conferred on those possessing the resources to traverse it. It is this restratification of social space that lends the postmodern city its heterotopic character. The postmodern city is the city made to the measure of the globalized world, fitted out with the infrastructure capable of enticing and satisfying the globals – with all the consequences this holds for the locals, increasingly defined in terms of their exclusion from a world in which the ability to overcome the limits of space has become paramount. The postmodern city and the entrepreneurial city are, in fact, one and the same.

The declaration of Leeds as a 'European' and a '24-hour' city at the end of the twentieth century was not, therefore, a merely truistic or tautological act. It signalled the fundamental change in space and time wrought by glocalization and the decomposition of modernity. By dint of its geographical location, Leeds has always been a European city. Its reflexive self-designation as such obliquely signals a change in the very meaning of location (Hall 1997; Lee 1997). It amounted to an attempt to convey the message, to those who need to hear it, that Leeds has more in common with, say, Barcelona than with its neighbouring city of Bradford. It therefore expressed a truth about glocalization, but one that diverges radically in meaning for rich and poor.[14] The connotations of the '24-hour city' point to precisely the same situation. The notion is presumably intended to suggest that the nocturnal city's former association with the shadowy forces of transgression has been overcome (Alvarez 1995; Creswell 1998; Schivelbusch 1988). It implies that the night-time city is increasingly to be regarded as a space of conspicuous consumption, redolent of continental European 'café-society' and the North American 'city that never sleeps' (Bianchini 1995; Comedia 1991; Heath 1997; Heath and Stickland 1997; Jencks 1996; Jones *et al.* 1999; Krietzman 1999; Lovatt and O'Connor 1995; Malbon 1998; Montgomery 1995; Warpole and Greenhalsh 1999). It also suggests, however, the sense in which urban time itself has been telescoped into a recurrent, episodic 24-hour period.[15] If the adoption of the 'European city' and '24-hour city' epithets by Leeds are symptomatic of glocalization, there are any number of visible signs of its human consequences.[16] Although widely acclaimed as a success story, the transformation of Leeds into a buoyant entrepreneurial city, fuelled by service-led economic growth (Tickell 1996; Leigh *et al.* 1994; Williams 1996; 1997), has always had to be qualified. Leeds was, from the start of its most recent economic transition, quite evidently a 'two-speed' city, riven by social polarization (Dalby 1993; Stillwell and Leigh 1996). The case for a more inclusive role for community groups in the entrepreneurial framework of the 'Leeds Initiative' (Haughton and Williams 1996) has been taken on board by the curiously entitled 'Community Involvement' programme (the title obliquely signalling that communities are no longer 'naturally' or inevitably a part of the cities to which they belong). The current situation, it seems, is no longer likely to provoke the kind of rioting described by Davies (1998), which the cynic might take as the reason for the attempt to patch up the cracks in the urban fabric in the first place. Such ameliorative measures as neighbourhood renewal (Leeds Initiative 2001; Social Exclusion Unit 2001) cannot, however, hope to remedy the underlying structural fault that polarizes the city's inhabitants along the dimensions of space and time, now emptied of the meaning once lent to them by modernity. The disintegration of modern space-time rebounds in the divergent experience of glocalization, and there is precious little reason to believe that stop-gap solutions will be able to remedy matters, when the underlying problem is the global system of consumerism itself.

Postscript

today, reality itself is hyperrealist

(Baudrillard 1993a, 74)

The year 2000 saw the completion of a major refurbishment of Leeds Town Hall, in part to improve its facilities and acoustics as an international concert venue. Four audience boxes, added to the gallery to increase the main hall's seating capacity, were actually included in the original plans, though not in the construction itself, upon completion in 1858: 'until now it had never become a reality' (Leeds City Council 2000, 8).

Notes

Introduction

1 A related, if subsidiary, issue concerns the compatibility of the work of these two authors. Some, such as Gane (1991, 4), have entirely spuriously implied that Bauman has 'disguised' his 'dependence on Baudrillard's work'. Others have failed to dwell on the resonances in much detail (Smith 1999; Beilharz 2000). I shall tend to point up certain similarities between their respective formulations, without intending to imply some kind of essential consonance.

2 In many senses, consumption has always been much in evidence in writings on cities. Yet until relatively recently, contemporary urban analysis has fought shy of any sustained engagement with consumption. This is particularly true if consumption is compared to production (Harvey 1985a; 1985b). Stretching the point only a little, one has to go back as far as Berry (1967) to find an explicit urban geography of consumption (a somewhat weary, revamped version appeared as Berry and Parr 1988).

1 Consumption controversies

1 This assertion taps into the distinction drawn by Lockwood (1964) between social and system integration, developed more recently by Archer (1996) and Mouzelis (1997).

2 The connection between postmodernity and consumption was first explicitly made by Jameson (1983), though in somewhat idiosyncratic terms. Bauman (1987) arguably offered the first thoroughgoing consideration in the English language. Baudrillard's work, dating from the late 1960s, had effectively explored the connection even before the term 'postmodern' gained currency. Featherstone (1991) represents another important contribution.

3 The situation is analogous to the way in which a 'first' only ever *follows* a 'second'. In commonsensical terms, a 'second' can only come *after* a 'first'. However, without a 'second', a 'first' is not a 'first' at all. It is not part of a series, and only becomes a 'first' with the arrival of a 'second'. A 'first' only ever appears in the future anterior: it *will have been*.

4 It should be noted that neither Bauman's nor Baudrillard's relationship to the 'postmodern' is straightforward. Baudrillard, for instance, has rarely used the term. Bauman, on the other hand, has made great play of the notion (cf. Bauman 1992a; 1993b; 1997a) – though, on 27 June 1999, he announced that he would no longer be using the term; in a seminar entitled 'Modern and post-modern adventures of work', given to the Department of Sociology, University of Leeds. This was not a denouncement. The concept has, for Bauman, simply

been wrung dry. Appropriately, his preferred alternative is now 'liquid modernity' (Bauman 2000).

5 Accordingly, we also need to dispense with many other unconvincing versions of the connection between postmodern and consumption, such as that resting on the misconception that postmodernist-cum-poststructuralist thought invites an interpretative, humanistic study of 'consumption as text' (Hirschman and Holbrook 1992). Such a conception mistakenly assumes, and promulgates, what Derrida (1981, 66) disparagingly refers to as 'a new 'idealism' ... of the text'.

6 Miller *et al.* (1998, 26) efface such differences, erroneously implying an essential compatibility between 'the work of Mary Douglas, Pierre Bourdieu and Daniel Miller' by speaking of them in the same breath.

7 For the sake of comprehensiveness, let us note that the relationship between income and consumption is *the* classic concern of macroeconomic theory. Keynes (1936, 96) suggested that the dependence of consumption on income was a stable linear function, reflecting a 'fundamental psychological law'. He held the marginal propensity to consume to be positive but less than one (that is, as income rises people will consume more but at a decreasing rate); and the marginal propensity to consume to be less than the average propensity to consume (reasoning that, were income levels to fall relative to recent levels, people would not lower their consumption standards proportionately; just as a sudden rise in income would not see a proportionate rise in consumption). The non-proportionality of consumption to disposable income that Keynes held to apply in the short run is, though, contradicted both by long-run data suggesting a proportional relationship between consumption expenditure and average income levels as these have risen over time (Kuznets 1946), and by cross-sectional evidence of differences in household incomes and expenditures. The history of macroeconomic thought on consumption consists almost entirely of attempts to redefine 'income' as something broader than 'current income' alone, in an attempt to resolve the apparent contradiction. The most notable examples are Duesenberry's (1949) 'relative-income theory'; Friedman's (1957) 'permanent-income theory'; and the 'life-cycle hypothesis' (Ando and Modigliani 1963; Modigliani and Brumberg 1955). For an examination of the departure from neo-classical economic consumer theory implicit in Keynesian economics, particularly Keynes' rejection of the deeply ingrained Benthamite doctrine of utilitarianism, see Drakopoulos (1992).

8 The phrase 'the politics of consumption' is frequently used to refer to a politics cutting across class interests or even divorced from capitalism (Mort 1989; Jackson 1993). However, the power of capitalism and the autonomy of the human subject remain amongst the most significant political issues connected to consumption (Hearn and Roseneil 1999; Daunton and Hilton 2001).

9 Despite common assertions to the contrary, Marx did not see production as crudely or unidirectionally determining consumption. He is often regarded as having taken a diametrically opposed stance to Adam Smith (1937, 625), who held that: 'Consumption is the sole end and purpose of all production and the interest of the producer ought to be attended to, only so far as it may be necessary for promoting that of the consumer'. (Cf. Sterne (1997, 95): 'it is the consumption of our products, as well as the manufactures of them, which gives bread to the hungry, circulates trades, – brings in money, and supports the value of our lands'.) Marx (1973, 99), in fact, regarded 'production, distribution, exchange and consumption ... [as] members of a totality, distinctions within a unity', offering a far more complex view of their interrelations and interdependencies.

10 This distinction also underlies so-called 'underconsumption theories' of the crises into which capitalism is periodically plunged (Lenin 1960; Luxembourg

1951). Such theories are generally regarded, today, as partial at best (Bleaney 1976; Bottomore 1985; O'Connor 1987). According to no less a theorist than Hilferding (1981, 241), ' "overproduction of commodities" and "underconsumption" tell us very little'.

11 Note that *productive consumption*, in this sense, occurs even in situations where the labour process is, in Marx's sense, *unproductive* (that is, unproductive of surplus value). Marx gives the example of a jobbing tailor patching a capitalist's trousers: such labour is unproductive because it is paid for out of the capitalist's income, and hence out of surplus value, generating no surplus value of its own. Nonetheless, raw materials – cloth, cotton thread, the tailor's labour power – are all used up in the process. Their *consumption* is 'productive' even if the tailor's *labour* is not.

12 It precludes or forecloses on 'other pleasures', Soper (2000) suggests. The problem Soper underplays, however, is that the possibility of any alternative has been rendered increasingly *unviable*. Specifically, the 'aesthetic spacing' (see Chapter 3) promoted by postmodernity is characterized by the fact that 'it does not choose as its points of reference and orientation the traits and qualities possessed by or ascribed to the objects of spacing, but the attributes of the spacing subject (like interest, excitement, satisfaction or pleasure)' (Bauman 1996a, 33). Thus, as Lyotard (1993a, 66) proposes, emancipation 'is no more situated as an alternative to reality, as an ideal set to conquer and force itself upon reality from outside' (cited in Bauman 1996a, 34). Consequently, 'the militant practice has been replaced by a defensive one, one that is easily assimilated by the "system" since it is now assumed that the latter contains all the bits and pieces from which the "emancipated self" will be eventually assembled' (Bauman 1996a, 34). Soper (2000) is right to point up what the system excludes. The difficulty lies, however, not so much in imagining an alternative as in the absence of the conditions in which it might be accommodated (Chapter 6).

13 It ought to be stressed that the 'consumer society' is an ineluctably Western phenomenon, at least in terms of its origins. Although American society has often been portrayed as the archetypal consumer society (cf. Jones 1965; Fox and Lears 1983; Glickman 1999), its European co-paternity should certainly not be forgotten. Nor should the significance of its translation into other cultural contexts be neglected: see, for example, Clammer's (1997) consideration of the Japanese city in terms of consumerism.

2 Everything you ever wanted to know about consumption (but were afraid to ask Baudrillard)

1 Essentially a technical elaboration of Bentham's doctrine of utilitarianism, the neo-classical economic theory of consumption, which emerged as part of the 'marginalist revolution' (in the work of Jevons in 1871, Walras in 1874, and anticipated by Gossen in 1854), initially assumed the possibility of a *cardinal* measure of utility. Marshall's partial equilibrium analysis of 1890 embodied similar assumptions. Modern utility theory is founded on an *ordinal* approach. Indifference analysis, initiated by Edgeworth in 1881, shorn of its last cardinal remnants by Pareto in 1906, and given its most rigorous treatment by Hicks and Allen (1934) is, however, in very much the same mould, as is the later 'revealed preference' approach (Samuelson 1948). Mishan (1961) provides an internal critique of such theories. Most recent developments (see Herden *et al.* 1999) work on the basis of *subjective expected utility* – an approach originating with Bernoulli in 1738, formally developed by von Neumann and Morgenstern (1947), and absorbed into econometric discrete-choice modelling approaches such as 'random utility theory' (Bates 1988). Much of the initial work in the 'applied' discipline of marketing aimed to open up the 'black box' of the utility-maximizing consumer

in order to add a dose of behavioural and psychological realism (Howard and Sheth 1969). Foxall (1988) provides a thorough, if, by now, dated account, whilst Belk (1995) reviews more recent contributions. Geographical work spanning economic and behavioural approaches includes Rushton (1969) and Bacon (1984); Shepherd and Thomas (1980) and Golledge and Stimson (1997) provide fuller reviews.

2 Consumption has always almost been regarded as being concerned with satisfying needs. It was once common, following Maslow (1954), to speak of a 'hierarchy of needs' – moving from basic physiological needs to higher, more refined ones. Although Maslow applied the term 'needs' across the board, a similar idea is evident in the distinction between 'authentic' and 'artificial' ('false') needs, and between 'necessities' and 'luxuries'. Attempting to overcome the difficulties of such hierarchical/dichotomous conceptions, Douglas and Isherwood (1996, xxi) rightly emphasise the duality of *all* consumer goods:

> As far as keeping a person alive is concerned, food and drink are needed for physical services; but as far as social life is concerned, they are needed for mustering solidarity, attracting support, requiting kindnesses, and this goes for the poor as well as the rich

– and for 'luxury' as well as 'basic' consumption (even if some goods are more 'observable', and hence 'positional' [Hirsch 1976], than others [Frank 1985]).

3 There will be cause to refine these rough-and-ready definitions in due course, but they will suffice for the time being.

4 As this example reveals, the sign is composed of a 'signifier' (its material element) and a 'signified' (conceptual element). Words are signs, composed of a signifier (e.g. 'cat' – the three letters 'c-a-t' materially inscribed on the page, or the corresponding verbal 'sound-image'); and a signified (the idea of a feline mammal). Signifier and signified are inseparable insofar as they are mutually constitutive: they form the two sides of a sign in the same way that the recto and verso form the two sides of a sheet of paper. Signifier and signified might, on occasion, appear to come undone, though this is a deceptive appearance. When addressed in a foreign tongue, for instance, one seemingly has access only to signifiers; though without their corresponding signifieds they no longer signify, thus losing their status *as* signifiers. Likewise, when a word (signifier) is 'on the tip of one's tongue', the signified also remains infuriatingly ungraspable. Elaborating on the floral example, Sturrock (1979, 7) notes that there 'can be no precise signified for a wreath because the language of flowers is too loose, at any rate in our culture; but equally there can be no wreath without a significance of some kind'.

5 Harvey (1989a, 287) also reads Baudrillard as claiming 'that Marx's analysis of commodity production is outdated because capitalism is now predominantly concerned with the production of signs, images and sign systems rather than with commodity systems themselves'. As Smith (1997, 308n) notes,

> this is not at all what Baudrillard is arguing. Baudrillard [1981, 146] is arguing that, *'the logic of the commodity and political economy is at the very heart of the sign … the structure of the sign is at the very heart of the commodity form'*.

Such readings of Baudrillard as Campbell's, Harvey's, and also Gottdeiner's (1994), are possible, but are neither true to the original nor, for that matter, particularly enlightening.

6　Baudrillard's (1981) *critique* of the 'political economy of the sign' does achieve some vital insights (Genosko 1994a; 1994b).

7　Noting the term's etymological roots (*sýmbolon* as distinct from *diábolon* [Müri 1931]), Luhmann (1988, 96) suggests that symbols

> presuppose the difference between unfamiliar and familiar and they operate in such a way as to enable the re-entry of this difference into the familiar. In other words, symbols represent the distinction between familiar and unfamiliar within the familiar world. They are forms of self-reference using the self-reference of form.

8　Cassirer (1955, 53) claims that 'in every case "symbolic form" is a condition either of the knowledge of meaning or of the human expression of meaning'. For Peirce (cited in Almeder 1980, 25), this involves a particular kind of meaning: a symbol is something like a sign that is arbitrarily linked to its object 'by means of an association of ideas or habitual connection' (one thinks of the cross as a symbol of faith in Jesus Christ, for example). This accords with Cassirer's (1955) sense in which 'spiritual meaning is attached to a concrete, material sign and intrinsically given to this sign' in a 'symbolic form' (cited in Panofsky 1991, 41). For Lacan, the symbolic relates not simply to 'meaning', nor even 'meaning of a particular kind'. It relates to nothing less than the linguistic constitution of the human subject (Easthope 1999). The subject – the 'I' by which I identify myself – amounts, in Lacan's conception, to a position granted by the symbolic order; but insofar as no signifier is ever adequate to what it claims to represent, this positioning incurs an inevitable cost (hence Lacan's reference to the subject's *manque à être* ('want to be'), rather than its 'being' as such). Julien (1994, 61) defines the Lacanian 'symbolic' as 'pure language, a combinatorial system for the inscription of places' – noting, crucially, that 'absence is the condition of symbolic presence'. Lacan's cybernetic analogy clarifies this: 'the minimal binary choice of 0 and 1 inscribes presence against a background of absence' (*ibid.*). Lacan (1977, 104) thus follows Hegel in stating that the 'symbol manifests itself first of all as the murder of the thing': language is never simply a naming of what is; it apparently creates something from nothing (*ex nihilo*), but does so only at the cost of losing something. The lack issuing from the symbolic is what drives the subject; what compels it into action in an attempt to overcome the absence that foregrounds its presence. Since this is an impossibility (the lack cannot be made good inasmuch as it is a *constitutive* lack), the subject, 'in search of unity, is fated to know only division' (Lapsley 1997, 187). Whatever the particular demands the consuming subject expresses (which might, for a while, be satisfied, in the manner in which a drink quenches thirst), they are but singular expressions of an interminable desire that is destined to remain insatiable (in the way in which thirst *as such* will always have been an unslaked thirst).

9　Following Barthes (1973), it is difficult to concede to Campbell that there is any special case involved here. It seems, rather, to be the general case – in much the same way in which poetry is not a special case of language but its general quality (Mukařovský 1964), and writing not a special case of language but its general condition (in the sense that language is ineluctably *graphematic*: Derrida 1973; 1976; 1978).

10　Gifts, in contemporary society, are undeniably of interest (Belk 1979; Cheal 1988; Miller 1993). Their relation to Mauss' sense of the gift is, though, far from straightforward (Berking 1999). For Carrier (1991; 1994), many instances of gift giving, conventionally understood, are explicitly ruled out of Mauss' formulation: put bluntly, Mauss is dealing with something else. The present discussion circumnavigates some important recent literature (notably Munn 1986; Strathern 1988), restricting itself to Mauss' own text.

11 The same principle works from the other direction, too. 'It is all right to send flowers to your aunt in the hospital, but never right to send the cash they are worth with a message to 'get yourself some flowers'' (Douglas and Isherwood 1996, 38).

12 One can, perhaps, buy *oneself* a gift, though 'treating oneself' in this way does not necessitate a commercial purchase. For more on 'self-gifts' see Mick and DeMoss (1993).

13 Gifts are sometimes regarded as the *opposite* of economic exchange. Belk (1979, 100), for instance, writes of the gift 'that (a) it is something voluntarily given and (b) that there is no expectation of compensation'. In Mauss' account of the gift, there is *every* expectation of 'compensation' or, more accurately, a 'counter-gift': one gives in order to receive in (re)turn.

14 'If humans are something "intrinsically", they are *social*' (Bauman and Tester 2001, 43).

15 Sahlins (1972, 169) writes that gift exchange involves and amounts to 'a kind of social contract for primitives', an aspect emphasized by Miller (1987). I would concur only to the extent that a corresponding systemic purpose is served by the obligation to reciprocate as is achieved by the system of possessive individualism (for which the notion of a 'social contract' is a euphemism). Imposing the notion of a 'social contract' willy-nilly on any society is misleading in that it fails to account for a seminal feature of modern society – that society is conceived as 'Society'; as an abstract, disembodied principle (*q.v.*).

16 To anticipate an argument yet to be made, Bataille's divergence from Mauss is evident here: 'In overall industrial development, are there not social conflicts and planetary wars?' asks Bataille (1988, 20), for whom war is a catastrophic expenditure of excess energy, an excess that inevitably outstrips society's ability to put it to purposeful, productive use (and which, if acknowledged, might be expended in less catastrophic, though equally useless, ways) (*q.v.*).

17 Simmel (1978) describes the lifeworld of modernity as a 'culture of things' (Frisby 1992, 119).

18 For a range of recent considerations of the relations between subjects and objects, see: Doel and Clarke (1999a); Kopytoff (1986); Latour (1991; 1992); Law (1986); Wildavsky (1991). For broader accounts dealing with the bracketing of the distinction between 'human' and 'non-human' actors, see Bijker (1995); Bingham (1996); Callon (1986); Murdoch (1997).

19 If this appears merely to *invert* social contract theory, suggesting a unilateral determinism in the opposite direction, rather than something more dialectical, it is worth emphasizing Althusser's debt to Lacan's (1977) conception of the 'mirror stage'. The 'mirror categories' of 'Society' and 'individual' imply not a faithful reflection of one in the other, but a process of misrecognition (*méconnaissance*) of oneself as an *in*dividual (that is, as a self-contained unit). This imaginary plenitude involves identification with an ideal ego – a reflection that is taken to be *in* the mirror but which is, in fact, *of* the mirror (Doel and Clarke 2002).

20 Land (1992, 45) notes that Bataille, like 'Freud ... is an energeticist'. He adds that, in the wake of 'Lacan and his semiological ilk one would never suspect Freud's leanings. I remain unconvinced, however, that the distinction between 'semiotic' and 'energeticist' modes of thought is as clear or dichotomous as Land contends (*q.v.*).

21 Marx failed to push this insight to its logical conclusion. If he occasionally glimpsed use-value as an anthropocentric form, he nonetheless cast use-value axiomatically in terms of a natural law, even though he managed to place the blame for that situation squarely on the shoulders of capitalism (*q.v.*; Baudrillard 1975, 1981, 1993a; Clarke and Doel, 2000).

22 Whether Bataille's (1988) discussion of Aztec sacrifice is appropriate as an exemplification of this understanding is a matter of dispute (see Duverger 1979; translated extract, Duverger 1989), though Richardson (1994), drawing on Todorov (1984), defends Bataille's account. Smith (1998) offers a reliable discussion of current knowledge of Aztec society.

23 Of course, this logic is entirely phallic: the other 'is always described in terms of deficiency or atrophy', just like 'the other side of the sex that alone holds a monopoly on value: the male sex' (Irigaray 1985, 69).

24 Despite the long history of the notion, which extends back at least as far as Adam Smith, the actual term 'consumer sovereignty' is of relatively recent vintage. According to Rothenberg (1968), the term was coined only in 1936, by the economist Hutt (1936).

25 Marx himself did not imply any such manipulation. His conception concerned the overall influence on society of the circuit of capital and the imperative to accumulate (Marx 1973, 99; cf. Harvey 1996, 62–68). For a variety of fuller discussion of Marx's conception of need, see Heller (1976); Fraser (1998); Leiss (1976); Soper (1981).

26 This is not to deny that the work of Galbraith (see Reisman 1980), along with that of Sweezy (1942), Baran (1957), and Baran and Sweezy (1966), has served to open up economic theory to new lines of development (Cowling 1982).

27 Veblen's is an explicitly gendered analysis. In the rigidly patriarchal late nineteenth-century society Veblen analysed, it was above all *men* of property who exhibited their pecuniary strength through conspicuous consumption, even exercising 'a distorting influence over the ideals of feminine beauty' (Storey 1999, 38). As Veblen (1994, 90) sardonically observed, the woman of the leisure class is, ideally, 'useless and expensive, and ... consequently valuable as evidence of pecuniary strength'. The male pronoun is used advisedly in this section.

28 It is a supreme irony, therefore, that Veblen's writings have been incorporated into economics as a minor footnote to marginal utility theory, as in Leibenstein's (1950) discussion of instances where the usual assumption of independent consumer preferences fails to hold. The so-called 'Veblen effect' produces an anomalous demand curve, where price is no longer inversely related to the quantity of a good demanded, since a higher cost will entice more purchases (from consumers eager to display their pecuniary strength conspicuously). It is significant, as Drakopoulos (1992) notes, that Keynes rejected outright the neo-classical model of demand because of its unrealistic assumptions regarding consumer preferences, as did the closely related work of Duesenberry (1949), whose approach has been likened to Veblen's (McCormick 1983). The devastation wrought on the neo-classical framework by allowing interdependent consumer preferences is demonstrated by Pollack (1970; 1976), who develops Duesenberry's work. These resonances are unsurprising, given that Veblen's work embodied a wholesale rejection of the line of economic thought running from Smith to Bentham to Marshall (Hamilton 1987). Veblen reserved his most vitriolic criticism for Bentham's utilitarianism. Marshall (1920, cited in Ackerman 1997, 153), who confined 'discussion of demand ... to an elementary analysis of a purely formal kind', claiming that the 'higher study of consumption must come after, and not before, the main body of economic analysis', also received short shrift from Veblen (though it is perhaps true that Marshall has received a bad press with respect to these particular remarks [Endres 1991]). It is worth noting that Veblen's ideas were anticipated by Rae (Edgell and Tilman 1991), and that the term 'conspicuous consumption' has since developed a cosmopolitan history of its own (Mason 1981; 1984).

29 With an irony that Veblen might have relished, leisure itself has not proved immune to the processes initially unleashed by the leisure class. Linder (1970), for

example, notes that economic growth promotes a tendency to economize on leisure time, as well as to use leisure time more efficiently, often directly through the use of consumer goods (Hirsch 1976; Johnson 1967): 'the consumer's free time is being sold to him [*sic.*]', as Baudrillard remarks (1998a, 153; cf. Rojek 1990).

30 Veblen (1994, 22) notes: 'In making use of the term "invidious" ... there is no intention to extol or depreciate, or to commend or deplore any of the phenomena which the word is used to characterize. The term is used in a technical sense. ... An invidious comparison is a process of valuation of persons in respect of worth'.

31 Though redolent of Bataille, Veblen lays a rather different emphasis on production. Productive enterprise is, for Veblen, a vital aspect of human nature. He considers 'the instinct of workmanship' an essential part of the human desire to seek 'in every act the accomplishment of some concrete, objective, impersonal end', which promotes 'a sense of the merit of serviceability or efficiency and of the demerit of futility, waste, or incapacity' (Veblen 1994, 9). Veblen usually manages to avoid overtly romanticizing such activity as honest toil: his view derives from the sense in which our creative powers represent a propensity to put things right; an impulse towards '"teleological" activity' (*ibid.*).

32 Veblen, like Simmel (1957), stressed the city as the locus of new tastes and fashions. Glennie and Thrift's (1992) discussion of 'new consumption discourses' circulating within the urban sphere of early eighteenth-century Europe highlights the longevity of this association.

33 Pre-modern societies are likened, by Bataille (1988, 117), to 'a body composed like all living organisms of nonhomogeneous parts, that is of a hierarchy of functions'. The military aristocracy, the clergy, and labour each carried out the tasks proper to their rank, receiving something from the other two estates in return. Protection, spiritual well-being, and the fruits of manual labour thus exchanged within an interlinked, divinely ordained social body, governed by sumptuary law, the doctrine of 'just price', and the prohibition of usury – earning interest on money-lending – by canon law (Tawney 1929). Such a society is a rigorously *organized* social body, a 'body with organs' (Deleuze and Guattari 1983), markedly different from the modern social body that Veblen sought to analyse.

34 Such concerns, of course, owed a great deal to the self-interested conviction that luxuries ought to be kept the preserve of an elite, and were, as Mandeville (1970) satirically pointed out, programmatically inimical to capitalist economic development (Appleby 1978; 1993; Berry 1994; Sekora 1977; Sombart 1967).

35 The late eighteenth-century consumer revolution, McKendrick *et al.* (1982) contend, ultimately spawned, or at least heralded, the subsequent nineteenth-century industrial revolution. Fine and Leopold (1990) vigorously dispute this, emphasising instead the dramatic nature of coincident changes in production and arguing that any kind of 'trickle-down' mechanism – operating, for instance, through the emulation by domestic servants of their employer's tastes – was severely constrained by the meagre financial resources available to servants, thus restricting the development of cultivated consumption practices. Fine and Leopold (1990) insist upon the class dimension of the issue, whereas McKendrick *et al.* (1982) deal with emulation in terms of social rank. Though he affords scant attention to this last point, Slater (1997) deploys the broader notion of a 'commercial revolution', inaugurated in the sixteenth century, in attempting to circumnavigate this chicken-and-egg dispute. He draws attention to Thirsk's (1978) discussion of the early-modern 'projects' – most famous for prompting Adam Smith's observations of the division of labour – which produced consumer goods and simultaneously provided the wages necessary for their acquisition.

36 Pleasure is undoubtedly a vital aspect of modern consumption. This presents
    Campbell (1987) with a specific challenge. He must attempt to negotiate the
    seeming disjuncture between the ascetic Protestant ethic, which Weber (1976) saw
    as the animating spirit of capitalism, and the seemingly incompatible hedonistic
    consumerist attitudes that this system eventually bred. To account for it,
    Campbell enlists the notion of a Romantic ethic, which he holds to be respon-
    sible for gestating a new form of 'autonomous imaginative hedonism', capable of
    sustaining the *insatiable* desires necessary for capitalism to reproduce itself. In
    effect, Campbell demonstrates the way in which economic success served, over
    time, to erode plain and simple living: a story neatly summed up in the saying
    that 'the Quakers came to Pennsylvania to do good and ended up doing well' (cf.
    Shi 1985; Peet 1997). Although Campbell considers a number of transformations
    in the nature of pleasure, he is nonetheless guilty of taking the link between
    consumption and pleasure somewhat at face value (Chapter 5).
37 One of the seminal features of modernity was the belief that order could be
    instated simply by expunging disorder. Yet as we have seen with respect to
    symbolic exchange, disorder can never be eradicated in this way. The order,
    meaningfulness or 'entropy' of a structuring system can only be maintained by
    unceasingly sucking 'negentropy' (negative entropy) from outside its established
    boundaries (Schrödinger's formulation), such that 'the system and its environ-
    ment [enter into] a network of constant and regular mutual relations' (Bauman
    1999a, 47). Such structuring systems are, as 'Rapoport put it in another cele-
    brated dictum, tiny "islands of order" in a sea of increasing disorder' (*ibid.*). The
    novelist Thomas Pynchon (1996) offers an entertaining story that revolves
    around meaning, entropy and the second law of thermodynamics (the propensity
    for a system to tend, irreversibly, towards a situation of maximum entropy).
38 Although we will let the issue speak for itself, it is worth stating that 'sign-value'
    and 'sign exchange-value' are synonymous in Baudrillard's lexicon: he employs
    the same concept in different forms according to context. Sign-value is, in effect,
    'non-economic' exchange-value, which is exactly what is highlighted by speaking
    of 'sign exchange-value' and 'economic exchange-value' in the same breath: both
    forms issue from self-referential systems, in a way that is most apparent, obvious,
    and familiar in the latter case. The two homologous forms of value simply apply
    to different fields (though the difference involved is a matter of social rather than
    natural fact). Reference simply to 'sign-value' accentuates the fact that there are
    not only two logics of value at play in the commodity (use-value and exchange-
    value), but three, constituting a composite 'commodity-sign'. The distinction in
    terminology does not amount to a conceptual distinction, though many appear
    to have been flummoxed on this score.
39 'What more could anyone possibly want?' is the rhetorical question on the lips of
    the consumer society.
40 In Lacan's celebrated (1979, 158) formulation, desire is 'the desire of the Other'.
    The enigmatic 'Other' equates to the 'symbolic order' that grants the subject a
    position from which to know itself in the first place. This is, moreover, the only
    possible position the subject could be granted, the only way in which the subject
    can be constituted: 'there is no Other of the Other'. Desire issues from the fact
    that the subject occupies a position granted by the symbolic, since the symbolic,
    by its very nature, can only ever achieve a presence that is foregrounded by an
    absence. In Lacan's (1979, 158) words, the subject is necessarily channelled
    through the 'defile of the signifier', such that the symbolic opens up a rent in the
    subject at one and the same moment that it provides it with a sense of presence.
    It is this constitutive gap that compels the subject towards the impossible task of
    achieving a completion it can never attain. Desire is thus a desire for an impos-
    sible unity. It is *necessarily* insatiable.

41 Baudrillard's words echo those of Lefebvre (1994, 60ff), from whom the suggestion that ours is a 'Bureaucratic Society of Controlled Consumption' originates. Heller (1984, 215) similarly notes that whereas the synthetic world-views of religion, philosophy (and, more recently, science) struggled to bring the 'partial syntheses' of everyday life (such as 'folk wisdom' and the like) under their purview, their declining authority has also witnessed a decline in such 'partial syntheses': 'those that do appear are organized and manipulated, and turned into ready-made consumer goods'.

42 Sheppard and Barnes (1990, 226) point out that this particular mechanism is a plausible but not strictly a necessary component of the reproduction process:

> The surplus that it produced at the end of the production period can be used in three different ways: workers' consumption, capitalists' consumption or reinvestment. It is logically possible, therefore, for capitalist consumption and reinvestment, say, to increase and thereby meet the new levels of effective demand that are required to sustain a higher rate of accumulation, while worker consumption remains constant.

Baudrillard's point, however, is broader than a technical argument about clearing effective demand (cf. Ahrne 1988; Gilliatt 1992; Jhally 1990).

3 Consumption and the city, modern and postmodern

1 If one were forced to single out *the* defining characteristic of modernity, it would perhaps be the sense in which it 'opposes itself to tradition' (Baudrillard 1987b). As Bauman (1993c, 592) notes, the term – which, in essence, means little more than 'current', 'of recent origin' – rose to dominate intellectual debate in the seventeenth century, though it had been in sporadic use since the fifth century: 'From the moment of its triumphant entry into public discourse, the idea of the modern tended to recast the old as antiquated, obsolete, out of date, about to be (deservedly) sunk into oblivion and replaced'. Modernity thus 'refuses to accept longevity as an entitlement to authority' (Bauman 1999b, 133).

2 Capitalism and modernity are neither coincident nor coterminous (Sayer 1991; Wrong 1992).

3 For Bauman (1991, 1), ambivalence, 'the possibility of assigning an object or an event to more than one category, is a language-specific disorder: a failure in the naming (segregating) function that language is meant to perform'. And for Douglas (1970, 48): 'Dirt is the by-product of [the] systematic ordering and classification of matter ... the reaction which condemns any object or idea likely to confuse or contradict cherished classifications'. Or as Freud (in 'Character and anal eroticism', 1973–1986, vol. 7, 213) succinctly put it: 'Dirt is matter in the wrong place'. Ambivalence is thus inevitable: 'Ambivalence is ... the *alter ego* of language, and its permanent companion – indeed, its normal condition' (Bauman 1991, 1). This understanding of classification and order, of ambivalence, dirt and disorder, is vital to what follows.

4 In fact, though it is a characteristic of reason to present itself as universal, the qualification '*European* reason' would perhaps be a preferable formulation in considering modernity (see Luhmann 1998, on the specificity of 'European rationality').

5 This process finds its initial model in the Christian Church. Whereas the ambivalence of the enchanted world was evident in 'the ancient cults where the basic intuition of a specificity of evil and death was still strong' (Baudrillard 1993a, 149), the Church sought to 'impose the pre-eminent principle of the Good

(God), reducing evil and death to a negative principle, dialectically subordinate to the other (the Devil)' (*ibid.*). In a certain sense, therefore, it was not that the religious basis of society eventually, and at a fairly late stage, became secularized (Chadwick 1975), but that it gradually lost its specificity within the sphere of religion.

6 The 'disenchantment of the world' equates to the 'culturalization of nature': the redefinition of what lies within the scope of human intervention. Modernity dramatically increased the range and span of the cultural thus defined.

7 Such contrasts are sometimes seen as being drawn too strongly. Munn (1992, 111), for instance, complains of the 'analytical perspective that views clock time as a "lifeless time"' (the embedded term comes from Jaques [1982]), noting that, when 'considered in the context of daily activity, clock time is quite alive, embodied in personal activity and experience ... [e]ndowed with potency and affect'. This is, of course, entirely true – but the danger lies in thereby dismissing the specificity of clock time. Clock time amounts to a form of disavowal, and so it is unsurprising that the practices associated with it should continually confirm what it seeks to repress.

8 This is not to imply that modernity precluded or excluded movements that detracted from its own ideal representation. To the contrary, the gradual increase in the gestation of a variety of (often nostalgic or utopian) characterizations of modernity as productive of a false, flawed or inauthentic social existence – from Marx and Engels to Ruskin, William Morris, Pugin and so on – bears witness to the extent to which modernity actively gestated visions of its own inadequacies and impossibilities or actively produced its own discontents (Bourne *et al.* 1987).

9 Durham (1998) offers a consideration of Baudrillard's use of the simulacrum in relation to other thinkers.

10 The 'culturalization of nature' initiated by modernity was ultimately to come full circle, 'to naturalize culture: cultural facts might be human products, but once produced they confront their erstwhile authors with all the unyielding and indomitable obstinacy of nature' (Bauman 1999a, x). The tale is one of the 'tragedy of culture' (Simmel 1968), first worked through in the social thought of the nineteenth century and culminating in Durkheim's conception of 'social facts'.

11 Thrift (2000b) puts forward what purports to be a case to the contrary.

12 Engels' position is nothing if not complex (Blanchard 1985). Thus Freenberg (1980, 118) insists that

> Marx and Engels write about urban life without any nostalgia for earlier communitarian phases in social development. For them the city is a place of increasing freedom and of public life accessible to the lower classes, even if under capitalism it is the place of ruthless exploitation, disease and demoralization as well.

The dialectical tensions thrown up by the city are equally complex in Simmel (1950a; 1957). Cf. Williams (1973).

13 'A question with no answer is a barrier that cannot be breached. In other words, it is questions with no answers that set the limits of human possibilities, describe the boundaries of human existence' (Kundera 1984a, 139).

14 These remarks are indebted to the lecture 'The city and the stranger' given by Zygmunt Bauman on 7 December 1994, under the auspices of the Centre for Cultural Studies at the University of Leeds.

15 For Giddens (1990, 28), modernity involved a *disembedding* process that removed social relations from 'the immediacies of context': the separation 'of time and space and their formation into standardised "empty" dimensions cut through the

connections between social activity and its "embedding" in the particularities of contexts of presence', providing 'the basis for their recombination in relation to social activity' and thus permitting a 'complex co-ordination ... across large tracts of time-space' (*ibid.*, 20, 19, 20). This 'stretching' of social systems, allowing for a scale of social activity unprecedented by those based on localized co-presence, was marked, as the case of the stranger demonstrates, by 'dialectical features, provoking opposing characteristics' (*ibid.*, 19). Giddens' account is directed against Parsons' (1951) view of the 'problem of order', which was predicated on an assumption of 'the boundedness of social systems, because it is defined as a question of integration – what holds the system together in the face of division of interest that would "set all against all"' (Giddens 1990, 14). The boundedness Parsons takes as given is what Giddens (*ibid.*) regards as calling for explanation:

> We should reformulate the question of order as a problem of how it comes about that social systems 'bind' space and time. The problem of order is ... one of *time-space distanciation* – the conditions under which time and space are organized so as to connect presence and absence,

with all the contradictory tensions this provokes.

16 On the 'slimy' ('*le visqueux*'), see Sartre (1957, 604–659); and Bauman (1999a, 109–124).

17 It might be ventured that it was this situation that underpinned the proliferation of popular institutions characterizing the nineteenth-century city, which arguably represented a distinctively modern means of overcoming the loss of pre-modern, 'organic' communities (Baernreither 1889; Morris 1983; Razzell and Wainwright 1973, see especially Letter XXI, 320–323). The significance of the fact that such associations eventually included consumer cooperation should not be overlooked (Cole 1944; Clarke and Purvis 1994; Furlough 1991; Furlough and Strikwerda 1999; Gurney 1996; Purvis 1990; 1998).

18 In the context of what follows, it might be noted that the term 'statistics' initially related to the categorization and enumeration of the inhabitants of the territory of the nation-*state* (Porter 1986).

19 Proust (1981) likens the random commun(icat)ion between two homosexuals observed by the narrator, Marcel, to that between an insect and a flower, the *transversal* nature of which is discussed in detail by Deleuze (1972) and Deleuze and Guattari (1983). This accords with Benjamin's (1973, 26) remark that the *flâneur* 'goes botanizing on the asphalt', and Wilson's (1995, 65) comment that the *flâneur* is 'the naturalist of this unnatural environment'.

20 The broadly simultaneous development of the cinema is also worth noting here (Charney and Schwartz 1995; Clarke 1997b; Friedberg 1993; Nava 1997; Williams 1982). According to Williams (1982, 82), 'the passive solitude of the moviegoer resembles the behaviour of department store shoppers who also submit to the reign of imagery with a strange combination of intellectual and physical passivity and emotional hyperactivity' (*q. v.*).

21 Cf. Wilson (1991, 2000).

22 Though the department store has been represented, in this context, as the most significant space of its kind, other developments included ladies-only eating establishments, tea rooms and the like (Wilson 1995; Scobey 1992). Glennie (1995) specifically cautions against overstating the significance of the department store, particularly when it is portrayed as vital to the development of a modern consumer culture. This retail form came relatively late to England compared to France and America, despite England's well established and diverse consumer

culture (Adburgham 1981; Alexander 1970; Davis 1966; Fraser 1981; Jefferys 1954; Mui and Mui 1989; Thrift and Glennie 1993; Winstanley 1983). The shopping mall has come to be regarded as the latter-day counterpart of the nineteenth-century department store. As Lehtonen and Mäenpää (1997, 136) note, the term derives from the 'playing alley' of the seventeenth-century London game of pall-mall. Today, the term is used less in the UK than in the USA, largely as a result of the proliferation of the form in North American development (Kowinski 1985; Lord 1985). Most current debate – much of it centring on the world's largest, the West Edmonton Mall – focuses on the appropriateness of different cultural methodologies; specifically, whether academic readings claiming semiotic or some other brand of theoretical rigour in fact amount to elitist renditions, which exclude the sort of ordinary engagements that more sociological qualitative methodologies might reveal (Jackson and Holbrook 1995; cf. Chaney 1990; Crawford 1992; Goss 1993; Hopkins 1991; Miller *et al.* 1998; Morris 1988; Shields 1989; 1994b; Williamson 1992). A recent move has been to leave the mall in search of other spaces of consumption (Crewe and Gregson 1997; 1998).

23 'Pleasure is this society's Special Offer. ... And it is female desire which makes us respond and take up the offer', states Coward (1984, 13; cf. Bowlby 1993; Radner 1995). It is a commonplace of psychoanalysis that feminine desire is of a different order than masculine desire: Lacan (1982, 152) characterizes the male orgasm as 'the *jouissance* of the idiot'. Cf. Mitchell (1975), Gallup (1982) and Williamson (1987); for an alternative perspective, see Weinbaum and Bridges (1979).

24 The two regimes of light identified by Schivelbusch (1988) – street-lighting, geared towards the securing of public order; and the commercial lighting of commodified display – neatly illustrate the proteophobic and proteophilic forces at work in the modern city. Their proximity to one another should not, however, be underestimated. Thus Oettermann (1997, 41) suggests that 'the panorama and panopticon are at the same time identical and antithetical' – both are 'schools of vision' – whilst Foucault (1977, 317) ponders whether 'Bentham [was] aware of the Panoramas that Baker was constructing at exactly the same period' that the Panopticon was being proposed.

25 I have returned the English translation of *instance* (as 'rule') to the original French, which better captures the meaning of this phrase. A suitable alternative might be 'tribunal'. Rendering *instance* as 'rule', though, is in danger of eliding the crucial difference between the *law* and the *rule*. The only other available English version (Baudrillard 1990a, 26) translates *instance* as 'institution', similarly obscuring the distinction Baudrillard is driving at here – between the asymptotic curvature of the norm and the strict margin of the law.

26 'Like the sign form, the commodity is a code managing the exchange of values' (Baudrillard 1981, 146). The specific modality of the code lies in its operational functioning as a means of designation and classification. For example, the 'object-cum-advertising system ... constitutes less a language, whose living syntax its lacks, than a set of significations. Impoverished yet efficient, it is basically a code. It does not structure the personality, but designates and classifies it' (Baudrillard 1996, 193). This coded form has become 'the fundamental social form of control, which works by infinitely dividing practices and responses' (Baudrillard 1993a, 62).

27 Perhaps nothing has been more misunderstood than the contrast between death and disappearance in today's world. 'Rather than a mortal mode of disappearance' (death pure and simple), what faces us today, proposes Baudrillard (1993c, 4), is 'a fractal mode of dispersal'. What dies is gone for good, whereas what disappears 'disappears into the void along a crooked path that only rarely

happens to intersect with other such paths. This is the pattern of the fractal'
(*ibid.*, 6). Or again, what disappears 'becomes an event in a cycle that may bring
it back many times' (Baudrillard 1990d, 92; Chapter 6).

28 As if to underscore the point, Manchester has recently opened Urbis, a museum
dedicated to the exhibition of city life, curiously turning the city in on itself.

4 Seduced and repressed: collective consumption revisited

1 In an interesting aside, Gilliatt (1992) explores how the gap between optimal and
actual public-service provision calls upon the consumer to perform the work
necessary to the smooth functioning of the system of provision, in the same way
that Gershuny (1978) has focused on the increasing significance of independent
production (such as DIY) within the household: the so-called 'self-service
economy'.

2 Castells' early work, notably *Monopolville* (Castells and Godard 1974), deployed
a conception of monopoly capitalism (cf. Baran and Sweezy 1966). Ironically, the
attack on collectivization launched by the New Right in the 1980s charged the
public sector with holding a 'monopoly' on the provision of public goods,
arguing that the consequent lack of competition was against the public interest.

3 The theory of public goods was developed as a branch of welfare economics, as
a normative or prescriptive theory designed to cope with the supply of, and
demand for, non-market goods (Buchanan 1968; Samuelson 1954; 1955). Public
goods are formally distinguished by:

> (i) 'non-rival' consumption (such that the marginal cost of an additional
> consumer is zero); and
> (ii) 'non-excludability' (provision for one amounts to provision for all, in the
> form of an 'undepletable' positive externality effect (Baumol and Oates
> 1975)).

The latter implies the underprovision of public goods (e.g. roads) relative to
private goods (e.g. cars) if provision is left entirely to market forces (the examples
are 'impure' but suitably illustrative). Such characteristics are *not* definitive of
most goods and services that, historically speaking, have been provided collectively (Steiner 1974); and there is significant dispute over the extent to which
conventional examples, such as lighthouses, conform to these criteria (Coase
1974). Such technical problems have seen the concept 'turned around by new
right economists, in whose hands it becomes a weapon to declare illegitimate all
forms of state intervention which do not neatly conform to the "collective
goods" model' (Dowding and Dunleavy 1996, 48–49). Malkin and Wildavsky
(1991) have called for the distinction to be abandoned altogether. On the geography of collective consumption, see Pinch (1985; 1989; 1996).

4 Harvey is fond of citing Engels' statement that the 'bourgeoisie has only one
solution to its housing problems: it moves them around'. Engels put his finger on
a 'solution' with far wider bearing than the 'housing problem' alone.

5 The terminology of 'cleavages' derived from work on the interests and identifications underlying party-political affiliations (Lipset and Rokkan 1967).

6 Saunders, like Castells, elides consumption and provision. So-called 'collective
consumption' typically concerns the collective *provision* of means of consumption that may well be *consumed* 'individually' (e.g. public transport). Likewise,
privately provided goods may be consumed 'collectively' (e.g. private education).
Some of the combinatorial possibilities this generates are detailed for education
by Bradford and Burdett (1989, table 8.1 on page 193), employing Savas' (1982)
three dimensions of arranger, producer and financer, each of which may be

public- or private-sector institutions. Otnes (1988) points to a further complexity, noting that 'collective' and 'social' (or 'socio-material') are frequently elided, despite the fact that 'individual' or 'private' consumption is itself ineluctably 'social'.

7 Abrams (1977) offers a review of the ideology of private consumption, albeit in a manner somewhat removed from the issues Saunders raises.

8 Capitalism has, plainly, always required consumption too. Much of the historical literature concerns the source of effective demand underpinning – and predating – the industrial revolution (McKendrick *et al.* 1982). The precession of the 'consumer revolution' might seem to discredit the hypothesis of the twentieth-century development of a specifically 'consumer society'. As we have argued, however, it does not undermine the coherence of arguments supporting a relatively recent switch from the (modern) primacy of work to the (postmodern) primacy of consumption in terms of their systemic role (Bauman 1998a).

9 Such a pattern clearly resonates with the social geography of class observed by Engels in 1844: 'I have never seen', records Engels (1973, 86), 'so systematic a shutting out of the working class from the thoroughfares, so tender a conceal-ment of everything that would affront the eye and the nerves of the bourgeoisie, as in Manchester'. 'The finest part of the arrangement', he wrote (*ibid.*, 85), was that members of the bourgeoisie 'can take the shortest road through the middle of all the labouring districts to their places of business without ever seeing that they are in the midst of the grimy misery that lurks to the right and the left'.

10 It is also worth noting the changed emphasis from taxes levied on labour to taxes levied on consumption that took place, in the United Kingdom at least, over the 1980s and 1990s.

11 In fact, 'membership' is a poor term for the kind of belonging neo-tribalism entails: 'one can hardly speak of membership ... considering the absence of tribal executive powers, the rudimentary nature of formalization and almost complete lack of control over the individual adherents' decisions to "join" or to "leave" and move to another tribe' (Bauman 1992c).

5 The meaning of lifestyle

1 Goss (1995) has sought to expose the instrumental rationality behind lifestyle-related marketing. In geodemographics, the basis of much direct marketing, geographical areas are classified according to the likely characteristics of their residents (and almost attributed with the agency of consumers themselves: 'Area X consumes high levels of tinned goods'). Although such areal classifications (in common with other, aspatial 'lifestyle' segmentation methods) are based on the characteristics of *consumers*, they are actually far more concerned with the allo-cation of *consumer goods* (and *advertising messages*) to areas/market segments than they are with 'lifestyle' as such. There is a common tendency to overplay the 'panoptic' nature of lifestyle marketing, overemphasizing its sinister, Big-Brother quality. Doel and Clarke (1999a) offer a somewhat more Kafkaesque reading.

2 It is not difficult to see why, in this context, a good deal of suspicion should surround the notion of 'lifestyle'. The sense in which the concept is intensely *ideo-logical*, rather than merely *chaotic*, is evident in the arena of health research (Blaxter 1990). There is, for instance, a well established correlation between 'lifestyle' and 'health' (Cox 1987; Cox *et al.* 1993). This cannot simply be dismissed as mere statistical sleight of hand. Nonetheless, increasing emphasis on this relation is far from scientifically neutral. It potentially diminishes the signifi-cance of calls to tackle poverty with *public* programmes by heightening the importance of *personal* ('lifestyle') changes – thus placing firmly in the court of the individual problems that are not wholly or even primarily individual in nature.

3 In the light of what follows, it is interesting that Sayer should (unconsciously?) phrase his assessment in terms of 'social *psychology*'.
4 The study of the 'correlation between society and the manner of making use of the body', as Maffesoli (1997, 35) puts it, originates with Mauss (1973) and is central not only to Foucault's work on discipline but also on the 'care of the self' (Foucault 1986; 1997).
5 The entry into, and eventual colonization of, the economic sphere by disciplinary power is not, perhaps, surprising. ' "Discipline" ', says Foucault (1977, 215),

> may be identified neither with an institution nor with an apparatus; it is a type of power, a modality for its exercise, comprising a whole set of instruments, techniques, procedures, levels of application, targets; it is a 'physics' or 'anatomy' of power, a technology.

As such, it has a momentum all of its own.
6 As Foucault (1980, 39) notes, there is no strict correlation 'between the two processes, global and local'. 'In England the same capillary modification of power occurred as in France. But there the person of the king ... was displaced within the system of political representations, rather than eliminated' (*ibid.*; cf. Hill 1955).
7 Disciplinary power has since become more or less synonymous with Bentham's nineteenth-century design for the 'all-seeing' architectural power-apparatus he named the 'panopticon' (Himmelfarb 1968). As Foucault (1980, 148) notes: 'Bentham didn't merely imagine an architectural design calculated to solve a specific problem, such as that of a prison, a school or a hospital. He proclaimed it as a veritable discovery, saying of it himself that it was "Christopher Columbus's egg" ' – something that is easy once one knows the trick. 'One important point should be noted: Bentham thought and said that his optical system was *the* great innovation needed for the easy and effective exercise of power. It has in fact been widely employed since the end of the eighteenth century' (*ibid.*).
8 The factory rules were one of the most obvious ways in which discipline infected the minutiae of everyday life. They frequently displayed only a tenuous connection to strictly economic concerns.

> Pernickety, often contradictory rules, prohibiting whistling or singing, setting penalties for having dirty hands but also for washing them ... make sense only when seen as an aid to drilling factory labourers into complete submission to their supervisors, and to extirpating the last shreds of the autonomy of the labourer's body.
>
> (Bauman 1983, 37)

9 See de Vries' (1993) 'Z-goods model' for a neo-classical economic formalization of this situation. Whatever its benefit in terms of clarity, such a formulation has the unfortunate effect of reducing an active social process to an individual decision and a matter of preference.
10 Cf. Harvey (1985c, 129): 'The closer we get to a stationary state (let alone actual decline), the more unhealthy the economy is judged to be. This translates into an ideology of growth ("growth is good") no matter what the environmental, human, or geopolitical consequences'.
11 As Veblen (1961b, 115) notes, Smith's image of an 'invisible hand' 'does not fall back on a meddling Providence who is to set human affairs straight when they are in danger of going askew'. Smith's view is of a Creator who has 'established the natural order to serve the ends of human welfare; and has very nicely

adjusted the efficient causes comprised in the natural order, including human aims and motives, to this work that they are to accomplish'. Smith did not, then, believe in 'path-dependence'; if things go wrong, then *left alone* they have an inbuilt, self-correcting tendency.

12 As Williams (1980, 173) makes clear, it 'is not true ... that with the coming of factory production large scale advertising became economically necessary'. Though the quantity of advertising increased rapidly with the industrial revolution, its form was initially continuous with earlier periods (broadly resembling what today would be termed 'classifieds'). It was the kind of 'puffery' associated with 'occasional and marginal products' (*ibid.*, 172) that provided the model for the 'full development of an organized system of commercial information and persuasion' that occurred during 'the half century between 1880 and 1930' (*ibid.*, 179), as part of the drive towards market control following the Great Depression of 1875–1894.

13 The labourer is, Marx (1967, 272–273) pointed out, 'free in the double sense that as a free individual he can dispose of his labour-power as his own commodity, and that, on the other hand, he has no other commodity for sale, i.e. he is rid of them, he is free of all the objects needed for the realization of his labour-power'. Consumption is characterized by a formidable chasing after objects; sometimes, as is evident in DIY tools or kitchen implements, objects that allow for the parodic re-realization of productive autonomy.

14 The consumer's body becomes 'an uneasy, poorly balanced mixture of love and horror' (Bauman 1983, 41), paralleling the earlier situation noted above (note 8). Bauman (*ibid.*) provides an exemplary instance: 'For many years now the two kinds of books most likely to win a place on the bestsellers list in the USA have been cookbooks, including the most eerie and exotic ones ... and diet books, prohibiting the consumption of practically everything the first category of books recommends'.

15 There is no need, therefore, to account for this chase after satisfaction in terms of the perennially unlikely scenario of the widespread manipulation and hoodwinking of consumers. Consumption is an entirely 'rational' strategy in the sense that it confronts, in the best way it can, an entirely irrational situation.

16 As a number of authors have observed, in a striking reversal of the former situation, a 'consumerist logic of identity' has begun to penetrate the world of work (Crompton 1996; du Gay 1996; Sennett 1998).

17 The situation Baudrillard describes clearly anticipates the focus of Beck's (1992, 21) Risk Society, where risk – 'defined as a *systematic way of dealing with hazards and insecurities induced and introduced by modernization itself* – automatically ensures 'not less but more modernity' (*ibid.*, 14). Modernity's assumption of a reflexive form (Beck *et al.* 1994) sees the 'logic of wealth production', as the prime characteristic of modern industrial society, 'gradually displaced by the logic of risk avoidance and risk management' (Bauman 1993b, 199). The upshot is a situation where no action or situation can be regarded as unambiguously dangerous or safe. Unlike 'danger', 'risk' belongs to the discourse of gambling, a 'discourse which does not sustain clear-cut opposition between success and failure, safety and danger ... thereby straddling the barricade which separates them in the discourse of 'order' which the term 'danger' comes from and represents' (*ibid.*, 200n).

18 The situation is the final, ironic twist of a process first unleashed by modernity. Pre-modern notions of 'fate' or 'fortune' (*fortuna*) conferred a degree of ontological security by taking a good deal of contingency out of play, rendering it external to human contrivance, beyond human intervention. Modernity broadened the boundaries of human powers, ushering in an 'altered perception of determination and contingency' (Giddens 1990, 34).

Only in early modern times did a new term (*riesgo, rischio,* risk) appear to indicate that unexpected results may be a consequence of our decisions, and not simply an aspect of cosmology, an expression of the hidden meanings of nature or ... God.

(Luhmann 1988, 96)

The 'discovery of "risk" ... adds another dimension to human experience' (*ibid.*), but where it initially held out the promise of increased control, it ultimately resolves itself in insecurity and uncertainty, as the unanticipated consequences of human decisions begin to outweigh those intended.

19 This also explains why, paradoxically, 'tradition is the talk of the postmodern town' (Bauman 1996b, 50) – why the progressive detraditionalization of modernity should give rise to an untimely *reanimation,* or *selective* reanimation, of 'tradition'. Despite its ostensible topic, therefore, the renewed discourse of tradition is essentially a discourse 'about absences and an unslaked thirst':

It is about what is missing and missed. First and foremost – it is about absent totality. Totality in space: a framed composition that would allow every brushstroke to bask in the glory of meaningful design. And totality in time: an unbroken thread that would keep every bead in place, and in its *right* place, as it is strung on the thread of time just after the one before and before the one that will come after. Tradition is such a missing totality. Or, rather, that entity that would make a totality out of fragments scattered in space and time: of the episodes otherwise devoid of connection, orphaned by the past and heirless.

Hence Giddens' (1994) reframing of postmodern society as 'post-traditional' society, which needs to be

understood as referring not to tradition running out of fashion, but to the *surplus* of traditions: the *excess* of readings of the past vying for acceptance, the absence of a single reading of history likely to command a universal or near-universal trust.

(Bauman 1999b, 133)

20 Confidence, Giddens (1990, 27) argues, derives from the kind of 'weak inductive knowledge' arising from the 'calculation of the reliability of likely future events' typical of the situated knowledge of pre-modern, local contexts. Whilst the 'context independence' of modern knowledge, which results from its basis in impersonal principles, underlies their capacity for distanciation, it simultaneously undermines localized bonds of familiarity, giving rise to increased insecurity. One thinks, for example, of the trust entailed in relations with strangers, as 'representatives of the unknown'.

21 Modernity formally opposed itself to tradition: modernity begins where tradition ends. Tradition thus supplied modernity with the alibi it needed to ensure its own identity and coherence. In actual fact, of course, modernity dissolved *and* (re)invented tradition in a single stroke (Hobsbawm and Ranger 1983). It employed a distinctively modern form of 'power-assisted' tradition in attempting to forge a stable social order in the crucible of the nation-state.

22 Giddens' (1994) conception only superficially resembles Becker and Murphy's (1988, 675–676) claim that 'people get addicted not only to alcohol, cocaine, and cigarettes but also to work, eating, music, television, their standard of living, other people, religion'. Giddens posits the prevalence of addiction not as a

universal condition but as an *historically specific* one. Becker's (1992) aim is to demonstrate that 'rational choice theory' copes well enough with addictions – a claim made by addicts of all kinds, of course. The approach places taste beyond investigation (Stigler and Becker 1977), redefines the consumer as a unit of production (Becker 1965; cf. Lancaster 1966a; 1966b; Muth 1966), and proceeds to interpret even the most powerful chemical addiction not as an indication that 'consumer preferences' have been distorted, but that knowledge of the 'technology' capable of delivering 'highs' has altered (tastes remaining as rational as ever). Fine (1997, 17) objects that addictions 'are simply understood and explained [away] on the basis of rational, optimizing individuals'.

23 Advertising, being the focus of the majority of claims that the consumer is duped into buying items that are not really needed (Packard 1957), has become a thorny topic. Much empirical research suggests that advertising works to promote brand loyalty or encourage brand-switching behaviour (Ehrenberg 1974), implying something far less sinister than the creation of false needs: everybody buys toilet-tissue anyway, advertising simply persuades one to buy brand Y rather than brand X (Chiplin and Sturgess 1981; Ehrenberg 1959; 1972). Whilst the latter explanation is the more convincing, it has led to an unreasonably narrow conviction that advertising is merely a form of information (one of the 'popular economic fallacies' unpicked by Mishan 1969). Advertising has proved problematic for economics, given that advertising's *raison d'être* contradicts the axioms of demand theory (Ackerlof 1970; Dixit and Norman 1978; Dorfman and Steiner 1954; Nelson 1974; Palda 1964; Schmalensee 1978), and also because of the difficulty of disentangling the effect of advertising on consumption from the relationship running in the opposite direction – the effect of consumption levels (sales) in determining advertising budgets and total advertising expenditure (Ashley *et al.* 1980; Blank 1962; Comanor and Wilson 1979; Cowling *et al.* 1975; Kaldor 1950; Lambin 1976; Nerlove and Arrow 1962; Taylor and Weiserbs 1964; Yang 1964). Advertising undoubtedly has a far broader influence on consumption than orthodox economics admits (Arriaga 1984), even if its manipulative role is a gross caricature.

24 Barthes (1990, 295) proposes that fashion possesses 'two durations: one strictly historical, the other what could be called *memorable*'. The latter explains the regular swings from one style to the next, the former, changes of far longer duration: 'in order for history to intervene in Fashion, it must modify its rhythm, which seems possible only with a history of very long duration' (*ibid.*, 296). An example is provided by the relatively recent change in men's fashions and consequent increased pace of change in men's fashion (Edwards 1997; Nixon 1992; 1996; Mort 1996).

25 'With myth', writes Lévi-Strauss (1963, 208), 'everything is possible'. It is in this sense that Baudrillard (1998a) invokes the term. In Lévi-Strauss' conception, myth might best be regarded as the inverse of poetry. Where the poetic phrase puts an idea into words in a way that cannot be bettered, and cannot be altered without loss, with myth it is precisely the opposite. Myth is a linguistic form that is resilient enough to survive its translation into many different, even diametrically opposed, forms. In other words, the purpose and meaning of a myth remains apparent through thick and thin. This is why myth is eminently suited to a *structural* analysis: the myth itself inheres in its structure and not in its elements (or again, the elements issue from the structure and not the other way round). It should be readily apparent why Baudrillard detects in consumption the mythic form.

26 'This negative discourse is the intellectual's second home' (Baudrillard 1998a, 196).

27 Hahnemann's principle: let like be treated with like. (In fact, the Latin motto is not to be found in Hahnemann's writings, which established the basis of modern homoeopathy, but appears as an anonymous side-note in Paracelsus' *Opera Omnia* [*c.*1490–1541, ed. 1658, vol. 1].)

28 Baudrillard is not guilty of a naive romanticism, of citing such conditions as some kind of a conscious protest against consumerism. He is attempting to explain a puzzling sociological datum: the peculiar co-incidence of anorexia *and* obesity within the midst of an affluent society.

6 Minimal utopia

1 This sense of foreboding is present in the very choice of the term 'obscene', the etymology of which implies 'ill omened'.

2 Detailed accounts of the history of Leeds are provided by Beresford (1988) and Fraser (1980); see also Morris (1990); Beresford and Jones (1967); Bateman (1986).

3 The widespread adoption of the term 'governance' signals the fact that governing the city is no longer seen as the sole prerogative of institutions of government (Rhodes 1996; Goodwin *et al.* 1993). The case of Leeds, as an example of a 'corporate city', is discussed in Haughton and Williams (1996). The corporatist model has its origins not in urban theory but in state theory; it 'is best understood not in terms of the representation of diffusely articulated interests (as the traditional liberal theory of the state would wish), but as a network of consultation, bargaining and compromise between functional "simulators"' – as a kind of *simulated politics* (Bauman 1982, 136; cf. Cawson 1986; Grant 1985; Harrison 1980). Updated information on the Leeds Initiative is available from the City Council's website: http://www.leeds.gov.uk/lcc/leedsini/inits.html.

4 One thinks, in particular, of the continuous testing of consumer markets (Kent 1989) and television audiences (Barwise and Ehrenberg 1988).

5 Harvey's is arguably the best-known work on the postmodern city. Equally prominent is the work of the putative 'LA School' (Davis 1990; 1998; Dear 2000; Dear *et al.* 1996; Scott and Soja 1996; Soja 1989; 1996; 2000). The focus on Los Angeles has prompted impatience in some quarters, particularly insofar as it portrays the city as a postmodern paradigm or archetype. A case in point is the labelling of the city – with an allusion to Benjamin's (1986) remarks on Paris – as 'capital of the twentieth century' (Scott and Soja 1986: a sobriquet that has also been claimed for New York; Wallock 1988). As an epicentre of urban analysis – though little of it geared explicitly towards consumption (cf. Venkatesh 1991) – Los Angeles at the end of the twentieth century is directly comparable to Chicago in the 1920s (Park *et al.* 1925). The attention lavished on LA has occasioned claims for alternative contenders for paradigmatic-city status (Nijman 2000) alongside calls for a more sober perspective (Amin and Graham 1997).

6 The fundamental scarcity of space implied by modernity accords to certain 'commonsensical notions of dimensionality based on extension' (Doel 1999, 185). Hyperbolic and certain fractal geometries (geometries involving fractional dimensions) depart from these assumptions to behave in precisely the manner suggested by Baudrillard's characterization of proliferation.

7 As if this were not enough, resistance also seems increasingly likely to be *criminalized*. Demonstrations that would probably have been seen as *legitimate* protests in western liberal democracies only a decade or so ago – such as the 'anti-capitalist demonstrations' mounted against the World Trade Organization in Seattle, November 1999, or the May Day events in London and elsewhere in 2001 – are increasingly deemed illegitimate and crushed by police powers (typically on the basis that such demonstrations stop 'ordinary people' going about their daily business: earning a crust, shopping, tourism, and so forth).

8 The self-serving discourse of neo-liberalism that underpins globalization has been aptly characterized by Peck and Tickell (1994) as the 'law of the jungle'. In fact, the forces of globalization do operate very much like external forces of nature. Bauman (1998b, 60) characterizes the globalized world as a 'vast expanse

of man-made wilderness (not the "natural" wilderness which modernity set out to conquer and tame; but to paraphrase Anthony Giddens' felicitous phrase, a "*manufactured* jungle" – the post-domestication wilderness, one that emerged *after* the conquest, and as its result)'. Gamble (2001, 170) laments the fact that globalization is typically treated 'as a natural phenomenon ... more akin to climate' than a 'contrivance of human agency', but fails to see that this situation is the direct result of the unintended consequences of the kind of modernist project he wishes to revive.

9 Any number of urban policy changes reflect this, such as the introduction of 'competitive bidding' for dwindling urban-regeneration funds in the British case. If the timing and extent of the introduction of competitive elements have varied by national context (DiGaetano and Klemanski 1993; Harding 1994; Lauria 1997; Stoker and Mossberger 1994; Stoker 1995), the underlying situation is remarkably similar.

10 Reference to the 'culture industries' in this context no longer recognizes the original critical connotations of the term (Adorno and Horkheimer 1972).

11 This emphasis on duality and division has been especially associated with work on 'world cities' (Sassen 1991; 1994; 1998), though it appears equally true of cities at lower tiers of the global urban hierarchy (Haughton and Williams 1996).

12 The basic insecurity of the consumer society often goes unremarked, despite its permanent presence as an ambient background warning. As Bauman (in correspondence with Smith [1999, 193–194]) notes, the 'Germans speak today of a *zwei-Drittel Gesellschaft*; it is no more fanciful to expect the coming of a one-third society'. The situation has been inscribed within the sociological lexicon:

> The term 'working class' belongs to the imagery of a society in which the tasks and functions of the better-off and worse-off are divided – different but *complementary*. ... The term 'lower class' belongs to the imagery of social mobility. ... The term 'underclass' belongs to the imagery of a society which is not all-embracing and comprehensive.
>
> (Bauman 1998a, 66)

The term evokes 'an image of a class of people who are beyond classes and outside hierarchy, with neither chance nor need of readmission; people without role, making no useful contribution to the lives of the rest, and in principle beyond redemption' (*ibid.*). There is absolutely no inevitability that the dividing line between the underclass and the majority will remain where it is.

13 As the high-profile example of 'gated communities' reveals, walls are today as likely to be built in order to keep people *out* as to keep them in (Flusty 1997).

14 Of all the myriad meanings conveyed by the 'European city' (Parker 1996), self-designation as such calls to mind the following statement: 'The European city, when released into the act of consumption, will consume itself interminably' (Barber 1995, 75).

15 Writing of the middle of last century, Waterhouse (1994, 55) remarked that 'Leeds was in many respects [in] the mid-to-late Thirties a schizophrenic city, half of it with its clogs firmly planted in the Edwardian and Victorian past, the other half determinedly modernistic'. Today, the pace of historical contractions has quickened, to the point where everything takes place within the space of twenty-four hours.

16 'Symptomatic' in the sense that they are amenable to a 'symptomatic reading', capable of revealing the 'unconscious' or 'problematic' of the city (Althusser and Balibar 1971).

Bibliography

Abelson, E. S. (1989) *When Ladies Go A-thieving: Middle-class Shoplifters in the Victorian Department Store*, Oxford University Press, Oxford.

Abercrombie, N., Hill, S. and Turner, B. (1986) *Sovereign Individuals of Capitalism*, Allen and Unwin, London.

Abrams, M. (1977) 'The ideology of private consumption', in Hirst, I. R. C. and Duncan Reekie, W. (eds) *The Consumer Society*, Tavistock, London, 117–127.

Ackerlof, G. A. (1970) 'The market for "lemons": quality uncertainty and the market mechanism', *Quarterly Journal of Economics*, 84(4), 488–500.

Ackerman, F. (1997) 'Foundations of economic theories of consumption', in Goodwin, N. R., Ackerman, F. and Kiron, D. (eds) *The Consumer Society*, Island Press, Washington DC, 149–158.

Acton, W. (1968) *Prostitution*, MacGibbon and Kee, London.

Adburgham, A. (1981) *Shops and Shopping, 1800–1914: Where and in What Manner the Well-dressed Englishwoman Bought Her Clothes*, Allen and Unwin, London.

Adorno, T. and Horkheimer, M. (1972) *The Dialectic of Enlightenment*, Continuum, New York.

Aglietta, M. (1979) *A Theory of Capitalist Regulation: The US Experience*, Verso, London.

Agnew, J.-C. (1993) 'Coming up for air: consumer culture in historical perspective', in Brewer, J. and Porter, R. (eds) *Consumption and the World of Goods*, Routledge, London, 19–39.

Ahrne, G. (1988) 'A labour theory of consumption', in Otnes, P. (ed.) *The Sociology of Consumption*, Solum Forlag, Oslo, 49–63.

Alexander, D. (1970) *Retailing in England During the Industrial Revolution*, Athlone, London.

Alliez, E. and Feher, M. (1989) 'Notes on the sophisticated city', in Feher, M. and Kwinter, S. (eds) *The Contemporary City: Zone Parts I and II*, Zone Books, New York, 41–55.

Almeder, R. (1980) *The Philosophy of Charles S. Peirce: A Critical Introduction*, Blackwell, Oxford.

Althusser, L. (1969) *For Marx*, New Left Books, London.

——(1972) *Politics and History: Montesque, Rousseau, Hegel and Marx*, New Left Books, London.

Althusser, L. and Balibar, E. (1971) *Reading Capital*, New Left Books, London.

Alvarez, A. (1995) *Night: An Exploration of Night Life, Night Language, Sleep and Dreams*, Jonathan Cape, London.

Amin, A. and Graham, S. (1997) 'The ordinary city', *Transactions of the Institute of British Geographers*, 22, 411–429.

Anderson, B. (1991) *Imagined Communities: Reflections on the Origin and Spread of Nationalism*, 2nd edn, Verso, London.

Ando, A. and Modigliani, F. (1963) 'The "life cycle" hypothesis of saving: aggregate implications and tests', *American Economic Review*, 3(1), 55–84.

Anon. (1839) *Northern Star*, 28 September 1839.

Anon. (1862) 'The rape of the glances', *The Saturday Review*, 1 February, 124–125.

Appadurai, A. (1990) 'Disjuncture and difference in the global cultural economy', in Featherstone, M. (ed.) *Global Culture: Nationalism, Globalization and Modernity*, Sage, London, 295–310.

Appleby, J. (1978) *Economic Thought in Seventeenth-century England*, Princeton University Press, Princeton.

——(1993) 'Consumption in early modern social thought', in Brewer, J. and Porter, R. (eds) *Consumption and the World of Goods*, Routledge, London, 162–173.

Archer, M. S. (1996) 'Social integration and system integration: developing the distinction', *Sociology*, 30(4), 679–699.

Ariés, P. (1962) *Centuries of Childhood: A Social History of Family Life*, Cape, London.

Arnold, J. (2001) 'Why consumer confidence matters', *BBC Online*, 25 September 2001, at: *http://news.bbc.co.uk/hi/english/business/newsid_1561000/1561162.stm*, accessed 1 October 2001.

Arriaga, P. (1984) 'On advertising: a Marxist critique', *Media, Culture and Society*, 6(1), 53–64.

Ashley, R., Grainger, C. W. J. and Schmalensee, R. (1980) 'Advertising and aggregate consumption: an analysis of causality', *Econometrica*, 48(5), 1149–1167.

Ashworth, G. and Voogd, H. (1990) *Selling the City: Marketing Approaches in Public Sector Urban Planning*, Belhaven, London.

Augé, M. (1995) *Non-places: Introduction to an Anthropology of Supermodernity*, Verso, London.

Bacon, R. W. (1984) *Consumer Spatial Behaviour: A Model of Purchasing Decisions Over Space and Time*, Clarendon Press, Oxford.

Baernreither, J. M. (1889) *English Associations of Working Men*, Sonnenschein, London.

Bagnall, G. (1996) 'Consuming the past', in Edgell, S., Hetherington, K. and Warde, A. (eds) *Consumption Matters: The Production and Experience of Consumption*, Blackwell, Oxford, 227–247.

Baker, F. (1988) 'Archaeology and the heritage industry', *Archaeological Review from Cambridge*, 7(2), 141–144.

Bakhtin, M. (1984) *Rabelais and His World*, Indiana University Press, Bloomington.

Ball, M. (1983) *Housing Policy and Economic Power: The Political Economy of Owner Occupation*, Methuen, London.

——(1985) 'Coming to terms with owner occupation', *Capital and Class*, 24, 15–44.

Baran, P. (1957) *The Political Economy of Growth*, Monthly Review Press, New York.

Baran, P. and Sweezy, P. (1966) *Monopoly Capital*, Monthly Review Press, New York.

Barber, S. (1995) *Fragments of the European City*, Reaktion, London.

Barlow, J. and Duncan, S. (1988) 'The use and abuse of housing tenure', *Housing Studies*, 3(4), 219–231.

Barnekov, T., Boyle, R. and Rich, D. (1989) *Privatism and Urban Policy in Britain and the United States*, Oxford University Press, Oxford.

Barret-Ducrocq, F. (1991) *Love in the Time of Victoria*, Verso, London.

Barthes, R. (1973) *Mythologies*, Paladin, London.

——(1990) *The Fashion System*, University of California Press, Berkeley.

Barwise, P. and Ehrenberg, A. S. C. (1988) *Television and its Audience*, Sage, London.

Bassett, K. (1993) 'Urban cultural strategies and urban regeneration: a case study', *Environment and Planning A*, 25, 1773–1788.

Bataille, G. (1985) 'The notion of expenditure', in *Visions of Excess: Selected Writings, 1927–1939*, ed. A. Stoekl, Manchester University Press, Manchester, 116–29.

——(1988) *The Accursed Share: An Essay on General Economy. Volume I. Consumption*, Zone Books, New York.

Bateman, M. (1986) 'Leeds: a study in regional supremacy', in Gordon, G. (ed.) *Regional Cities in the UK, 1890–1980*, Harper and Row, London, 99–115.

Bates, J. J. (1988) 'Stated preference techniques and the analysis of consumer choice', in Wrigley, N. (ed.) *Store Choice, Store Location and Market Analysis*, Routledge, London, 187–202.

Baudelaire, C. (1970) *Paris Spleen*, New Directions Books, New York.

Baudrillard, J. (1975) *The Mirror of Production*, Telos, St Louis.

——(1981) *For a Critique of the Political Economy of the Sign*, Telos, St Louis.

——(1983) *In the Shadow of the Silent Majorities*, Semiotext(e), New York.

——(1987a) 'When Bataille attacked the metaphysical principle of economy', *Canadian Journal of Political and Social Theory*, 11(3), 57–62.

——(1987b) 'Modernity', *Canadian Journal of Political and Social Theory*, 11(3), 63–72.

——(1987c) *Forget Foucault*, Semiotext(e), New York.

——(1988) *America*, Verso, London.

——(1990a) *Fatal Strategies*, Pluto, London.

——(1990b) *Seduction*, Macmillan, London.

——(1990c) *Revenge of the Crystal: Selected Writings on the Modern Object and its Destiny, 1968–1983*, eds P. Foss and J. Pefanis, Pluto, London.

——(1990d) *Cool Memories*, Verso, London.

——(1993a) *Symbolic Exchange and Death*, Sage, London.

——(1993b) *Baudrillard Live: Selected Interviews*, ed. M. Gane, Routledge, London.

——(1993c) *The Transparency of Evil: Essays on Extreme Phenomena*, Verso, London.

——(1994a) *Simulacra and Simulation*, University of Michigan Press, Ann Arbor.

——(1994b) *The Illusion of the End*, Polity, Cambridge.

——(1995a) 'Symbolic exchange: taking theory seriously. An interview with Jean Baudrillard by Roy Boyne and Scott Lash', *Theory, Culture and Society*, 12, 79–95.

——(1995b) 'The virtual illusion: or the automatic writing of the world', *Theory, Culture and Society*, 12, 97–107.

——(1996) *The System of Objects*, Verso, London.

——(1998a) *The Consumer Society: Myths and Structures*, Sage, London.

——(1998b) 'The end of the millenium or the countdown', *Theory, Culture and Society*, 15(1), 1–9.

——(1998c) *Paroxism: Interviews with Philippe Petit*, Verso, London.

——(2001) 'Utopia: the smile of the Cheshire Cat', in Genosko, G. (ed.) *The Uncollected Baudrillard*, Sage, London, 59–60.

Bauman, Z. (1972) 'Culture, values and science of society', *The University of Leeds Review*, 15(2), 185–203.

——(1982) *Memories of Class: The Pre-history and After-life of Class*, Routledge and Kegan Paul, London.

——(1983) 'Industrialism, consumerism and power', *Theory, Culture and Society*, 1(3), 32–43.

——(1987) *Legislators and Interpreters: On Modernity, Post-modernity and Intellectuals*, Polity, Cambridge.

——(1988a) 'Disappearing into the desert', *Times Literary Supplement*, December, 16–22.

——(1988b) *Freedom*, Open University Press, Milton Keynes.

——(1990a) *Thinking Sociologically*, Blackwell, Oxford.

——(1990b) 'Modernity and ambivalence', in Featherstone, M. (ed.) *Global Culture: Nationalism, Globalization and Modernity*, Sage, London, 143–169.

——(1991) *Modernity and Ambivalence*, Polity, Cambridge.

——(1992a) *Intimations of Postmodernity*, Routledge, London.

——(1992b) *Mortality, Immortality and other Life Strategies*, Polity Press, Cambridge.

——(1992c) 'Survival as a social construct', in Featherstone, M. (ed.) *Cultural Theory and Cultural Change*, Sage, London, 1–36.

——(1993a) 'The sweet scent of decomposition', in Rojek, C. and Turner, B. S. (eds) *Forget Baudrillard?*, Routledge, London, 22–46.

——(1993b) *Postmodern Ethics*, Blackwell, Oxford.

——(1993c) 'Modernity', in Krieger, J., Kahler, M., Nzongola-Ntalaja, G., Stallins, B. B. and Weir, M. (eds) *The Oxford Companion to Politics of the World*, Oxford University Press, Oxford, 592–596.

——(1994) 'Desert spectacular', in Tester, K. (ed.) *The Flâneur*, Routledge, London, 138–157.

——(1995) *Life in Fragments: Essays in Postmodern Ethics*, Blackwell, Oxford.

——(1996a) 'From pilgrim to tourist – or a short history of identity', in Hall, S. and du Gay, P. (eds) *Questions of Cultural Identity*, Sage, London, 18–36.

——(1996b) 'Morality in the age of contingency', in Heelas, P., Lash, S. and Morris, P. (eds) *Detraditionalization*, Blackwell, Oxford, 49–58.

——(1997a) *Postmodernity and its Discontents*, Polity, Cambridge.

——(1997b) 'The haunted house', *New Internationalist*, April, 24–25.

——(1998a) *Work, Consumerism and the New Poor*, Open University Press, Milton Keynes.

——(1998b) *Globalization: The Human Consequences*, Polity, Cambridge.

——(1999a) *Culture as Praxis*, new edn, Sage, London.

——(1999b) *In Search of Politics*, Polity, Cambridge.

——(1999c) 'Modern adventures of procrastination', *Parallax*, 5(1), 3–6.

——(2000) *Liquid Modernity*, Polity, Cambridge.

——(2001a) *The Individualized Society*, Polity, Cambridge.

——(2001b) *Community: Seeking Safety in an Insecure World*, Polity, Cambridge.

——(2001c) 'Consuming life' *Journal of Consumer Culture*, 1(1), 9–29.

——(2001d) 'On globalization: or globalization for some, localization for some others', in Beilharz, P. (ed.) *The Bauman Reader*, Blackwell, Oxford, 298–311.

——(2002) 'Scene and obscene', in Araeen, R., Cubitt, S. and Sardar, Z. (eds) *The Third Text Reader on Art, Culture and Theory*, Continuum, London, 279–290.

Bauman, Z. and Tester, K. (2001) *Conversations with Zygmunt Bauman*, Polity, Cambridge.

Baumol, W. and Oates, W. (1975) *The Theory of Environmental Policy*, Prentice-Hall, Englewood Cliffs.

Beauregard, R. A. (1995) 'Theorizing the global–local connection', in Knox, P. L. and Taylor, P. J. (eds) *World Cities in a World-System*, Cambridge University Press, Cambridge, 232–248.

Beck, U. (1992) *Risk Society: Toward a New Modernity*, Sage, London.

——(2000) *What is Globalization?*, Polity, Cambridge.

Beck, U., Giddens, A. and Lash, S. (1994) 'Preface', in *Reflexive Modernization: Politics, Tradition and Aesthetics in the Modern Social Order*, Polity Press, Cambridge, i–viii.

Becker, G. S. (1965) 'A theory of the allocation of time', *Economic Journal*, 75, 493–517.

——(1992) 'Habits, addictions, and traditions', *Kyklos*, 45(3), 327–46.

Becker, G. S. and Murphy, K. M. (1988) 'A theory of rational addiction', *Journal of Political Economy*, 96(4), 675–700.

Beier, A. L. (1985) *Masterless Men: The Vagrancy Problem in England, 1500–1640*, Methuen, London.

Beilharz, P. (2000) *Zygmunt Bauman: Dialectic of Modernity*, Sage, London.

Belk, R. W. (1979) 'Gift-giving behaviour', in Sheth, J. E. (ed.) *Research in Marketing, Volume 2*, JAI Press, Greenwich, 95–126.

——(1995) 'Studies in the new consumer behaviour', in Miller, D. (ed.) *Acknowledging Consumption: A Review of New Studies*, Routledge, London, 58–95.

Bell, D. (1976) *The Cultural Contradictions of Capitalism*, 2nd edn, Heinemann, London.

Bell, W. (1958) 'Social choice, lifestyles and suburban residence', in Dobringer, W. (ed.) *The Suburban Community*, Putnam, New York, 225–247.

Benjamin, W. (1973) *Charles Baudelaire: A Lyric Poet in the Era of High Capitalism*, New Left Books, London.

——(1975a) 'On some motifs in Baudelaire', in *Illuminations*, Fontana, London, 155–200.

——(1975b) 'The work of art in the age of mechanical reproduction', in *Illuminations*, Fontana, London, 219–253.

——(1985) 'Central Park', *New German Critique*, 34, 32–58.

——(1986) 'Paris, capital of the nineteenth century', in *Reflections: Essays, Aphorisms, Autobiographical Writings*, Schocken, New York, 146–162.

——(1999) *The Arcades Project*, Belknap Press of Harvard University Press, Cambridge MA.

Bennett, A. (1999) 'Subcultures or neo-tribes? Rethinking the relationship between youth, style and musical taste', *Sociology*, 33(3), 599–617.

Bennington, G. (1995) 'Introduction to economics I: because the world is round', in Gill, C. B. (ed.) *Bataille: Writing the Sacred*, Routledge, London, 46–57.

Benson, J. (1994) *The Rise of Consumer Society in Britain, 1880–1980*, Longman, London.

Benson, S. P. (1986) *Counter Cultures: Saleswomen, Managers, and Customers in American Department Stores, 1890–1940*, University of Illinois Press, Chicago.

Beresford, M. (1988) *East End, West End: The Face of Leeds during Urbanisation, 1684–1842*, The Thoresby Society, Leeds.

Beresford, M. W. and Jones, G. R. J. (eds) (1967) *Leeds and its Region*, Leeds Local Executive Committee of the British Association for the Advancement of Science, Leeds.

Berking, H. (1999) *Sociology of Giving*, Sage, London.

Berman, M. (1983) *All that is Solid Melts into Air: The Experience of Modernity*, Verso, London.

Bermingham, A. and Brewer, J. (eds) (1995) *The Consumption of Culture: Image, Object, Text*, Routledge, London.

Berry, B. J. L. (1967) *Geography of Market Centers and Retail Distribution*, Prentice-Hall, Englewood Cliffs.

Berry, B. J. L. and Parr, J. (1988) *Market Centres and Retail Location: Theory and Applications*, Prentice-Hall, Englewood Cliffs.

Berry, C. J. (1994) *The Idea of Luxury: A Conceptual and Historical Investigation*, Cambridge University Press, Cambridge.

Bianchi, M. (ed.) (1998) *The Active Consumer: Novelty and Surprise in Consumer Choice*, Routledge, London.

Bianchini, F. (1995) 'Night cultures, night economies', *Planning Practice and Research*, 10(2), 121–126.

Bianchini, F., Dawson, J. and Evans, R. (1992) 'Flagship projects in urban regeneration', in Healy, P., Davoudi, S., O'Toole, M., Tavsanoglu, S., O'Toole, M. and Usher, D. (eds) *Rebuilding the City: Property-led Urban Regeneration*, Spon, London, 245–255.

Bianchini, F. and Parkinson, M. (1993) *Cultural Policy and Urban Regeneration: The West European Example*, Manchester University Press, Manchester.

Bijker, W. E. (1995) *Of Bicycles, Bakelites and Bulbs: Towards a Theory of Sociotechnical Change*, MIT Press, Cambridge MA.

Bingham, N. (1996) 'Object-ions: from technological determinism towards geographies of relations', *Environment and Planning D: Society and Space*, 14, 635–658.

Blackwell, R. D. and Talarzyk, W. W. (1983) 'Lifestyle retailing', *Journal of Retailing*, 59(4), 7–27.

Blanchard, M. E. (1985) *In Search of the City: Engels, Baudelaire, Rimbaud*, ANMA Libri, Saragota.

Blank, D. M. (1962) 'Cyclical behaviour of national advertising', *Journal of Business*, 35, 14–27.

Blau, P. (1964) *Exchange and Social Power*, John Wiley, New York.

Blaxter, M. (1990) *Health and Lifestyles*, Tavistock, London.

Bleaney, M. (1976) *Underconsumption Theories: A History and Critical Analysis*, Lawrence and Wishart, London.

Bloch, M. (1962) *Feudal Society. Volume One*, Clarendon Press, Oxford.

Blomley, N. (1996) ' "I'd like to dress her all over": masculinity, power and retail space', in Wrigley, N. and Lowe, M. (eds) *Retailing, Capital and Consumption: Towards the New Retail Geography*, Longman, London, 238–256.

Blumin, S. M. (1989) *The Emergence of the Middle Class: Social Experience in the American City, 1760–1900*, Cambridge University Press, Cambridge.

Bogdanor, V. (1999) 'Devolution, decentralisation or disintegration?', *Political Quarterly*, 70(2), 185–194.

Bordo, S. (1992) 'Anorexia nervosa: psychopathology as the crystallization of culture', in Curtin, D. W. and Heldke, A. M. (eds) *Cooking, Eating, Thinking: Transformative Philosophies of Food*, Indiana University Press, Indiana, 28–55.

Boseley, S. (2001), 'Tobacco firm to profit from cancer genes', *Guardian*, Monday 12 November, 1.

Bottomore, T. (1985) *Theories of Modern Capitalism*, Unwin Hyman, London.

Bourdieu, P. (1977) *Outline of a Theory of Practice*, Cambridge University Press, Cambridge.

——(1984) *Distinction: A Social Critique of the Judgement of Taste*, Routledge, London.

——(1987) 'Marginalia', in Schrift, A. (ed.) *The Logic of the Gift*, Routledge, London.

Bourne, B., Eichler, U. and Herman, D. (eds) (1987) *Modernity and its Discontents*, Spokesman, Nottingham.

Bowlby, R. (1985) *Just Looking: Consumer Culture in Dreisser, Gissing and Zola*, Methuen, London.

——(1987) 'Modes of shopping: Mallarmé at the Bon Marché', in Armstrong, N. and Tennenhouse, L. (eds) *The Ideology of Conduct: Essays on Literature and the History of Sexuality*, Methuen, London, 185–205.

——(1993) *Shopping with Freud*, Routledge, London.

——(2001) *Carried Away: The Invention of Modern Shopping*, Faber and Faber, London.

Boyer, P. (1990) *Tradition as Truth and Communication*, Cambridge University Press, Cambridge.

Boyer, R. (1990) *The Regulation School: A Critical Introduction*, Columbia University Press, New York.

Boyle, M. (1995) 'The politics of urban entrepreneurialism in Glasgow', *Geoforum*, 25, 452–469.

Bradford, M. G. (1995) 'Diversification and division in the English education system: towards a post-Fordist model?', *Environment and Planning A*, 27, 1595–1612.

Bradford, M. G. and Burdett, F. (1989) 'Privatization, education and the North/South divide', in Lewis, J. and Townsend, A. (eds) *The North/South Divide: Regional Change in Britain in the 1980s*, Paul Chapman, London, 192–212.

Braudel, F. (1981) *The Structures of Everyday Life*, Collins, London.

Breen, T. H. (1988) '"Baubles of Britian": the American and Consumer Revolution of the eighteenth century', *Past and Present*, 119, 73–104.

Briggs, A. (1968) *Victorian Cities*, Penguin, Harmondsworth.

Buchanan, I. (2000) *Michel de Certeau: Cultural Theorist*, Sage, London.

Buchanan, J. M. (1968) *The Demand and Supply of Public Goods*, Rand McNally, Chicago.

Buck-Morss, S. (1986) 'The *flâneur*, the sandwichman and the whore: the politics of loitering', *New German Critique*, 39, 99–140.

——(1989) *The Dialectics of Seeing: Walter Benjamin and the Arcades Project*, MIT Press, Cambridge MA.

Burroughs, W. S. (1959) *The Naked Lunch*, Grove Press, New York.

——(1988) *The Western Lands*, Picador, Basingstoke.

Burrows, R. and Butler, T. (1989) 'Middle mass and the pit: a critical review of Peter Saunders' sociology of consumption', *Sociological Review*, 37, 338–364.

Burrows, R. and Loader, B. (eds) (1994) *Towards a Post-Fordist Welfare State?*, Routledge, London.

Butler, R. (1999) *Jean Baudrillard: The Defence of the Real*, Sage, London.

Butler, T. (1997) *Gentrification and the Middle classes*, Aldershot, Ashgate.

Calinescu, M. (1987) *Five Faces of Modernity: Modernism, Avant Garde, Decadence, Kitch, Postmodernism*, Duke University Press, Durham NC.

Calle, S. (1999) *Double Game*, Violette Editions, London.

Calle, S./Baudrillard, J. (1988) *Suite Venitienne/Please Follow Me*, Bay Press, Seattle.

Calligaris, C. (1994) 'Memory lane: a vindication of urban life', *Critical Quarterly*, 36(4), 56–70.

Callon, M. (1986) 'Some elements of a sociology of translation: domestication of the scallops and the fishermen of St Brieuc Bay', in Law, J. (ed.) *Power, Action and Belief: A New Sociology of Knowledge?*, Routledge and Kegan Paul, London, 196–232.

Calvino, I. (1974) *Invisible Cities*, Harcourt Brace, London.

Camhi, L. (1993) 'Stealing femininity: department store kleptomania as sexual disorder', *Differences*, 5, 26–50.

Campbell, C. (1987) *The Romantic Ethic and the Spirit of Modern Consumerism*, Blackwell, Oxford.

——(1995) 'The sociology of consumption', in Miller, D. (ed.) *Acknowledging Consumption: A Review of New Studies*, Routledge, London, 96–126.

——(1996) 'Detraditionalization, character and the limits to agency', in Heelas, P., Lash, S. and Morris, P. (eds) *Detraditionalization*, Blackwell, Oxford, 149–169.

——(1997a) 'When the meaning is not a message: a critique of the consumption as communication thesis', in Nava, M., Blake, A., McRury, I. and Richards, B. (eds) *Buy this Book: Studies in Advertising and Consumption*, Routledge, London, 340–351.

——(1997b) 'Romanticism, introspection and consumption: a response to Professor Holbrook', *Consumption, Markets and Culture*, 1(2), 165–173.

Carrier, J. (1991) 'Gifts, commodities and social relations: a Maussian view of exchange', *Sociological Forum*, 6(1), 119–136.

——(1994) *Gifts and Commodities: Exchange and Western Capitalism since 1700*, Routledge, London.

Carroll, J. (1977) *Puritan, Paranoid, Remissive: A Sociology of Modern Culture*, Routledge and Kegan Paul, London.

Carter, E. (1984) 'Alice in the consumer wonderland', in McRobbie, A. and Nava, M. (eds) *Gender and Generation*, Macmillan, London, 185–214.

Carter, P. (1987) *The Road to Botany Bay: An Essay in Spatial History*, Chicago University Press, Chicago.

Cassirer, E. (1955) *The Philosophy of Symbolic Forms, Volume 2: Mythical Thought*, Yale University Press, New Haven.

Castells, M. (1976a) 'Is there an urban sociology?', in Pickvance, C. (ed.) *Urban Sociology: Critical Essays*, Tavistock, London, 33–59.

——(1976b) 'Theory and ideology in urban sociology', in Pickvance, C. (ed.) *Urban Sociology: Critical Essays*, Tavistock, London, 60–84.

——(1977) *The Urban Question: A Marxist Approach*, Edward Arnold, London.

——(1978) *City, Class and Power*, Macmillan, London.

——(1984) *The City and the Grassroots*, Edward Arnold, London.

——(1989) *The Informational City: Information Technology, Economic Restructuring and the Urban/Regional Process*, Blackwell, Oxford.

——(1996) *The Information Age: Economy, Society and Culture. Volume 1: The Rise of the Network Society*, Blackwell, Oxford.

Castells, M. and Godard, F. (1974) *Monopolville*, Mouton, Paris.

Castells, M. and Godard, F. (1974) *Monopolville*, Mouton, Paris.

Castells, M., Coh, L. and Kwok, R. Y.-W. (1990) *The Shek Kipp Mei Syndrome: Economic Development and Public Housing in Hong Kong and Singapore*, Pion, London.

Cawson, A. (1986) *Corporatism and Political Theory*, Blackwell, Oxford.

Céline, L.-F. (1988) *Journey to the End of the Night*, Calder, London.

de Certeau, M. (1984) *The Practice of Everyday Life*, University of California Press, Berkeley.

de Certeau, M., Giard, L. and Mayol, P. (1998) *The Practice of Everyday Life. Volume 2: Living and Cooking*, University of Minnesota Press, Minneapolis.

Chadwick, O. (1975) *The Secularization of the European Mind in the Nineteenth Century*, Cambridge University Press, Cambridge.

Chaney, D. (1983) 'The department store as cultural form', *Theory, Culture and Society*, 1(3), 22–31.

——(1990) 'Subtopia in Gateshead: the MetroCentre as cultural form', *Theory, Culture and Society*, 7(4), 49–68.

——(1996) *Lifestyles*, Routledge, London.

Charney, L. and Schwartz, V. (1995) 'Introduction', in Charney, L. and Schwartz, V. (eds) *Cinema and the Invention of Modern Life*, University of California Press, Berkeley, 1–12.

Cheal, D. (1988) *The Gift Economy*, Routledge, London.

Chiplin, B. and Sturgess, B. (1981) *The Economics of Advertising*, Holt, Rinehart and Winston, London.

Christopherson, S. (1994) 'The fortress city: privatized spaces, consumer citizenship', in Amin, A. (ed.) *A Post-Fordist Reader*, Blackwell, Oxford, 409–427.

Clammer, J. (1997) *Contemporary Urban Japan: A Sociology of Consumption*, Blackwell, Oxford.

Clarke, D. B. (1995) 'The limits to retail capital', in Wrigley, N. and Lowe, M. (eds) *Retailing, Capital and Consumption: Towards the New Retail Geography*, Longman, London, 284–301.

——(1997a) 'Consumption and the city, modern and postmodern', *International Journal of Urban and Regional Research*, 21(2), 218–237.

——(1997b) 'Previewing the cinematic city', in Clarke, D. B. (ed.) *The Cinematic City*, Routledge, London, 1–18.

——(1998) 'Consumption, identity and space-time', *Consumption, Markets and Culture*, 2(3), 233–258.

——(2000) 'Space, knowledge and consumption', in Bryson, J. R., Daniels, P. W., Henry, N. D. and Pollard, J. S. (eds) *Knowledge, Space, Economy*, Routledge, London, 209–225.

Clarke, D. B. and Purvis, M. (1994) 'Dialectics, difference and the geographies of consumption', *Environment and Planning A*, 26, 1091–1109.

Clavel, P. (1986) *The Progressive City: Planning and Participation, 1969–1984*, Rutgers University Press, New Brunswick.

Coase, R. H. (1974) 'The lighthouse in economics', *Journal of Law and Economics*, 17, 357–376.

Cobbe, H. (ed.) (1979) *Cook's Voyages and Peoples of the Pacific*, British Museum, London.

Cole, G. D. H. (1944) *A Century of Co-operation*, Co-operative Union, Manchester.

Comanor, W. S. and Wilson, T. A. (1979) 'The effect of advertising on competition: a survey', *Journal of Economic Literature*, 17, 453–476.

Comedia (1991) *Out of Hours: A Study of Economic, Social and Cultural Life in Twelve Town Centres in the UK*, Comedia/Gulbenkian Foundation, London.

Cook, I. and Crang, P. (1996) 'The world on a plate: culinary culture, displacement and geographical knowledges', *Journal of Material Culture*, 1(2), 131–153.

Cooke, P. (1988) 'Modernity, postmodernity and the city', *Theory, Culture and Society*, 5, 472–492.

Corbett, G. (2000) 'Women, body image and shopping for clothes', in Baker, A. (ed.) *Serious Shopping: Essays in Psychotherapy and Consumerism*, Free Association, London, 114–132.

Corbridge, S., Thrift, N. J. and Martin, R. (eds) (1994) *Money, Power and Space*, Blackwell, Oxford.

Corrigan, P. (1997) *The Sociology of Consumption: An Introduction*, Sage, London.

Corrigan, P. and Sayer, D. (1981) 'How the law rules: variations on some themes in Karl Marx', in Fryer, B., Hunt, A., McBarnet, D. and Moorhouse, B. (eds) *Law, State and Society*, Croom Helm, London.

——(1985) *The Great Arch: English State Formation as Cultural Revolution*, Blackwell, Oxford.

Coward, R. (1984) *Female Desire: Women's Sexuality Today*, Paladin, London.

Cowling, K. (1982) *Monopoly Capitalism*, Macmillan, London.

Cowling, K., Cable, J., Kelly, M. and McGuinness, T. (1975) *Advertising and Economic Behaviour*, Macmillan, London.

Cox, B. D. (1987) *The Health and Lifestyle Survey*, Cambridge University Press, Cambridge.

Cox, B. D., Huppert, F. A. and Whichelow, M. J. (eds) (1993) *The Health and Lifestyle Survey: Seven Years On*, Cambridge University Press, Cambridge.

Cox, K. (1993) 'The local and the global in the new urban politics: a critical view', *Environment and Planning D: Society and Space*, 11, 433–448.

——(1995) 'Globalization, competition and the politics of local economic development', *Urban Studies*, 32, 213–225.

Cox, K. and Mair, A. (1988) 'Locality and community in the politics of local economic development', *Annals of the Association of American Geographers*, 78, 137–146.

Crawford, M. (1992) 'The world in a shopping mall', in Sorkin, M. (ed.) *Variations on a Theme Park: The New American City and the End of Public Space*, Noonday Press, New York, 3–30.

Cresswell, T. (1996) *In Place, Out of Place: Geography, Ideology and Transgression*, University of Minnesota Press, Minneapolis.

——(1997) 'Imagining the nomad: mobility and the postmodern primitive', in Benko, G. and Strohmayer, U. (eds) *Space and Social Theory: Interpreting Modernity and Postmodernity*, Blackwell, Oxford, 360–379.

——(1998) 'Night discourse: producing/consuming meaning on the street', in Fyfe, N. (ed.) *Images of the Street: Planning, Identity and Control in Public Space*, Routledge, London, 268–279.

Crewe, L. and Gregson, N. (1997) 'The bargain, the knowledge, and the spectacle: making sense of consumption in the space of the car boot sale', *Environment and Planning D: Society and Space*, 15, 87–112.

——(1998) 'Tales of the unexpected: exploring car boot sales as marginal spaces of contemporary consumption', *Transactions of the Institute of British Geographers*, 23, 39–54.

Crompton, R. (1996) 'Consumption and class analysis', in Edgell, S., Hetherington, K. and Warde, A. (eds) *Consumption Matters: The Production and Experience of Consumption*, Blackwell, Oxford, 113–132.

——(1998) *Class and Stratification: An Introduction to Current Debates*, 2nd edn, Polity, Cambridge.

Cross, B. (1993) *Time and Money: The Making of a Consumer Culture*, Routledge, London.

Crossick, G. and Jaumain, S. (eds) (1999) *Cathedrals of Consumption*, Ashgate, Aldershot.

Crozier, M. (1964) *The Bureaucratic Phenomenon*, University of Chicago Press, Chicago.

Dalby, S. (1993) 'City of rich and poor', *Financial Times*, 23 January, 32.

Daunton, M. and Hilton, M. (eds) (2001) *The Politics of Consumption: Material Culture and Citizenship in Europe and America*, Berg, Oxford.

Davidoff, L. and Hall, C. (1987) *Family Fortunes: Men and Women of the English Middle Class, 1780–1850*, University of Chicago Press, Chicago.

Davies, N. (1998) *Dark Heart: The Shocking Truth about Hidden Britain*, Vintage, London.

Davis, C. (1992) 'The Protestant Ethic and the comic spirit of capitalism', *British Journal of Sociology*, 43(3), 421–442.

Davis, D. (1966) *Fairs, Shops and Supermarkets: A History of English Shopping*, University of Toronto Press, Toronto.

Davis, J. (1989) 'From "rookeries" to "communities": race, poverty and policing in London, 1850–1985', *History Workshop*, 27, 66–85.

Davis, M. (1990) *City of Quartz: Excavating the Future in Los Angeles*, Verso, London.

——(1998) *Ecology of Fear: Los Angeles and the Imagination of Disaster*, Picador, London.

Deakin, N. and Edwards, J. (1993) *The Enterprise Culture and the Inner City*, Routledge, London.

Dear, M. (2000) *The Postmodern Urban Condition*, Blackwell, Oxford.

Dear, M. and Flusty, S. (1998) 'Postmodern urbanism', *Annals of the Association of American Geographers*, 88, 50–72.

Dear, M., Schockman, H. E. and Hise, G. (eds) (1996) *Rethinking Los Angeles*, Sage, Thousand Oaks.

Deas, I. and Ward, K. G. (2000) 'From the "new localism" to the "new regionalism"? The implications of Regional Development Agencies for city/regional relations', *Political Geography*, 19(3), 273–292.

Debord, G. (1977) *Society of the Spectacle*, Black and Red, Detroit.

Deleuze, G. (1972) *Proust and Signs*, George Braziller, New York.

——(1992) 'Postscript on the societies of control', *October*, 59, 3–7.

Deleuze, G. and Guattari, F. (1983) *Anti-Oedipus: Capitalism and Schizophrenia*, Athlone, London.

Derrida, J. (1973) *Speech and Phenomena, and Other Essays on Husserl's Theory of Signs*, Northwestern University Press, Evanston.

——(1976) *Of Grammatology*, Johns Hopkins University Press, Baltimore.

——(1978) *Writing and Difference*, Routledge, London.

——(1981) *Positions*, University of Chicago Press, Chicago.

——(1992) *Given Time: I Counterfeit Money*, University of Chicago Press, Chicago.

——(1994) *Specters of Marx: The State of the Debt, the Work of Mourning, and the New International*, Routledge, London.

de Vries, J. (1993) 'Between purchasing power and the world of goods: understanding the household economy in early modern Europe', in Brewer, J. and Porter, R. (eds) *Consumption and the World of Goods*, Routledge, London, 85–132.

The Dictionary of Human Geography (2000) eds Johnstone, R.J., Gregory, D., Pratt, G. and Watts, M., Blackwell, Oxford.

A Dictionary of Marxist Thought (1987) eds Bottomore, T., Harris, L., Kiernan, V. G. and Miliband, R., Blackwell, Oxford.

DiGaetano, A. and Klemanski, J. (1993) 'Urban regimes in comparative perspective: the politics of urban development in Britain', *Urban Review Quarterly*, 29, 54–83.

Dixit, A. and Norman, V. (1978) 'Advertising and welfare', *Bell Journal of Economics*, 9(1), 1–17.

Doel, M. A. (1996) 'A hundred thousand lines of flight: a machinic introduction to the nomadic thought and scrumpled geographies of Gilles Deleuze and Félix Guattari', *Environment and Planning D: Society and Space*, 14, 421–439.

——(1999) *Poststructuralist Geographies: The Diabolical Art of Spatial Science*, Edinburgh University Press, Edinburgh.

Doel, M. A. and Clarke, D. B. (1995) 'Transpolitical geography', *Geoforum*, 25, 505–524.

——(1998) 'Transpolitical urbanism: suburban anomaly and ambient fear', *Space and Culture*, 1(2), 13–36.

——(1999a) 'Dark panopticon. Or, Attack of the Killer Tomatoes', *Environment and Planning D: Society and Space*, 17, 427–450.

——(1999b) 'Virtual worlds: simulation, suppletion, seduction, and simulacra', in May, J., Crang, M. and Crang, P. (eds) *Virtual Geographies: Bodies, Space and Relations*, Routledge, London, 261–283.

——(2000) 'Cultivating ambivalence: the unhinging of economy and culture', in Cook, I., Crouch, D., Naylor, S. and Ryan, J. (eds) *Cultural Turns/Geographical Turns: Perspectives on Cultural Geography*, Prentice-Hall, Harlow, 214–233.

——(2002) 'Lacan: the movie', in Creswell, T. and Dixon, D. (eds) *Engaging Film: Geographies of Mobility and Identity*, Rowman and Littlefield, Lanham, 69–93.

Domosh, M. (1996) 'The feminized retail landscape: gender ideology and consumer culture in nineteenth-century New York', in Wrigley, N. and Lowe, M. (eds) *Retailing, Capital and Consumption: Towards the New Retail Geography*, Longman, London, 257–270.

Dorfman, R. and Steiner, P. O. (1954) 'Optimal advertising and optimal quality', *American Economic Review*, 44(5), 826–836.

Dostoevsky, F. (1991) *Notes from the Underground*, and *The Gambler*, Oxford University Press, Oxford.

Douglas, M. (1970) *Purity and Danger: An Analysis of Concepts of Pollution and Taboo*, Penguin, Harmondsworth.

——(1996) *Thought Styles: Critical Essays on Good Taste*, Sage, London.

Douglas, M. and Isherwood, B. (1996) *The World of Goods: Towards an Anthropology of Consumption*, new edn, Routledge, London.

Dowding, K. and Dunleavy, P. (1996) 'Production, disbursement and consumption: the modes and modalities of goods and services', in Edgell, S., Hetherington, K. and Warde, A. (eds) *Consumption Matters: The Production and Experience of Consumption*, Blackwell, Oxford, 36–65.

Dowling, R. (1993) 'Femininity, place and commodities: a retail case study', *Antipode*, 25(4), 295–319.

Drakopoulos, S. A. (1992) 'Keynes' economic thought and the theory of consumer behaviour', *Scottish Journal of Political Economy*, 39(3), 318–336.

Duesenberry, J. S. (1949) *Income, Saving and the Theory of Consumer Behaviour*, Harvard University Press, Cambridge MA.

du Gay, P. (1996) *Consumption and Identity at Work*, Sage, London.

Duke, V. and Edgell, S. (1984) 'The political economy of cuts in Britain and consumption sectoral cleavages', *International Journal of Urban and Regional Research*, 10, 177–201.

Dumont, L. (1977) *From Mandeville to Marx: The Genesis and Triumph of Economic Ideology*, University of Chicago Press, Chicago.

——(1986) *Essays on Individualism: Modern Ideology in Anthropological Perspective*, University of Chicago Press, Chicago.

Dunleavy, P. (1979) 'The urban basis of political alignment: social class, domestic property ownership, and state intervention in consumption processes', *British Journal of Political Science*, 9, 409–444.

——(1980) *Urban Political Analysis: The Politics of Collective Consumption*, Macmillan, London.

——(1986) 'The growth of sectoral cleavages and the stabilization of state expenditures', *Environment and Planning D: Society and Space*, 4, 129–44.

Durham, S. (1998) *Phantom Communities: The Simulacrum and the Limits of Postmodernism*, Stanford University Press, Stanford.

Durkheim, E. (1953) *The Elementary Forms of the Religious Life*, Allen and Unwin, London.

——(1972) *Emile Durkheim: Selected Writings*, ed. A. Giddens, Cambridge University Press, Cambridge.

Duverger, C. (1979) *La Fleur Létale: Economie du Sacrifice Aztèque*, Seuil, Paris.

——(1989) 'The meaning of sacrifice', in Feher, M. (ed.) *Fragments for a History of the Human Body: Part Three*, Zone Books, New York, 366–385.

Dyer, A. W. (1997) 'Prelude to a theory of *homo absurdus*: variations on themes from Thorstein Veblen and Jean Baudrillard', *Cambridge Journal of Economics*, 21, 45–53.

Dyer, C. (1989) 'The consumer and the market in the later Middle Ages', *Economic History Review*, 42, 305–327.

Easthope, A. (1999) *The Unconscious*, Routledge, London.

Edgell, S. and Tilman, R. (1991) 'John Rae and Thorstein Veblen on conspicuous consumption: a neglected intellectual relationship', *History of Polititcal Economy*, 23, 731–743.

Edgell, S. Hetherington, K. and Warde, A. (eds) (1996) *Consumption Matters: The Production and Experience of Consumption*, Oxford, Blackwell.

Edwards, T. (1997) *Men in the Mirror: Men's Fashion, Masculinity and Consumer Society*, Cassell, London.

Ehrenberg, A. S. C. (1959) 'The pattern of consumer purchases', *Applied Statistics*, 8, 26–41.

——(1972) *Repeat-buying: Theory and Applications*, 2nd edn, North Holland Press, Amsterdam.

——(1974) 'Repetitive advertising and the consumer', *Journal of Advertising Research*, 14(2), 25–34.

Eisinger, P. K. (1981) *The Rise of the Entrepreneurial State: State and Local Economic Development Policy in the United States*, Wisconsin University Press, Madison.

Elgin, D. (1981) *Voluntary Simplicity*, HarperCollins, New York.

Elias, N. (1991) *The Society of Individuals*, Oxford, Blackwell.

Ellin, N. (1996) *Postmodern Urbanism*, Oxford, Blackwell.

——(ed.) (1997) *Architecture of Fear*, Princeton Architectural Press, New York.

Endres, A. M. (1991) 'Marshall's analysis of economizing behaviour with particular reference to the consumer', *European Economic Review*, 35, 333–341.

Engels, F. (1973) *The Condition of the Working Class in England in 1844: From Personal Observation and Authentic Sources*, Progress Publishers, Moscow.

England, P. (1993) 'The separative self: androcentric bias in neoclassical assumptions', in Ferber, M. A. and Nelson, J. A. (eds) *Beyond Economic Man: Feminist Theory and Economics*, University of Chicago Press, Chicago, 37–53.

Erlich, D., Guttman, I., Schenback, P. and Mills, J. (1957) 'Postdecision exposure to relevant information', *Journal of Abnormal and Social Psychology*, 54, 98–102.

Etzoni, A. (1998) 'Voluntary simplicity: characterization, select psychological implications and societal consequences', *Journal of Economic Psychology*, 19, 619–643.

Evans, R. and Harding, A. (1997) 'Regionalisation, regional institutions and economic development', *Policy and Politics*, 25(1), 19–29.

Ewen, S. (1975) *Captains of Consciousness: Advertising and the Social Roots of the Consumer Culture*, McGraw-Hill, New York.

——(1990) 'Marketing dreams: the political elements of style', in Tomlinson, A. (ed.) *Consumption, Identity and Style: Marketing, Meanings, and the Packaging of Pleasure*, Routledge, London, 41–56.

Fainstein, S., Gordon, I. and Harloe, M. (eds) (1993) *The Divided City*, Blackwell, Oxford.

Falk, P. (1994) *The Consuming Body*, Sage, London.

——(1997a) 'The scopic regimes of shopping', in Falk, P. and Campbell, C. (ed.) *The Shopping Experience*, Sage, London, 177–185.

——(1997b) 'The genealogy of advertising', in Sulkunen, P., Holmwood, J., Radner, H. and Schulze, G. (eds) *Constructing the New Consumer Society*, St Martin's Press, New York, 81–107.

Feagin, J. (1988) *Free Enterprise City*, Rutgers University Press, New Brunswick.

Featherstone, M. (1991) *Consumer Culture and Postmodernism*, Sage, London.

Febvre, L. (1968) *Le Problème de l'Incroyance, au XVIᵉ Siècle*, Gallimard, Paris.

Ferguson, P. P. (1993) 'The *flâneur*: urbanization and its discontents', in Nash, S. (ed.) *Home and its Dislocations in Nineteenth-century France*, State University of New York Press, Albany, 45–64.

Fine, B. (1997) 'Playing the consumption game', *Consumption, Markets and Culture*, 1(1), 7–29.

Fine, B. and Harris, L. (1976) 'State expenditure in advanced capitalism: a critique', *New Left Review*, 98, 97–112.

Fine, B. and Leopold, E. (1990) 'Consumerism and the Industrial Revolution', *Social History*, 15(2), 151–179.

Firat, A. F. and Venkatesh, A. (1995) 'Liberatory postmodernism and the reenchantment of consumption', *Journal of Consumer Research*, 22, 239–267.

Flusty, S. (1997) 'Building paranoia', in Ellin, N. (ed.) *Architecture of Fear*, Princeton Architectural Press, New York.

Forrest, R. and Murie, A. (1988) *Selling the Welfare State: The Privatization of Public Housing*, Routledge, London.

Forrest, R., Murie, A. and Williams, P. (1990) *Home Ownership: Differentiation and Fragmentation*, Unwin Hyman, London.

Foucault, M. (1974) *The Order of Things: An Archaeology of the Human Sciences*, Routledge, London.

——(1977) *Discipline and Punish: The Birth of the Prison*, Vintage, New York.

——(1980) *Power/Knowledge: Selected Interviews and Other Writings by Michel Foucault, 1972–1977*, ed. C. Gordon, Harvester Press, Brighton.

——(1986) *The History of Sexuality, Volume 3. The Care of the Self*, Pantheon, New York.

——(1989) 'Space, knowledge and power', in *Foucault Live: Collected Interviews, 1961–1984*, ed. S. Lotringer, Semiotext(e), New York.

——(1997) 'Subjectivity and truth', in *The Essential Works of Michel Foucault, 1954–1984. Volume 1: Ethics.*, ed. P. Rabinow, Penguin, Harmondsworth, 87–92.

Fox, R. W. and Lears, T. J. J. (eds) (1983) *The Culture of Consumption: Critical Essays in American History, 1880–1980*, Pantheon, New York.

Foxall, G. R. (1988) *Consumer Behaviour: A Practical Guide*, Routledge, London.

Frank, R. H. (1985) 'The demand for unobservable and other nonpositional goods', *American Economic Review*, 75, 101–116.

——(1999) *Luxury Fever: Why Money Fails to Satisfy in an Era of Excess*, Free Press, New York.

Frank, R. H. and Cook, P. J. (1996) *The Winner-take-all Society: Why the Few at the Top Get so much more*, Penguin, Harmondsworth.

Fraser, D. (ed.) (1980) *A Modern History of Leeds*, Manchester University Press, Manchester.

Fraser, I. (1998) *Hegel and Marx: The Concept of Need*, Edinburgh University Press, Edinburgh.

Fraser, W. (1981) *The Coming of the Mass Market, 1850–1914*, Hamish Hamilton, London.

Freenberg, A. (1980) 'The political economy of social space', in Woodward, K. (ed.) *The Myths of Information: Technology and Postindustrial Culture*, Routledge, London, 111–124.

Freud, S. (1955) *Civilization and its Discontents*, Hogarth Press, London.

——(1973–1986) *The Standard Edition of the Complete Psychological Works of Sigmund Freud*, 24 vols, Hogarth Press, London.

Friedberg, A. (1993) *Window Shopping: Cinema and the Postmodern*, University of California Press, Berkeley.

Friedman, M. (1957) *A Theory of the Consumption Function*, Princeton University Press, Princeton.

Frisby, D. (1984) *Georg Simmel*, Routledge, London.

——(1992) *Simmel and Since: Essays on Georg Simmel's Social Theory*, Routledge, London.

——(1994) 'The *flâneur* in social theory', in Tester, K. (ed.) *The Flâneur*, Routledge, London, 81–110.

Frow, J. (1987) 'Accounting for tastes: some problems of Bourdieu's sociology of culture', *Cultural Studies*, 1, 59–73.

Fukuyama, F. (1992) *The End of History and the Last Man*, Free Press, New York.

Furlough, E. (1991) *Consumer Cooperation in Modern France: The Politics of Consumption*, Cornell University Press, Ithaca NY.

Furlough, E. and Strikwerda, C. (eds) (1999) *Consumers Against Capitalism? Consumer Co-operation in Europe, North America and Japan, 1840–1990*, Rowman and Littlefield, Lanham.

Gabriel, Y. and Lang, T. (1995) *The Unmanageable Consumer*, Sage, London.

Galbraith, J. K. (1958) *The Affluent Society*, Riverside Press, Cambridge.

——(1967) *The New Industrial State*, Penguin, Harmondsworth.

Gallup, J. (1982) *Feminism and Psychoanalysis: The Daughter's Seduction*, Macmillan, London.

Gamble, A. (2001) 'Political economy', in Philo, G. and Miller, D. (eds) *Market Killing: What the Free Market Does and What Social Scientists Can Do About It*, Longman, London, 170–175.

Gane, M. (1991) *Baudrillard: Critical and Fatal Theory*, Routledge, London.

Gans, H. J. (1968) 'Urbanism and suburbanism as ways of life', in Pahl, R. E. (ed.) *Readings in Urban Sociology*, Pergamon, Oxford, 95–118.

——(1995) *The War Against the Poor: The Underclass and Antipoverty Policy*, Basic Books, New York.

Garmaniko, E. and Purvis, J. (1983) *The Public and the Private*, Heinemann, London.

Genosko, G. (1992) 'The struggle for an affirmative weakness: de Certeau, Lyotard, and Baudrillard', *Current Perspectives in Social Theory*, 12, 179–194.

——(1994a) *Baudrillard and Signs: Signification Ablaze*, Routledge, London.

——(1994b) 'Bar games: Baudrillard's table of conversions', *Semiotic Inquiry*, 14(3), 105–120.

Gershuny, J. (1978) *After Industrial Society: The Emerging Self-service Economy*, Macmillan, London.

Giddens, A. (1990) *The Consequences of Modernity*, Polity, Cambridge.

——(1994) 'Living in a post-traditional society', in Beck, U., Giddens, A., and Lash, S., (eds) *Reflexive Modernization: Politics, Tradition and Aesthetics in the Modern Social Order*, Polity Press, Cambridge, 56–109.

Gilliatt, S. (1992) 'Consumers at work: the private life of public provision', *British Journal of Sociology*, 43(2), 239–265.

Gilloch, G. (1996) *Myth and Metropolis: Walter Benjamin and the City*, Polity, Cambridge.

Glaeser, E. L., Kolko, J. and Salz, A. (2001) 'Consumer city', *Journal of Economic Geography*, 1, 27–50.

Glennie, P. D. (1995) 'Consumption within historical studies', in Miller, D. (ed.) *Acknowledging Consumption: A Review of New Studies*, Routledge, London, 164–203.

Glennie, P. D. and Thrift, N. J. (1992) 'Modernity, urbanism and modern consumption', *Environment and Planning D: Society and Space*, 10, 423–443.

——(1993) 'Modern consumption: theorising commodities *and* consumers', *Environment and Planning D: Society and Space*, 10, 423–443.

——(1996) 'Consumption, shopping and gender', in Wrigley, N. and Lowe, M. (eds) *Retailing, Capital and Consumption: Towards the New Retail Geography*, Longman, London, 221–237.

Glickman, L. B. (1999) 'Bibliographic essay', in Glickman, L. B. (ed.) *Consumer Society in American History: A Reader*, Cornell University Press, Ithaca NY, 399–414.

Goetz, E. G. and Clarke, S. E. (1993) *The New Localism: Comparative Politics in a Global Era*, Sage, London.

Goetz, E. G. and Sydney, M. (1994) 'Revenge of the property owners: community development and the politics of property', *Journal of Urban Affairs*, 16(4), 319–334.

Gold, J. and Ward, S. V. (eds) (1994) *Place Promotion: The Use of Publicity and Marketing to Sell Towns and Regions*, John Wiley, Chichester.

Golledge, R. G. and Stimson, R. (1997) *Spatial Behavior: A Geographic Perspective*, Guilford, New York.

Goodwin, M. (1997) 'Housing', in Pacione, M. (ed.) *Britain's Cities: Geographies of Division in Urban Britain*, Routledge, London, 203–217.

Goodwin, M., Duncan, S. and Halford, S. (1993) 'Regulation theory, the local state and the transition of urban politics', *Environment and Planning D: Society and Space*, 11, 67–88.

Goss, J. (1993) 'The "magic of the mall": an analysis of form, function and meaning in the contemporary retail built environment', *Annals of the Association of American Geographers*, 83, 18–47.

——(1995) ' "We know who you are and we know where you live": the instrumental rationality of geodemographic systems', *Economic Geography*, 71(2), 171–198.

——(1998) 'Disquiet on the waterfront: reflections on nostalgia and utopia in the urban archetypes of festival marketplaces', *Urban Geography*, 17(3), 221–247.

——(1999) 'Once upon a time in the commodity world: an unofficial guide to Mall of America', *Annals of the Association of American Geographers*, 89, 45–75.

Gottdeiner, M. (1994) 'The system of objects and the commodification of everyday life: the early Baudrillard', in Kellner, D. (ed.) *Baudrillard: A Critical Reader*, Blackwell, Oxford, 25–40.

Gough, I. (1975) 'State expenditure in advanced capitalism', *New Left Review*, 92, 53–92.

——(1979) *The Political Economy of the Welfare State*, Macmillan, London.

Gouldner, A. (1960) 'The norm of reciprocity', *American Sociological Review*, 25, 161–178.

Graham, B., Ashworth, G. J. and Tunbridge, J. E. (2000) *A Geography of Heritage: Power, Culture and Society*, Arnold, London.

Grant, W. (ed.) (1985) *The Political Economy of Corporatism*, St Martin's Press, New York.

Gray, R. (1981) *The Aristocracy of Labour in Nineteenth Century Britain, c.1850–1900*, Macmillan, London.

Green, N. (1990) *The Spectacle of Nature: Landscape and Bourgeois Culture in Nineteenth-Century France*, Manchester University Press, Manchester.

Gregory, C. A. (1982) *Gifts and Commodities*, Academic Press, London.

Gregory, D. (1994) *Geographical Imaginations*, Blackwell, Oxford.

Griffiths, R. (1995) 'Cultural strategies and new modes of urban intervention', *Cities*, 12(4), 253–265.

Gurney, P. (1996) *Co-operative Culture and the Politics of Consumption in England, 1870–1930*, Manchester University Press, Manchester.

Habermas, J. (1983) 'Modernity: an incomplete project', in Foster, H. (ed.) *The Anti-Aesthetic: Essays on Postmodern Culture*, Bay Press, Seattle, 3–15.

——(1985) *The Philosophical Discourse of Modernity*, Polity Press, Cambridge.

——(1989) *The New Conservatism*, Polity Press, Cambridge.

Hall, T. (1995) ' "The second industrial revolution": cultural reconstructions of industrial regions', *Landscape Research*, 20, 112–123.

——(1997) '(Re)placing the city: cultural relocation and the city as centre', in Westwood, T. and Williams, S. (eds) *Imagining Cities: Scripts, Signs, Memories*, Routledge, London, 202–218.

Hall, T. and Hubbard, P. (1996) 'The entrepreneurial city: new urban politics, new urban geographies?', *Progress in Human Geography*, 20(2), 153–174.

——(eds) (1998) *The Entrepreneurial City: Geographies of Politics, Regime and Representation*, John Wiley, Chichester.

Hallsworth, A. G. (1992) *The New Geography of Consumer Spending: A Political Economy Approach*, Belhaven Press, London.

Hamilton, D. B. (1987) 'Institutional economics and consumption', *Journal of Economic Issues*, 21, 1531–1554.

Hamnett, C. (1989) 'Consumption and class in contemporary Britain', in Hamnett, C., McDowell, L. and Sarre, P. (eds) *The Changing Social Structure*, Sage, London, 199–243.

Hannertz, U. (1990) 'Cosmopolitans and locals in world culture', *Theory, Culture and Society*, 7, 237–351.

Hannigan, J. A. (1995) 'The postmodern city: a new urbanization?', *Current Sociology*, 43(1), 151–217.

——(1998) *Fantasy City: Pleasure and Profit in the Postmodern Metropolis*, Routledge, London.

Harding, A. (1992) 'Property interests and urban growth coalitions in the UK: a brief encounter', in Healy, P., Davoudi, S., O'Toole, M., Tavsanoglu, S. and Usher, D. (eds) *Rebuilding the City: Property-led Urban Regeneration*, Spon, London, 224–232.

——(1994) 'Urban regimes and growth machines: towards a cross-national research agenda', *Urban Affairs Quarterly*, 29(3), 356–382.

Harland, R. (1987) *Superstructuralism: The Philosophy of Structuralism and Post-structuralism*, Routledge, London.

Harloe, M. (1984) 'Sector and class: a critical comment', *International Journal of Urban and Regional Research*, 8, 228–237.

Harloe, M. and Paris, C. (1984) 'The decollectivisation of consumption: housing and local government finance in England and Wales, 1979–81', in Szelenyi, I. (ed.) *Cities in Recession*, Sage, London, 70–98.

Harrison, M. (1986) 'Consumption and urban theory: an alternative approach based on the social division of welfare', *International Journal of Urban and Regional Research*, 10, 232–242.

Harrison, R. J. (1980) *Pluralism and Corporatism*, Allen and Unwin, London.

Harvey, D. (1973) *Social Justice and the City*, Edward Arnold, London.

——(1982) *The Limits to Capital*, Blackwell, Oxford.

——(1985a) *The Urbanization of Capital. Studies in the History and Theory of Capitalist Urbanization, Volume 1*, Blackwell, Oxford.

——(1985b) *Consciousness and the Urban Experience. Studies in the History and Theory of Capitalist Urbanization, Volume 2*, Blackwell, Oxford.

——(1985c) 'The geopolitics of capitalism', in Gregory, D. and Urry, J. (eds) *Social Relations and Spatial Structures*, Macmillan, London, 128–163.

——(1987) 'Flexible accumulation through urbanization: reflections on postmodernism in the American city', *Antipode*, 19, 260–286.

——(1989a) *The Condition of Postmodernity: An Enquiry into the Origins of Cultural Change*, Blackwell, Oxford.

——(1989b) 'From managerialism to entrepreneurialism: the transformation in urban governance in late capitalism', *Geografiska Annaler*, 71B(1), 3–17.

——(1992) 'Social justice, postmodernism and the city', *International Journal of Urban and Regional Research*, 16, 588–601.

——(1996) *Justice, Nature and the Geography of Difference*, Blackwell, Oxford.

——(2000) *Spaces of Hope*, Edinburgh University Press, Edinburgh.

Haug, W. F. (1986) *Critique of Commodity Aesthetics: Appearance, Sexuality and Advertising in Capitalist Society*, Polity, Cambridge.

Haughton, G. and Williams, C. C. (1996) 'Leeds: a case of second city syndrome?', in Haughton, G. and Williams, C. C. (eds) *Corporate City? Partnership, Participation and Partition in Urban Development in Leeds*, Avebury, Aldershot, 3–17.

Hearn, J. and Roseneil, S. (eds) (1999) *Consuming Cultures: Power and Resistance*, Macmillan, London.

Heath, J. (2001) 'The structure of hip consumerism', *Philosophy and Social Criticism*, 27(6), 1–17.

Heath, T. (1997) 'The twenty-four hour city concept: a review of initiatives in British cities', *Journal of Urban Design*, 2(2), 193–204.

Heath, T. and Stickland, R. (1997) 'The twenty-four hour city concept', in Oc, T. and Tiesdell, S. (eds) *Safer City Centres: Reviving the Public Realm*, Paul Chapman, London, 170–183.

Hebdige, D. (1979) *Subculture: The Meaning of Style*, Methuen, London.

Heelas, P. (1996) *The New Age Movement: The Celebration of Self and the Sacralization of Modernity*, Blackwell, Oxford.

Heffernan, M. (1994) 'On geography and progress: Turgot's *Plan d'un ouvrage sur la géographie politique* (1751), and the origins of modern progressive thought', *Political Geography*, 13(4), 328–343.

Hefner, R. (1977) 'Baudrillard's noble anthropology: the image of symbolic exchange in political economy', *Sub Stance*, 17, 105–113.

Heimel, C. (n.d., *c.*1995) 'How to be creative', in *Godhaven II*, Godhaven Ink, Leeds.

Heller, A. (1976) *The Theory of Need in Marx*, Allison and Busby, London.

——(1984) *Everyday Life*, Routledge and Kegan Paul, London.

Herden, G., Knoche, K., Seidl, C. and Trockel, W. (eds) (1999) *Mathematical Utility Theory: Utility Functions, Models and Applications in the Social Sciences*, Springer-Verlag, New York.

Hewison, R. (1987) *The Heritage Industry*, Butterworth, London.

Hicks, J. and Allen, R. J. (1934) 'A reconsideration of the theory of value. Part II: a mathematical theory of individual demand functions', *Economica*, 1(2), 196–219.

Hilferding, R. (1981) *Finance Capital: A Study of the Latest Phase of Capitalist Development*, Routledge, London.

Hill, C. (1955) *The English Revolution 1640*, Lawrence and Wishart, London.

Himmelfarb, G. (1968) 'The haunted house of Jeremy Bentham', in Himmelfarb, G. (ed.) *Victorian Minds*, Weidenfeld and Nicolson, London, 32–81.

Hirsch, F. (1976) *The Social Limits to Growth*, Harvard University Press, Cambridge MA.

Hirschman, A. O. (1977) *The Passions and the Interests: Political Arguments for Capitalism before its Triumph*, Princeton University Press, Princeton.

Hirschman, E. C. and Holbrook, M. B. (eds) (1981) *Symbolic Consumer Behaviour*, Association for Consumer Research, Ann Arbor.

Hirschman, E. C. and Holbrook, M. B. (1992) *Postmodern Consumer Research: The Study of Consumption as Text*, Sage, London.

Hobsbawm, E. J. (1964) *Labouring Men*, Weidenfeld and Nicolson, London.

Hobsbawm, E. J. and Ranger, T. (1983) (eds) *The Invention of Tradition*, Cambridge University Press, Cambridge.

Holbrook, M. (1997) 'Romanticism, introspection and the roots of experiential consumption: Morris the epicurean', *Consumption, Markets and Culture*, 1(2), 97–163.

Holt, D. B. (1997) 'Distinction in America? Recovering Bourdieu's theory of tastes from its critics', *Poetics*, 25, 93–120.

Hopkins, J. S. P. (1991) 'West Edmonton Mall: landscapes of myth and elsewhere-ness', *Canadian Geographer*, 34, 2–17.

Howard, J. A. and Sheth, J. N. (1969) *The Theory of Buyer Behaviour*, John Wiley, New York.

Howard, M. C. and King, J. E. (1975) *The Political Economy of Marx*, Longman, Harlow.

Howes, D. (ed.) (1996) *Cross-cultural Consumption: Global Markets, Local Realities*, Routledge, London.

Hubert, H. and Mauss, M. (1964) *Sacrifice: Its Nature and Function*, Cohen and West, London.

Hudson, K. (1983) *The Archaeology of the Consumer Society: The Second Industrial Revolution in Britain*, Heinemann, London.

Hunt, A. (1996a) *Governance of the Consuming Passions: A History of Sumptuary Law*, Macmillan, London.

——(1996b) 'The governance of consumption: sumptuary laws and shifting forms of regulation', *Economy and Society*, 25(3), 410–427.

Hutt, W. H. (1936) *Economists and the Public: A Study of Competition and Opinion*, Jonathan Cape, London.

Huyssen, A. (1986) *After the Great Divide: Modernism, Mass Culture, Postmodernism*, Macmillan, London.

Irigaray, L. (1985) *This Sex Which Is Not One*, Cornell University Press, Ithaca NY.

Iwata, O. (1997) 'Attitudinal and behavioural correlates of voluntary simplicity lifestyles', *Social Behaviour and Personality*, 25, 233–240.

Jackson, P. (1993) 'Towards a cultural politics of consumption', in Bird, J., Curtis, B., Putnam, T., Robinson, G. and Tickner, L. (eds) *Mapping the Futures: Local Cultures, Global Change*, Routledge, London, 207–228.

——(1999) 'Commodity culture: the traffic in things', *Transactions of the Institute of British Geographers*, 24, 95–108.

Jackson, P. and Holbrook, B. (1995) 'Multiple meanings: shopping and the cultural politics of identity', *Environment and Planning A*, 27, 1913–1930.

Jameson, F. (1983) 'Postmodernism and consumer society', in Foster, H. (ed.) *The Anti-aesthetic: Essays on Postmodern Culture*, Bay Press, Port Townsend, 111–125.

Jaques, E. (1982) *The Form of Time*, Crane Russak, New York.

Jefferys, J. B. (1954) *Retail Trading in Great Britain, 1850–1950*, Cambridge University Press, Cambridge.

Jencks, C. (1996) 'The city that never sleeps', *New Statesman and Society*, 28 June, 26–28.

Jessop, B. (1994) 'The transition to post-Fordism and the Schumpeterian welfare state', in Burrows, R. and Loader, B. (eds) *Towards a Post–Fordist Welfare State* Routledge, London 13–37.

——(1997) 'The entrepreneurial city: re-imagining localities, redesigning economic governance, or restructuring capital?', in Jewson, N. and McGregor, D. (eds) *Transforming Cities: Contested Governance and New Spatial Divisions*, Routledge, London, 28–41.

——(1998) 'The narrative of enterprise and the enterprise of narrative: place marketing and the entrepreneurial city', in Hall, T. and Hubbard, P. (eds) *The Entrepreneurial City: Geographies of Politics, Regime and Representation*, John Wiley, Chichester, 77–99.

Jhally, S. (1990) *The Codes of Advertising: Fetishism and the Political Economy of Meaning in the Consumer Society*, Routledge, London.

John, P. and Whitehead, A. (1997) 'The renaissance of English regionalism in the 1990s', *Policy and Politics*, 25(1), 7–18.

Johnson, H. G. (1967) *Money, Trade and Economic Growth*, Harvard University Press, Cambridge MA.

Jonas, A. (1994) 'The scale politics of spatiality', *Environment and Planning D: Society and Space*, 12, 257–264.

Jones, G. and Smith, M. (2001) 'Britain "needs you to shop"', *Daily Telegraph*, 28 September, 1.

Jones, M. and MacLeod, G. (1999) 'Towards a regional renaissance? Reconfiguring and rescaling England's economic governance', *Transactions of the Institute of British Geographers*, 24, 295–313.

Jones, P., Hillier, D. and Turner, D. (1999) 'Towards the "24 hour city"', *Town and Country Planning*, 68(5), 164–165.

Jones, P. d'A. (1965) *The Consumer Society: A History of American Capitalism*, Penguin, Harmondsworth.

Jowett, B. (1871) *The Dialogues Of Plato*, Clarendon Press, Oxford.

Jowitt, B. (1992) *New World Disorder: The Leninist Extinction*, University of California Press, Berkeley.

Judd, D. R. and Ready, R. L. (1986) 'Entrepreneurial cities and the new politics of economic development', in Peterson, P. E. and Lewis, C. W. (eds) *Reagan and the Cities*, The Urban Institute Press, Washington DC, 209–247.

Julien, P. (1994) *Jaques Lacan's Return to Freud: The Real, the Symbolic and the Imaginary*, NYU Press, New York.

Kaldor, N. (1950) 'The economic aspects of advertising', *Review of Economic Studies*, 58(1), 1–27.

Kanter, R. M. (1995) *World Class: Thriving Locally in the Global Economy*, Simon and Schuster, New York.

Katona, G. (1964) *The Mass Consumption Society*, McGraw-Hill, New York.

Kearns, G. and Philo, C. (eds) (1993) *Selling Places: The City as Cultural Capital, Past and Present*, Pergamon, London.

Keat, R., Whitley, N. and Abercrombie, N. (eds) (1994) *The Authority of the Consumer*, Routledge, London.

Keating, M. (1997) 'The invention of regions: political restructuring and territorial government in Western Europe', *Environment and Planning C: Government and Policy*, 15, 383–398.

Kellner, D. (1989) *Jean Baudrillard: from Marxism to Postmodernism and Beyond*, Polity, Cambridge.

Kent, R. A. (1989) *Continuous Consumer Market Measurement*, Edward Arnold, London.

Kern, S. (1983) *The Culture of Time and Space, 1880–1918*, Harvard University Press, Cambridge MA.

Keynes, J. M. (1936) *The General Theory of Employment, Interest and Money*, Macmillan, London.

Kopytoff, I. (1986) 'The cultural biography of things: commoditization as process', in Appadurai, A. (ed.) *The Social Life of Things: Commodities in Cultural Perspective*, Cambridge University Press, Cambridge, 64–91.

Kowinski, W. S. (1985) *The Malling of America: An Inside Look at the Great Consumer Paradise*, William Morrow, New York.

Kracauer, S. (1937) *Offenbach and the Paris of his Time*, Constable, London.

Kreitzman, L. (1999) *The 24 Hour Society*, Profile Books, London.

Kundera, M. (1984a) *Immortality*, Faber and Faber, London.

——(1984b) *The Unbearable Lightness of Being*, Faber and Faber, London.

——(1996) *The Book of Laughter and Forgetting*, Faber and Faber, London.

Kuspit, D. (1990) 'The contradictory character of postmodernism', in Silverman, H. J. (ed.) *Postmodernism, Philosophy and the Arts*, Routledge, London, 53–68.

Kuznets, S. (1946) *National Product since 1869*, National Bureau of Economic Research, New York.

Lacan, J. (1977) *Ecrits: A Selection*, Tavistock/Routledge, London.

——(1979) *The Four Fundamental Concepts of Psycho-analysis*, Penguin, Harmondsworth.

——(1982) *Feminine Sexuality: Jacques Lacan and the 'Ecole Freudienne'*, Macmillan, London.

Laermans, R. (1993) 'Learning to consume: early department stores and the shaping of modern consumer culture, 1860–1914', *Theory, Culture and Society*, 10, 79–102.

Lake, R. (1990) 'Urban fortunes: the political economy of place: a commentary', *Urban Geography*, 11, 179–184.

Lambin, J. J. (1976) *Advertising, Competition and Market Conduct in Oligopoly over Time*, Elsevier, Amsterdam.

Lancaster, B. (1995) *The Department Store: A Social History*, Leicester University Press, Leicester.

Lancaster, K. (1966a) 'Change and innovation in the technology of consumption', *American Economic Review*, 56, 14–23.

——(1966b) 'A new approach to consumer theory', *Journal of Political Economy*, 74, 132–157.

Land, N. (1992) *The Thirst for Annihilation: Georges Bataille and Virulent Nihilism*, Routledge, London.

Landry, C. and Bianchini, F. (1995) *The Creative City*, Comedia, London.

Lapsley, R. (1997) 'Mainly in cities and at night: some notes on cities and film', in Clarke, D. B. (ed.) *The Cinematic City*, Routledge, London, 186–208.

Lasch, C. (1980) *The Culture of Narcissism: American Life in the Age of Diminishing Expectations*, Abacus, London.

——(1984) *The Minimal Self: Psychic Survival in Troubled Times*, Picador, London.

Latour, B. (1991) 'Technology is society made durable', in Law, J. (ed.) *A Sociology of Monsters: Essays on Power, Technology and Domination*, Routledge, London, 103–132.

——(1992) 'Where are the missing masses? The sociology of a few mundane artifacts', in Bijker, W. E. and Law, J. (eds) *Shaping Technology/Building Society: Studies in Sociotechnical Change*, MIT, Cambridge MA, 225–258.

Laumann, E. O. and House, J. S. (1970) 'Living room styles and social attributes: the patterning of material artifacts in a modern urban community', *Sociology and Social Research*, 54(3), 321–342.

Lauria, M. (ed.) (1997) *Reconstructing Urban Regime Theory*, Sage, London.

Law, J. (1986) 'On the methods of long-distance control: vessels, navigation and the Portuguese route to India', in Law, J. (ed.) *Power, Action and Belief: A New Sociology of Knowledge?*, Routledge and Kegan Paul, London, 234–263.

Lawrence, J. C. (1992) 'Geographical space, social space, and the realm of the department store', *Urban History*, 19, 64–83.

Laws, G. (1995) 'Social justice and urban politics', *Urban Geography*, 15, 603–612.

Leach, W. R. (1984) 'Transformations in a culture of consumption: women and department stores, 1890–1925', *Journal of American History*, 71, 319–342.

Lears, J. (1989) 'Beyond Veblen: rethinking consumer culture in America', in Bronner, S. J. (ed.) *Consuming Visions: Accumulation and Display of Goods in America, 1880–1920*, Norton, New York, 73–97.

Le Bon, G. (1982) *The Crowd: A Study of the Popular Mind*, Fraser Publishing Company, Burlington.

Lechner, F. J. (1991) 'Simmel on social space', *Theory, Culture and Society*, 8(3), 192–202.

Lee, M. J. (1992) *Consumer Culture Reborn: The Cultural Politics of Consumption*, Routledge, London.

——(1997) 'Relocating location: cultural geography, the specificity of place, and the city habitus', in McGuigan, J. (ed.) *Cultural Methodologies*, Sage, London, 126–141.

——(2000) 'Applause for new look Town Hall', *Leeds: The Newspaper of the City of Leeds*, 9, 8.

Leeds Initiative (1999) *The Vision for Leeds: A Strategy for Sustainable Development, 1999–2009*, Leeds Initiative, Leeds.

——(2001) *A Neighbourhood Renewal Strategy for Leeds*, Leeds Initiative, Leeds.

Lefebvre, H. (1994) *Everyday life in the Modern World*, Transaction Publishers, London.

——(1996) *Writings on Cities*, eds E. Kofman and E. Lebas, Blackwell, Oxford.

Le Goff, J. (1980) *Time, Work and Culture in the Middle Ages*, University of Chicago Press, Chicago.

Lehtonen, T.-K. and Mäenpää, P. (1997) 'Shopping in the East Centre Mall', in Falk, P. and Campbell, C. (ed.) *The Shopping Experience*, Sage, London, 136–165.

Leibenstein, H. (1950) 'Bandwagon, snob and Veblen effects in the theory of consumers' demand', *Quarterly Journal of Economics*, 44(2), 183–207.

Leigh, C., Stillwell, J. and Tickell, A. (1994) 'The West Yorkshire economy: breaking with tradition', in Haughton, G. and Whitney, D. (eds) *Reinventing a Region: Restructuring in West Yorkshire*, Avebury, Aldershot, 61–90.

Leiss, W. (1976) *The Limits to Satisfaction: An Essay on the Problem of Needs and Commodities*, University of Toronto Press, Toronto.

Leiss, W., Kline, S. and Jhally, S. (1986) *Social Communication in Advertising: Persons, Products and Images of Well-being*, Methuen, London.

Leitner, H. (1990) 'Cities in pursuit of economic growth: the local state as entrepreneur', *Political Geography Quarterly*, 9, 146–170.

Lemire, B. (1990) 'Reflections on the character of consumerism, popular fashion and the English market in the eighteenth century', *Material History Bulletin*, 31, 65–70.

Lenin, V. I. (1960) 'On the so-called market question', in *Collected Works: Volume 1, 1893–1894*, Foreign Languages Publishing House, Moscow, 75–125.

Lévi-Strauss, C. (1960) 'On manipulated sociological models', *Bijdragen tot de taal-, land- en volkenkunde*, 116, 45–54.

——(1963) *Structural Anthropology*, Basic Books, New York.

——(1969) *The Elementary Structures of Kinship*, Beacon Press, Boston MA.

——(1987) *Introduction to the Work of Marcel Mauss*, Routledge and Kegan Paul, London.

Levinas, E. (1981) *Otherwise than Being, or Beyond Essence*, Martinus Nijhoff, The Hague.

Ley, D. (1996) *The New Middle Class and the Remaking of the Central City*, Oxford University Press, Oxford.

Leyshon, A. and Thrift, N. J. (1997) *Money/Space: Geographies of Monetary Transformation*, Routledge, London.

Lieberman, R. (1993) 'Shopping disorders', in Massumi, B. (ed.) *The Politics of Everyday Fear*, University of Minnesota Press, Minneapolis, 245–265.

Linder, S. B. (1970) *The Harried Leisure Class*, Columbia University Press, New York.

Lipset, S. M. and Rokkan, S. E. (1967) 'Cleavage, party systems, and voter alignments: an introduction', in Lipset, S. M. and Rokkan, S. E. (eds) *Party, System and Voter Alignments*, Free Press, New York, 1–64.

Lipietz, A. (1986) 'New tendencies in the international division of labour: regimes of accumulation and modes of regulation', in Scott, A. J. and Storper, M. (eds) *Production, Work, Territory: The Geographical Anatomy of Industrial Capitalism*, Unwin Hyman, London, 16–39.

Lockwood, D. (1964) 'Social integration and system integration', in Zollschan, G. and Hirsch, H. (eds) *Explorations in Social Change*, Routledge, London, 244–257.

Lofland, L. H. (1973) *A World of Strangers: Order and Action in Urban Public Space*, Basic Books, New York.

Logan, J. R. and Molotch, H. L. (1987) *Urban Fortunes: The Political Economy of Place*, University of California Press, Berkeley.

Logan, J. R. and Swanstrom, T. (1990) *Beyond the City Limits: Urban Policy and Economic Restructuring in Comparative Perspective*, Temple University Press, Philadelphia.

Lojkine, J. (1976) 'Contribution to a Marxist theory of urbanization', in Pickvance, C. (ed.) *Urban Sociology: Critical Essays*, Tavistock, London, 119–146.

Lord, J. D. (1985) 'The malling of the American landscape', in Dawson, J. A. and Lord, J. D. (eds) *Shopping Centre Development: Policies and Prospects*, Croom Helm, Beckenham, 209–225.

Lovatt, A. and O'Connor, J. (1995) 'Cities and the night-time economy', *Planning Practice and Research*, 10(2), 127–133.

Lovejoy, A. O. (1961) *The Great Chain of Being: A Study of the History of an Idea*, Harvard University Press, Cambridge MA.

Lovering, J. (1995) 'Creating discourses rather than jobs: the crisis in the cities and the transition fantasies of intellectuals and policy makers', in Healey, P., Cameron, S., Davoudi, S., Graham, S. and Madani-Pour, A. (eds) *Managing Cities: The New Urban Context*, John Wiley, Chichester, 109–126.

——(1997) 'Global restructuring and local impact', in Paddison, R. (ed.) *Britain's Cities: Geographies of Division in Urban Britain*, Routledge, London, 63–87.

Lowe, D. (1995) *The Body in Late-capitalist USA*, Duke University Press, Durham NC.

Lowe, S. (1986) *Urban Social Movements: The City after Castells*, Macmillan, London.

Lowenthal, D. (1996) *The Heritage Crusade and the Spoils of History*, Viking, London.

Luhmann, N. (1979) *Trust and Power: Two Works by Niklas Luhmann*, John Wiley, Chichester.

——(1988) 'Familiarity, confidence, trust: problems and alternatives', in Gambetta, D. (ed.) *Trust: Making and Breaking Co-operative Relations*, Blackwell, Oxford.

——(1996) 'Complexity, structural contingencies and value conflicts', in Heelas, P., Lash, S. and Morris, P. (eds) *Detraditionalization*, Blackwell, Oxford, 59–71.

——(1998) *Observations on Modernity*, Stanford University Press, Stanford.

Lukás, G. (1971) *History and Class Consciousness: Studies in Marxist Dialectics*, Merlin, London.

Luke, T. W. (1996) 'Identity, meaning and globalization: deterritorialization in post-modern space-time compression', in Heelas, P., Lash, S. and Morris, P. (eds) *Detraditionalization*, Blackwell, Oxford, 109–133.

——(1998) 'Running flat out on the road ahead: nationality, sovereignty and territoriality in the world of the information superhighway', in O'Tuathail, G. and Dalby, S. (eds) *Rethinking Geopolitics*, Routledge, London, 274–294.

Lunt, P. K. and Livingstone, S. M. (1992) *Mass Consumption and Personal Identity: Everyday Economic Experience*, Open University Press, Milton Keynes.

Lury, C. (1996) *Consumer Culture*, Polity, Cambridge.

Luxembourg, R. (1951) *The Accumulation of Capital*, Routledge and Kegan Paul, London.

Lyons, M. (1998) 'Neither chaos, nor stark simplicity: a comment on "A new look at gentrification"', *Environment and Planning A*, 30, 367–370.

Lyotard, J.-F. (1984) *The Postmodern Condition: A Report on Knowledge*, Manchester University Press, Manchester.

——(1992) *The Postmodern Explained to Children: Correspondence 1982–1985*, Turnaround, London.

——(1993a) *Moralités Postmodernes*, Galilée, Paris.

——(1993b) *Libidinal Economy*, Indiana University Press, Bloomington.

MacClancy, J. (1992) *Consuming Culture*, London, Paul Chapman.

McCormick, K. (1983) 'Duesenberry and Veblen: the demonstration effect revisited', *Journal of Economic Issues*, 17, 1125–1129.

McCracken, G. (1990) *Culture and Consumption: New Approaches to the Symbolic Character of Consumer Goods and Activities*, Indiana University Press, Bloomington.

Macfarlane, A. (1978) *The Origins of English Individualism: The Family, Property and Social Transition*, Blackwell, Oxford.

McKendrick, N., Brewer, J. and Plumb, J. H. (1982) *The Birth of a Consumer Society: The Commercialization of Eighteenth-century England*, Europa, London.

MacLeod, G. (1999) 'Place, politics and "scale dependence": exploring the structuration of Euro-regionalism', *European Urban and Regional Studies*, 6(3), 231–253.

MacPherson, C. B. (1962) *The Political Theory of Possessive Individualism*, Blackwell, Oxford.

McRobbie, A. (1992) 'The *Passagenwerk*, and the place of Walter Benjamin in cultural studies: Benjamin, cultural studies, Marxist theories of art', *Cultural Studies*, 6, 147–169.

——(1997) 'Bridging the gap: feminism, fashion and consumption', *Feminist Review*, 55, 73–89.

Maffesoli, M. (1996) *The Time of the Tribes: The Decline of Individualism in Mass Society*, Sage, London.

——(1997) 'The return of Dyonisus', in Sulkunen, P., Holmwood, J., Radner, H. and Schultze, G. (eds) *Constructing the New Consumer Society*, Macmillan, London, 21–37.

Malbon, B. (1998) 'The club. Clubbing: consumption, identity and the spatial practices of every-night life', in Skelton, T. and Valentine, G. (eds) *Cool Places: Geographies of Youth Culture*, Routledge, London, 266–286.

Malinowski, B. (1978) *Argonauts of the Western Pacific: An Account of Native Enterprise and Adventure in the Archepelagos of Melanesian New Guinea*, Routledge and Kegan Paul, London.

Malkin, J. and Wildavsky, A. (1991) 'Why the traditional distinction between public and private goods should be abandoned', *Journal of Theoretical Politics*, 3, 355–379.

Mandeville, B. (1970) *The Fable of the Bees*, Penguin, Harmondsworth.

Marcuse, H. (1964) *One-dimensional Man: Studies in the Ideology of Advanced Industrial Society*, Sphere, London.

Marcuse, P. and van Kempen, R. (eds) (2000) *Globalizing Cities: A New Spatial Order?*, Blackwell, Oxford.

Marshall, A. (1920) *Principles of Economics*, 8th edn, Macmillan, London.

Martin, R. (ed.) (1999) *Money and the Space Economy*, John Wiley, London.

Marx, K. (1967) *Capital: A Critique of Political Economy. Volume I*, International Publishers, New York.

——(1973) *Grundrisse: Foundations of the Critique of Political Economy*, Penguin, Harmondsworth.

Maslow, A. (1954) *Motivation and Personality*, Harper and Row, New York.

Mason, R. (1981) *Conspicuous Consumption: A Study of Exceptional Consumer Behaviour*, Gower, Farnborough.

——(1984) 'Conspicuous consumption: a literature review', *European Journal of Marketing*, 18(3), 26–39.

Massey, D. (1991) 'Flexible sexism', *Environment and Planning D: Society and Space*, 9, 31–57.

Mauss, M. (1973) 'Techniques of the body', *Economy and Society*, 2(1), 70–88.

——(1990) *The Gift: The Form and Reason for Exchange in Archaic Societies*, Routledge, London.

Mawson, J. (1983) 'Organising for economic growth: the formulation of local authority economic development policies in West Yorkshire', in Young, K. and

Mason, C. (eds) *Urban Economic Development: New Roles and Relationships*, Macmillan, London, 79–105.

Mawson, J. and Hall, S. (2000) 'Joining it up locally? Area regeneration and holistic government in England', *Regional Studies*, 34(1), 67–74.

Mayer, M. (1992) 'The shifting local political system in European cities', in Dunford, M. and Kafkalas, G. (eds) *Cities and Regions in the New Europe: Global/Local Interplay and Spatial Development Strategies*, Belhaven, London, 255–278.

——(1995) 'Urban governance in post-Fordist cities', in Healey, P., Cameron, S., Davoudi, S., Graham, S. and Madani-Pour, A. (eds) *Managing Cities: The New Urban Context*, John Wiley, Chichester, 231–249.

Mellenkopf, J. H. (1983) *The Contested City*, Princeton University Press, Princeton.

Mellenkopf, J. H. and Castells, M. (eds) (1991) *Dual City: Restructuring New York*, Russell Sage Foundation, New York.

Melville, H. (1977) *Bartleby, the Scrivener*, The Pittsburgh Bibliophiles, Pittsburgh.

Mennell, S. (1985) *All Manner of Foods: Eating and Taste in England and France from the Middle Ages to the Present*, Blackwell, Oxford.

Merrifield, A. and Swyngedouw, E. (eds) (1997) *The Urbanization of Injustice*, New York University Press, New York.

Meyrowitz, J. (1985) *No Sense of Place: The Impact of Electronic Media on Social Behaviour*, Oxford University Press, Oxford.

Mick, D. G. and DeMoss, M. (1993) 'Self-gifts: phenomenological insights from four contexts', *Journal of Consumer Research*, 17, 322–332.

Miller, D. (1985) *Artifacts as Categories: A Study of Ceramic Variability in Central India*, Cambridge University Press, Cambridge.

——(1987) *Material Culture and Mass Consumption*, Blackwell, Oxford.

——(1988) 'Appropriating the state on the council estate', *Man*, 23, 353–72.

——(ed.) (1993) *Unwrapping Christmas*, Clarendon Press, Oxford.

——(1995) 'Consumption as the vanguard of history: a polemic by way of an introduction', in Miller, D. (ed.) *Acknowledging Consumption: A Review of New Studies*, Routledge, London, 1–57.

——(1997) 'Consumption and its consequences', in Mackay, H. (ed.) *Consumption and Everyday Life*, Sage, London, 14–50.

——(2000) 'Virtualism: the culture of political economy', in Cook, I., Crouch, D., Naylor, S. and Ryan, J. (eds) *Cultural Turns/Geographical Turns: Perspectives on Cultural Geography*, Prentice-Hall, London, 196–213.

Miller, D., Jackson, P., Thrift, N. J., Holbrook, B. and Rowlands, M. (1998) *Shopping, Place and Identity*, Routledge, London.

Miller, M. (1981) *The Bon Marché: Bourgeois Culture and the Department Store, 1869–1920*, Princeton University Press, London.

Miller, P. and Rose, N. (1997) 'Mobilizing the consumer: assembling the sign of consumption', *Theory, Culture and Society*, 14(1), 1–36.

Mingione, E. (1981) *Social Conflict and the City*, Oxford, Blackwell.

Mishan, E. J. (1961) 'Theories of consumption: a cynical view', *Economica*, 28(109), 1–11.

——(1969) *Twenty-one Popular Economic Fallacies*, Penguin, Harmondsworth.

Mitchell, J. (1975) *Psychoanalysis and Feminism*, Penguin, Harmondsworth.

Modigliani, F. and Brumberg, R. E. (1955) 'Utility analysis and the consumption function: an interpretation of cross-sectional data', in Kurihara, K. K. (ed.) *Post-Keynsian Economics*, Allen and Unwin, London, 388–436.

Mohun, S. (1977) 'Consumer sovereignty', in Green, F. and Nore, P. (eds) *Economics: An Anti-text*, Macmillan, London, 57–75.

Mollenkopf, J. and Castells, M. (eds) (1991) *Dual City: Restructuring New York*, Russell Sage Foundation, New York.

Molotch, H. (1976) 'The city as a growth machine', *American Journal of Sociology*, 82(2), 309–332.

——(1988) 'Strategies and constraints of growth elites', in Cummings, S. (ed.) *Business Elites and Urban Development*, State University of New York Press, Albany.

Molotch, H. and Logan, J. R. (1985) 'Urban dependencies: new forms of use and exchange in US cities', *Urban Affairs Quarterly*, 21(2), 143–169.

Montgomery, J. (1995) 'Urban vitality and the culture of cities', *Planning Practice and Research*, 10(2), 101–109.

Moore Jr, W. B. (1978) *Injustice: The Social Basis of Obedience and Revolt*, Macmillan, London.

Moorhouse, H. F. (1978) 'The Marxist theory of the labour aristocracy', *Social History*, 3(1), 481–490.

——(1983) 'American automobiles and workers' dreams', *Sociological Review*, 31, 403–426.

Morris, M. (1988) 'Things to do with shopping centres', in Sheridan, S. (ed.) *Grafts: Feminist Cultural Criticism*, Verso, London, 193–225.

Morris, R. J. (1983) 'Voluntary associations and British urban elites, 1780–1850: an analysis', *The History Journal*, 26, 95–118.

——(1990) *Class, Sect and Party: the Making of the British Middle Class, Leeds, 1820–1850*, Manchester University Press, Manchester.

Mort, F. (1989) 'The politics of consumption', in Hall, S. and Jacques, M. (eds) *New Times: The Changing Face of Politics in the 1990s*, Lawrence and Wishart, London, 160–172.

——(1996) *Cultures of Consumption: Masculinities and Social Space in Late Twentieth-century Britain*, Routledge, London.

Mouzelis, N. (1997) 'Social and system integration: Lockwood, Habermas, Giddens', *Sociology*, 31(1), 111–119.

Muckerji, C. (1983) *Graven Images: Patterns of Modern Materialism*, Columbia University Press, New York.

Mui, H.-C. and Mui, L. H. (1989) *Shops and Shopkeeping in Eighteenth-century England*, Routledge, London.

Mukařovský, J. (1964) 'Standard language and poetic language', in Garvin, P. L. (ed.) *A Prague School Reader*, Georgetown University Press, Washington DC, 17–30.

Munn, N. (1986) *The Fame of Gawa: A Symbolic Study of Value Transformation in a Massim (Papua New Guinea) Society*, Cambridge University Press, Cambridge.

——(1992) 'The cultural anthropology of time: a critical essay', *Annual Review of Anthropology*, 21, 93–123.

Murdoch, J. (1997) 'Inhuman/nonhuman/human: actor-network theory and the prospects for a nondualistic and symmetrical perspective on nature and society', *Environment and Planning D: Society and Space*, 15, 731–756.

Müri, W. (1931) *Symbolon: Wort- und sachgeschichtliche Studie*, Beilage zum Jahresbericht über das städtische Gymnasium Bern, Bern.

Muth, R. (1966) 'Household production and consumer demand functions', *Econometrica*, 34, 281–302.

Myers, J. and Gutman, J. (1974) 'Life style: the essence of social class', in Wells, W. (ed.) *Life Style and Psychographics*, American Marketing Association, Chicago, 235–256.

Nava, M. (1992) *Changing Cultures: Feminism, Youth and Consumerism*, Sage, London.

——(1997) 'Modernity's disavowal: woman, the city and the department store', in Falk, P. and Campbell, C. (ed.) *The Shopping Experience*, Sage, London, 56–91.

Nelson, P. (1974) 'Advertising as information', *Journal of Political Economy*, 81, 729–745.

Nerlove, M. and Arrow, K. (1962) 'Optimal advertising policy under dynamic conditions', *Economica*, 29, 129–142.

Nickel, D. R. (1995) 'The progressive city? Urban redevelopment in Minneapolis', *Urban Affairs Review*, 30(3), 255–377.

Nijman, J. (2000) 'The paradigmatic city', *Annals of the Association of American Geographers*, 90, 135–145.

Nixon, S. (1992) 'Have you got the look? Masculinities and shopping', in Shields, R. (ed.) *Lifestyle Shopping: The Subject of Consumption*, Routledge, London, 149–169.

——(1996) *Hard Looks: Masculinities, Spectatorship and Contemporary Consumption*, UCL Press, London.

Norris, C. and Armstrong, C. (1999) *The Maximum Surveillance Society: The Rise of CCTV*, Berg, Oxford.

Noys, B. (2000) *Georges Bataille: A Critical Introduction*, Pluto, London.

O'Brien, R. (1992) *Global Financial Integration: The End of Geography?*, Chatham House/Pinter, London.

O'Brien, S. and Ford, R. (1988) 'Can we at last say goodbye to social class?', *Journal of the Market Research Society*, 3, 289–332.

O'Connor, J. (1973) *The Fiscal Crisis of the State*, St Martins Press, New York.

——(1987) *The Meaning of Crisis: A Theoretical Introduction*, Blackwell, Oxford.

Oettermann, S. (1997) *The Panorama: History of a Mass Medium*, Zone Books, New York.

Offe, C. (1996) *Modernity and the State: East, West*, Polity, Cambridge.

Ogborn, M. (1993) 'Ordering the city: surveillance, public space and the reform of urban policy in England, 1835–56', *Political Geography*, 12(6), 505–21.

Ohmae, K. (1995) *The End of the Nation State: The Rise of Regional Economies*, Free Press, New York.

Olalquiaga, C. (1992) *Megalopolis: Contemporary Cultural Sensibilities*, Minnesota University Press, Minneapolis.

Otnes, P. (1988) 'Housing consumption: collective systems services', in Otnes, P. (ed.) *The Sociology of Consumption*, Solum Forlag, Oslo, 119–138.

Packard, V. (1957) *The Hidden Persuaders*, Penguin, Harmondsworth.

Paddison, R. (1993) 'City marketing, image reconstruction and urban restructuring', *Urban Studies*, 30(2), 339–350.

Pahl, R. (1975) *Whose City? And Further Essays on Urban Society*, Penguin, Harmondsworth.

——(1978) 'Castells and collective consumption', *Sociology*, 12, 309–315.

——(1995) *After Success: Fin-de-Siècle Anxiety and Identity*, Polity, Cambridge.

Palda, K. S. (1964) *The Measurement of Cumulative Advertising Effects*, Prentice-Hall, London.

Panofsky, E. (1991) *Perspective as Symbolic Form*, Zone Books, New York.

Paolucci, G. (2001) 'The city's continuous cycle of consumption: towards a new definition of the power over time?', *Antipode*, 33, 647–659.

Park, R. E., Burgess, E. W. and McKenzie, R. (1925) *The City: Suggestions of Investigation of Human Behaviour in the Urban Environment*, Chicago University Press, Chicago.

Parker, S. (1996) 'Cities of light, cities of dread: the European metropolis and the conflicts of modernity', *Contemporary European History*, 5(1), 139–151.

Parkinson, M. R. and Judd, D. R. (eds) (1990) *Leadership in Urban Regeneration*, Sage, London.

Parry, J. (1986) '*The Gift*, the Indian gift and "the Indian gift" ', *Man*, 21, 453–473.

Parsons, T. (1951) *The Social System*, Free Press, Glencoe.

Pasdermadjian, H. (1954) *The Department Store: Its Origins, Evolution, and Economics*, Newman Books, London.

Pawlett, W. (1997) 'Utility and excess: the radical sociology of Bataille and Baudrillard', *Economy and Society*, 26(1), 95–125.

Peck, J. A. (1995) 'Moving and shaking: business elites, state localism and urban privatism', *Progress in Human Geography*, 19, 16–46.

Peck, J. A. and Tickell, A. (1994) 'Jungle law breaks out: neo-liberalism and global–local disorder', *Area*, 26(4), 317–26.

Peet, R. (1986) 'World capitalism and the destruction of regional cultures', in Johnston, R. J. and Taylor, P. J. (eds) *A World in Crisis? Geographical Perspectives*, Blackwell, Oxford, 175–199.

——(1997) 'The cultural production of economic form', in Lee, R. and Wells, J. (eds) *Geographies of Economies*, Arnold, London, 37–46.

Pefanis, J. (1991) *Heterology and the Postmodern: Bataille, Baudrillard, and Lyotard*, Duke University Press, Durham NC.

Pennance, F. G. (1977) 'The market in and consumption of housing', in Hirst, I. R. C. and Duncan Reekie, W. (eds) *The Consumer Society*, Tavistock, London, 145–166.

Perec, G. (1988) *Life: A User's Manual*, Harvill, London.

Peterson, P. (1981) *City Limits*, Chicago University Press, Chicago.

Philo, C. (2000) 'More words, more worlds: reflections on the "cultural turn" and human geography', in Cook, I., Crouch, D., Naylor, S. and Ryan, J. (eds) *Cultural Turns/Geographical Turns: Perspectives on Cultural Geography*, Prentice-Hall, London, 26–53.

Pickvance, C. (1976) 'On the study of urban social movements', in Pickvance, C. (ed.) *Urban Sociology: Critical Essays*, Tavistock, London, 198–218.

——(1977) 'Marxist approaches to the study of urban politics', *International Journal of Urban and Regional Research*, 1, 218–255.

Pinch, S. (1985) *Cities and Services: The Geography of Collective Consumption*, Routledge, London.

——(1989) 'Collective consumption', in Wolch, J. and Dear, M. (eds) *The Power of Geography: How Territory Shapes Social Life*, Unwin Hyman, London, 41–60.

——(1996) *Worlds of Welfare: Understanding the Changing Geographies of Welfare Provision*, Routledge, London.

Pollack, R. (1970) 'Habit formation and dynamic demand functions', *Journal of Political Economy*, 78, 745–763.

——(1976) 'Interdependent preferences', *American Economic Review*, 66, 309–320.

Porter, T. M. (1986) *The Rise of Statistical Thinking, 1820–1900*, Princeton University Press, Princeton.

Poulantzas, N. (1973) *Political Power and Social Classes*, New Left Books, London.

Pratt, G. (1986) 'Against reductionism: the relations of consumption as a mode of social structuration', *International Journal of Urban and Regional Research*, 10, 377–400.

Preteceille, E. (1981) 'Collective consumption, the state and the crisis of capitalist society', in Harloe, M. and Lebas, E. (eds) *City, Class and Capital: New Developments in the Political Economy of Cities and Regions*, Edward Arnold, London, 1–16.

——(1986) 'Collective consumption, urban segregation, and social classes', *Environment and Planning D: Society and Space*, 4, 145–154.

Preteceille, E. and Terrail, J.-P. (1985) *Capitalism, Consumption and Needs*, Blackwell, Oxford.

Proust, M. (1981) *Remembrance of Things Past*, Chatto and Windus, London.

Punter, J. (1990) 'The privatization of the public realm', *Planning Practice and Research*, 5(3), 17–21.

Purvis, M. (1990) 'The development of co-operative retailing in England and Wales, 1851–1901: a geographical study', *Journal of Historical Geography*, 16, 314–331.

——(1998) 'Societies of consumers and consumer societies: co-operation, consumption and politics in Britain and continental Europe c.1850–1920', *Journal of Historical Geography*, 24(2), 147–169.

Pynchon, T. (1996) *The Crying of Lot 49*, Vintage, London.

Radner, H. (1995) *Shopping Around: Feminine Culture and the Pursuit of Pleasure*, Routledge, London.

Rappaport, E. D. (1995) ' "A new era of shopping": the promotion of women's pleasure in London's West End, 1909–1914', in Charney, L. and Schwartz, V. (eds) *Cinema and the Invention of Modern Life*, University of California Press, Berkeley, 130–155.

Razzell, P. E. and Wainwright, R. W. (eds) (1973) *The Victorian Working Class: Selections from Letters to the Morning Chronicle*, Cass, London.

Redfern, P. (1997a) 'A new look at gentrification: 1. Gentrification and domestic technologies', *Environment and Planning A*, 29, 1275–1296.

——(1997b) 'A new look at gentrification: 2. A model of gentrification', *Environment and Planning A*, 29, 1335–1354.

——(1998) 'Letter to the Editor. Neither chaos nor stark simplicity – a reply to Lyons', *Environment and Planning A*, 30, 2075–2077.

Reekie, G. (1993) 'Changes in the Adamless Eden: the spatial and sexual transformation of the Brisbane department store, 1930–90', in Shields, R. (ed.) *Lifestyle Shopping: The Subject of Consumption*, Routledge, London, 170–193.

——(1993) *Temptations: Sex, Selling and the Department Store*, Allen and Unwin, London.

Reisman, D., with Glazer, N. and Reuel, D. (1950) *The Lonely Crowd: A Study of the Changing American Character*, Yale University Press, New Haven.

Reisman, D. A. (1980) *Galbraith and Market Capitalism*, New York University Press, New York.

Relph, E. (1976) *Place and Placelessness*, Pion, London.

Rhodes, R. (1996) 'The new governance: governing without government', *Political Studies*, XLIV, 652–667.

Ricardo, D. (1951) *The Principles of Political Economy and Taxation*, Cambridge University Press, Cambridge.

Richards, T. S. (1991) *The Commodity Culture of Victorian England: Advertising and Spectacle, 1851–1914*, Verso, London.

Richardson, M. (1994) *Georges Bataille*, Routledge, London.

Ritzer, G. (1998) 'Introduction', in Baudrillard, J., *The Consumer Society: Myths and Structures*, Sage, London, 1–24.

——(1999) *Enchanting a Disenchanted World: Revolutionizing the Means of Consumption*, Sage, London.

Roberts, J. (1999) 'Philosophizing the everyday: the philosophy of praxis and the fate of cultural studies', *Radical Philosophy*, 98, 16–29.

Roberts, M. J. D. (1988) 'Public and private in early nineteenth-century London: the Vagrant Act of 1822 and its enforcement', *Social History*, 13, 273–94.

Roberts, S. M. and Schein, R. H. (1993) 'The entrepreneurial city: facilitating urban development in Syracuse, New York', *Professional Geographer*, 45, 21–33.

Robertson, R. (1992) *Globalization: Social Theory and Global Culture*, Sage, London.

——(1995) 'Glocalization: time-space and homogeneity-heterogeneity', in Featherstone, M., Lash, S. and Robertson, R. (eds) *Global Modernities*, Sage, London, 24–45.

Robins, K. (1991) 'Prisoners of the city: whatever could a postmodern city be?', *New Formations*, 15, 1–22.

——(1995) 'Collective emotion and urban culture', in Healey, P., Cameron, S., Davoudi, S., Graham, S. and Madani-Pour, A. (eds) *Managing Cities: The New Urban Context*, John Wiley, Chichester, 45–61.

Robinson, J. (1980) 'Introduction', in Walsh, V. and Gram, H. (eds) *Classical and Neoclassical Theories of General Equilibrium: Historical Origins and Mathematical Structure*, Oxford University Press, Oxford, xi–xvi.

Robson B. T., Bradford, M. G., Deas, I., Hall, E., Harrison, E., Parkinson, M., Evans, R., Garside, P., Harding, A. and Robinson, F. (1994) *Assessing the Impact of Urban Policy*, HMSO, London.

Rojek, C. (1990) 'Baudrillard and leisure', *Leisure Studies*, 9, 7–20.

Rothenberg, J. (1968) 'Consumer sovereignty', in Sills, D. L. (ed.) *International Encyclopedia of the Social Sciences*, vol. 3, Free Press, New York, 326–335.

Rule, J. (1987) 'The property of skill in the period of manufacture', in Joyce, P. (ed.) *The Historical Meanings of Work*, Cambridge University Press, Cambridge, 99–118.

Rushton, G. (1969) 'Analysis of behavior by revealed space preference', *Annals of the Association of American Geographers*, 59, 391–400.

Ryan, J. (1994) 'Women, modernity and the city', *Theory, Culture and Society*, 11, 35–64.

Sack, R. D. (1988) 'The consumer's world: place as context', *Annals of the Association of American Geographers*, 78, 642–664.

——(1993) *Place, Modernity and the Consumer's World: A Relational Framework for Geographical Analysis*, Johns Hopkins University Press, Baltimore.

Saegert, S. (1980) 'Masculine cities and feminine suburbs: polarized ideas, contradictory realities', *Signs: Journal of Women in Culture and Society*, 5(1) (supplement: 'Women and the American city') 96–111.

Sahlins, M. (1972) *Stone Age Economics*, Tavistock, London.

Sampson, P. (1994) 'Postmodernity', in Sampson, P., Samuel, V. and Sugden, C. (eds) *Faith and Modernity*, Regnum, Oxford, 29–57.

Samuel, R. (1992) 'Mechanization and hand labour in industrializing Britain', in Berlanstein, L. (ed.) *The Industrial Revolution and Work in Nineteenth-century Europe*, Routledge, London, 26–40.

Samuelson, P. (1948) 'Consumer theory in terms of revealed preferences', *Economica*, 15(60), 243–253.

——(1954) 'The pure theory of public expenditure', *Review of Economics and Statistics*, 36, 387–389.

——(1955) 'Diagrammatic exposition of a theory of public expenditure', *Review of Economics and Statistics*, 37, 360–366.

Sartre, J.-P. (1957) *Being and Nothingness*, Methuen, London.

Sassen, S. (1991) *The World City: New York, London, Tokyo*, Princeton University Press, Princeton.

——(1994) *Cities in a World Economy*, Pine Forge Press, London.

——(1998) *Globalization and its Discontents*, Free Press, New York.

Saunders, P. (1979) *Urban Politics: A Sociological Interpretation*, Hutchinson, London.

——(1984a) 'Beyond housing classes: the sociological significance of private property rights in means of consumption', *International Journal of Urban and Regional Research*, 8, 202–225.

——(1984b) 'Rethinking local politics', in Boddy, M. and Fudge, C. (eds) *Local Socialism?*, Macmillan, London, 22–48.

——(1985) 'Space, the city and urban sociology', in Gregory, D. and Urry, J. (eds) *Social Relations and Spatial Structures*, Macmillan, London, 67–89.

——(1986a) *Social Theory and the Urban Question*, 2nd edn, Hutchinson, London.

——(1986b) 'Comment on Dunleavy and Preteceille', *Environment and Planning D: Society and Space*, 4, 155–163.

——(1988) 'The sociology of consumption: a new research agenda', in Otnes, P. (ed.) *The Sociology of Consumption*, Solum Forlag, Oslo, 141–156.

——(1989) *Social Class and Stratification*, Tavistock, London.

——(1990) *A Nation of Homeowners*, Unwin Hyman, London.

de Saussure, F. (1959) *Course in General Linguistics*, The Philosophical Library, New York.

Savage, M. and Warde, A. (1993) *Urban Sociology, Capitalism and Modernity*, Macmillan, London.

Savage, M., Barlow, J., Dickens, P. and Fielding, T. (1992) *Property, Bureaucracy and Culture: Middle Class Formation in Contemporary Britain*, Routledge, London.

Savas, E. S. (1982) *Privatising the Public Sector*, Chatham House, Chatham.

Sawyer, M. C. (1989) *The Challenge of Radical Political Economy: An Introduction to the Alternatives to Neo-classical Economics*, Harvester Wheatsheaf, Hemel Hempstead.

Sayer, A. (1984) *Method in Social Science: A Realist Approach*, Hutchinson, London.

Sayer, D. (1991) *Capitalism and Modernity: An Excursus on Marx and Weber*, Routledge, London.

Schama, S. (1988) *The Embarassment of Riches: An Interpretation of Dutch Culture in the Golden Age*, University of California Press, Berkeley.

Schivelbusch, W. (1988) *Disenchanted Night: The Industrialisation of Light in the Nineteenth Century*, University of California Press, Berkeley.

——(1993) *Tastes of Paradise: A Social History of Spices, Stimulants, and Intoxicants*, Vintage, London.

Schor, J. (1998) *The Overspent American: Upscaling, Downshifting, and the New Consumer*, Basic Books, New York.

Schmalensee, R. (1978) 'A model of advertising and product quality', *Journal of Political Economy*, 86, 485–503.

Schuetz, A. (1944) 'The stranger: an essay in social psychology', *American Journal of Sociology*, 49(6), 499–507.

Schwartz, H. (1989) 'The three-body problem and the end of the world', in Feher, M. (ed.) *Fragments for a History of the Human Body: Part Two*, Zone Books, New York, 406–465.

Scitovsky, T. (1976) *The Joyless Economy: An Inquiry into Human Satisfaction and Consumer Dissatisfaction*, Oxford University Press, Oxford.

Scobie, D. (1992) 'Anatomy of the promenade: the rise of bourgeois sociability in nineteenth-century New York', *Social History*, 17, 203–227.

Scott, A. and Soja, E. (1986) 'Los Angeles: capital of the twentieth century', *Environment and Planning D: Society and Space*, 4, 249–254.

——(eds) (1996) *The City: Los Angeles and Urban Theory at the End of the Twentieth Century*, University of California Press, Berkeley.

Sekora, J. (1977) *Luxury: The Concept in Western Thought, Eden to Smollett*, Johns Hopkins University Press, Baltimore.

Sennett, R. (1970) *The Uses of Disorder: Personal Identity and City Life*, Knopf, New York.

——(1998) *The Corrosion of Character: The Personal Consequences of Work in the New Capitalism*, Norton, London.

Shammas, C. (1990) *The Pre-industrial Consumer in England and America*, Oxford University Press, Oxford.

Shepherd, I. D. H. and Thomas, C. J. (1980) 'Urban consumer behaviour', in Dawson, J. A. (ed.) *Retail Geography*, Croom Helm, Beckenham, 18–94.

Sheppard, E. and Barnes, T. J. (1990) *The Capitalist Space Economy: Geographical Analysis after Ricardo, Marx and Sraffa*, Unwin Hyman, London.

Sherratt, A. G. (1981) 'Plough and pastoralism: aspects of the secondary products revolution', in Hammond, N., Hodder, I. and Isaac, G. (eds) *Pattern of the Past: Essays in Honour of David Clarke*, Cambridge University Press, Cambridge, 261–305.

——(1998) 'The human geography of Europe: a prehistoric perspective', in Butlin, R. A. and Dodgshon, R. A. (eds) *An Historical Geography of Europe*, Clarendon Press, Oxford, 1–25.

Shi, D. E. (1985) *The Simple Life: Plain Living and High Thinking in American Culture*, Oxford University Press, Oxford.

Shichar, D. (1995) *Punishment for Profit: Private Prisons/Public Concerns*, Sage, London.

Shields, R. (1989) 'Social spatialization and the built environment: the West Edmonton Mall', *Environment and Planning D: Society and Space*, 7, 147–164.

——(1992) 'The individual, consumption cultures and the fate of community', in Shields, R. (ed.) *Lifestyle Shopping: The Subject of Consumption*, Routledge, London, 99–113.

——(1994a) 'Fancy footwork: Walter Benjamin's notes on *flânerie*', in Tester, K. (ed.) *The Flâneur*, Routledge, London, 61–80.

——(1994b) 'The logic of the mall', in Riggins, S. H. (ed.) *The Socialness of Things: Essays on the Socio-semiotics of Objects*, Mouton de Gruyter, New York, 203–229.

Sibley, D. (1981) *Outsiders in Urban Society*, Blackwell, Oxford.

——(1995) *Geographies of Exclusion: Society and Difference in the West*, Routledge, London.

Simmel, G. (1950a) 'Faithfulness and gratitude', in Wolff, K. (ed.) *The Sociology of Georg Simmel*, Free Press, New York, 379–395.

——(1950b) 'The metropolis and mental life', in Wolff, K. (ed.) *The Sociology of Georg Simmel*, Free Press, New York, 324–339.

——(1950c) 'The stranger', in Wolff, K. (ed.) *The Sociology of Georg Simmel*, Free Press, New York, 402–408.

——(1957) 'Fashion', *American Journal of Sociology*, 62(6), 541–558.

——(1968) 'On the concept and the tragedy of culture', in *Georg Simmel: The Conflict in Modern Culture and Other Essays*, Teachers' College Press, New York, 27–46.

——(1978) *The Philosophy of Money*, Routledge, London.

——(1997) 'The sociology of space', in Frisby, D. and Featherstone, M. (eds) *Simmel on Culture*, Sage, London, 137–170.

Slater, D. (1997) *Consumer Culture and Modernity*, Polity, Cambridge.

Smith, A. (1937) *An Inquiry into the Nature and Causes of the Wealth of Nations*, Modern Library, New York.

Smith, A. D. (1986) *The Ethnic Origins of Nations*, Blackwell, Oxford.

Smith, D. (1999) *Zygmunt Bauman: Prophet of Postmodernity*, Polity, Cambridge.

Smith, M. E. (1998) *The Aztecs*, Blackwell, Oxford.

Smith, M. P. and Feagin, J. R. (eds) (1987) *The Capitalist City: Global Restructuring and Community Politics*, Blackwell, Oxford.

Smith, M. P. and Tardanico, R. (1987) 'Urban theory reconsidered: production, reproduction and collective action', in Smith, M. P. and Feagin, J. R. (eds) *The Capitalist City: Global Restructuring and Community Politics*, Blackwell, Oxford, 87–110.

Smith, N. (1992) 'Geography, difference and the politics of scale', in Doherty, J., Graham, E. and Malek, M. (eds) *Postmodernism and the Social Sciences*, Macmillan, London, 57–79.

——(1996) *The New Urban Frontier: Gentrification and the Revanchist City*, Routledge, London.

Smith, R. G. (1997) 'The end of geography and radical politics in Baudrillard's philosophy', *Environment and Planning D: Society and Space*, 15, 305–320.

Sobel, M. (1981) *Lifestyle and Social Structure: Concepts, Definitions, Analyses*, Academic Press, New York.

Social Exclusion Unit (2001) *A New Commitment to Neighbourhood Renewal: National Strategy Action Plan*, Home Office, London.

Soja, E. (1989) *Postmodern Geographies: The Reassertion of Space in Critical Social Theory*, Verso, London.

——(1996) *Thirdspace: Journeys to Los Angeles and Other Real-and-Imagined Places*, Blackwell, Oxford.

——(2000) *Postmetropolis: Critical Studies of Cities and Regions*, Blackwell, Oxford.

Sombart, W. (1967) *Luxury and Capitalism*, University of Michigan Press, Ann Arbor.

Soper, K. (1981) *On Human Needs*, Harvester Press, Brighton.

——(2000) 'Other pleasures: the attractions of post-consumerism', in Panitch, L. and Leys, C. (eds) *Socialist Register 2000*, Merlin, Rendlesham, 115–132.

Stedman Jones, G. (1983) *The Languages of Class: Studies in English Working Class History, 1832–1982*, Cambridge University Press, Cambridge.

Steiner, P. O. (1974) 'Public expenditure budgeting', in *The Economics of Public Finance*, Brookings Institution, Washington DC, 345–371.

Stigler, G. and Becker, G. S. (1977) 'De gustibus non est disputandum', *American Economic Review*, 67, 76–90.

Stillwell, J. and Leigh, C. (1996) 'Exploring the geographies of social polarisation in Leeds', in Haughton, G. and Williams, C. C. (eds) *Corporate City? Partnership, Participation and Partition in Urban Development in Leeds*, Avebury, Aldershot, 59–178.

Sterne, L. (1997) *The Life and Opinions of Tristram Shandy, Gentleman*, Penguin, Harmondsworth.

Stoker, G. (1995) 'Urban regime theory and urban politics', in Judge, D., Stoker, G. and Wolman, H. (eds) *Theories of Urban Politics*, Sage, London, 54–71.

Stoker, G. and Mossberger, K. (1994) 'Urban regime theory in comparative perspective', *Environment and Planning C: Policy and Politics*, 12, 195–212.

Stone, C. L. (1989) *Regime Politics: Governing Atlanta, 1946–1988*, University Press of Kansas, Lawrence.

Stone, G. P. (1954) 'City shoppers and urban identification: observations on the social psychology of city life', *American Journal of Sociology*, 60(1), 36–45.

Storey, J. (1999) *Consumer Culture and Everyday Life*, Arnold, London.

Strathern, M. (1988) *The Gender of the Gift*, University of California Press, Berkeley.

——(1999) *Property, Substance and Effect*, Athlone, London.

Sturrock, J. (1979) 'Introduction', in Sturrock, J. (ed.) *Structuralism and Since: from Lévi-Strauss to Derrida*, Oxford University Press, Oxford, 1–18.

Sweezy, P. (1942) *The Theory of Capitalist Development*, Monthly Review Press, New York.

Swyngedouw, E. (1989) 'The heart of the place: the resurrection of locality in an age of hyperspace', *Geografiska Annaler*, 71B, 31–42.

——(1997) 'Neither global nor local: "glocalization" and the politics of scale', in Cox, K. (ed.) *Spaces of Globalization: Reasserting the Power of the Local*, Guilford Press, New York, 137–166.

Tagg, J. (1996) 'This city which is not one', in King, A. D. (ed.) *Re-presenting the City: Ethnicity, Capital and Culture in the 21st-century Metropolis*, Macmillan, London, 179–182.

Tawney, R. H. (1929) *Religion and the Rise of Capitalism: A Historical Study*, John Murray, London.

Taylor, L. D. and Weiserbs, D. (1964) 'Advertising and the aggregate consumption function', *American Economic Review*, 62, 642–656.

Taylor, P. J. (1996) 'What's modern about the modern world-system? Introducing ordinary modernity through world hegemony', *Review of International Political Economy*, 3(2), 260–286.

Tester, K. (ed.) (1994) *The Flâneur*, Routledge, London.

Tewdwr-Jones, M. and McNeill, D. (2000) 'The politics of city-regional planning and governance: reconciling the national, regional and urban in the competing voices of institutional restructuring', *European Urban and Regional Studies*, 7(2), 119–134.

Theret, B. (1982) 'Collective means of consumption, capital accumulation and the urban question', *International Journal of Urban and Regional Research*, 6, 345–371.

Thirsk, J. (1978) *Economic Policy and Projects: The Development of a Consumer Society in Early-modern England*, Clarendon Press, Oxford.

Thompson, A. (1977) 'Public and private consumption', in Hirst, I. R. C. and Duncan Reekie, W. (eds) *The Consumer Society*, Tavistock, London, 37–50.

Thompson, E. P. (1963) *The Making of the English Working Class*, Methuen, London.

——(1967) 'Time, work discipline, and industrial capitalism', *Past and Present*, 38, 56–97.

Thompson, G. F. (1997) 'Where goes economics and the economies?', *Economy and Society*, 26(4), 599–610.

Thrift, N. J. (1996) *Spatial Formations*, Sage, London.

——(2000a) 'Introduction: dead or alive', in Cook, I., Crouch, D., Naylor, S. and Ryan, J. (eds) *Cultural Turns/Geographical Turns: Perspectives on Cultural Geography*, Prentice-Hall, London, 1–6.

——(2000b) ' "Not a straight line but a curve", or, cities are not mirrors of modernity', in Bell, D. and Haddour, A. (eds) *City Visions*, Prentice-Hall, London, 233–263.

Thrift, N. J. and Glennie, P. D. (1993) 'Historical geographies of urban life and modern consumption', in Kearns, G. and Philo, C. (eds) *Selling Places: The City as Cultural Capital, Past and Present*, Pergamon, London, 33–48.

Tickell, A. (1996) 'Taking the initiative: the Leeds financial centre', in Haughton, G. and Williams, C. C. (eds) *Corporate City? Partnership, Participation and Partition in Urban Development in Leeds*, Avebury, Aldershot, 103–118.

——(2003) 'Cultures of money', in Anderson, K., Domosh, M., Pile, S. and Thrift, N. J. (eds) *The Handbook of Cultural Geography*, Sage, London.

Tiersten, L. (1993) 'Redefining consumer culture: recent literature on consumption and the bourgeoisie in western Europe', *Radical History Review*, 57, 116–159.

Todorov, T. (1984) *The Conquest of America: the Question of the Other*, Harper and Row, New York.

Tönnies, F. (1955) *Community and Society*, Harper and Row, New York.

Trudgill, E. (1973) 'Prostitution and paterfamilias', in Dyos, H. J. and Wolff, M. (eds) *The Victorian City: Images and Realities. Volume 2*, Routledge, London, 693–706.

Turner, B. (1987) 'A note on nostalgia', *Theory, Culture and Society*, 4(1), 147–156.

——(1988) *Status*, Open University Press, Milton Keynes.

Turner, V. W. (1969) *The Ritual Process: Structure and Anti-structure*, Routledge, London.

Turock, I. (1992) 'Property-led regeneration: panacea or placebo?', *Environment and Planning A*, 24, 361–379.

Vanderbeck, R. and Johnson, J. (2000) 'That's the only place where you can hang out: urban young people and the space of the mall', *Urban Geography*, 21(1), 5–25.

Vaneigem, R. (2001) *The Revolution of Everyday Life*, Rebel Press, London.

Veal, A. J. (1993) 'The concept of lifestyle: a review', *Leisure Studies*, 12, 233–52.

Veblen, T. (1961a) 'Why is economics not an evolutionary science?', in *The Place of Science in Modern Civilization and other Essays*, Russell and Russell, New York, 56–81.

——(1961b) 'The preconceptions of economic science II', in *The Place of Science in Modern Civilization and other Essays*, Russell and Russell, New York, 114–147.

——(1975) *The Theory of Business Enterprise*, Augustus M. Kelly, New York.

——(1994) *The Theory of the Leisure Class: An Economic Study in the Evolution of Institutions*, Dover, New York.

Veltz, P. (2000) 'European cities in the world economy', in Bagnasco, A. and Le Galès, P., (eds) *Cities in Contemporary Europe*, Cambridge University Press, Cambridge, 33–47.

Venkatesh, A. (1991) 'Changing consumption patterns: the transformation of Orange County since World War II', in Kling, R., Olin, S. and Poster, M. (eds) *Posturban California: The Transformation of Orange County since World War II*, University of California Press, Berkeley, 142–164.

Vidler, A. (1993) 'The explosion of space: architecture and the filmic imaginary', *Assemblage*, 21, 44–59.

Virilio, P. (1991) *The Aesthetics of Disappearance*, Semiotext(e), New York.

——(1994) *The Vision Machine*, BFI, London.

von Neumann, J. and Morgenstern, O. (1947) *Theory of Games and Economic Behaviour*, 2nd edn, Princeton University Press, Princeton.

Walkowitz, J. (1977) 'The making of an outcast group: prostitution and working women in nineteenth-century Plymouth and Southampton', in Vinicus, M. (ed.) *A Widening Sphere: Changing Roles of Victorian Women*, Methuen, London.

——(1980) *Prostitution and Victorian Society: Women, Class and the State*, Cambridge University Press, Cambridge.

——(1992) *City of Dreadful Delight: Narratives of Sexual Danger in Late Victorian England*, Virago, London.

Wallerstein, I. (1974) *The Modern World-system: Capitalist Agriculture and the Origins of the European World-economy in the Sixteenth Century*, Academic Press, New York.

——(1980) *The Modern World-system II: Mercantilism and the Consolidation of the European World-economy, 1600–1750*, Academic Press, New York.

Wallock, L. (1988) 'New York City: capital of the twentieth century', in Wallock, L. (ed.) *New York: Culture Capital of the World, 1940–1965*, Rizzoli, New York, 17–52.

Walsh, V. and Gram, H. (1980) *Classical and Neoclassical Theories of General Equilibrium: Historical Origins and Mathematical Structure*, Oxford University Press, Oxford.

Ward, K. G. (1997) 'Coalitions in urban regeneration: a regime approach', *Environment and Planning A*, 29, 1493–1506.

Warde, A. (1990) 'Production, consumption and social change: reservations regarding Peter Saunders' sociology of consumption', *International Journal of Urban and Regional Research*, 14, 228–248.

——(1992) 'Notes on the relationship between production and consumption', in Burrows, R. and Marsh, C. (eds) *Consumption and Class: Divisions and Change*, Macmillan, London, 15–31.

——(1994) 'Consumers, identity and belonging: reflections of some theses of Zygmunt Bauman', in Keat, R., Whiteley, N. and Abercrombie, N. (eds) *The Authority of the Consumer*, Routledge, London, 58–74.

——(1996) 'Afterword: the future of the sociology of consumption', in Edgell, S., Hetherington, K. and Warde, A (eds) *Consumption Matters: The Production and Experience of Consumption*, Blackwell, Oxford, 302–312.

——(1997) *Consumption, Food and Taste: Culinary Antinomies and Commodity Culture*, Sage, London.

Warpole, K. and Greenhalsh, L. (1999) *The Richness of Cities: Urban Policy in a New Landscape*, Comedia/Demos, London.

Waterhouse, K. (1994) *City Lights: A Street Life*, Hodder and Stoughton, London.

Watson, S. (1991) 'Guilding the smokestacks: the new symbolic representations of deindustrialized regions', *Environment and Planning D: Society and Space*, 9, 59–70.

Weatherhill, L. (1988) *Consumer Behaviour and Material Culture in Britain, 1660–1760*, Routledge, London.

Weber, M. (1946) 'Class, status, party', in Bendix, R. and Lipset, S. M. (eds) *Class, Status and Power: A Reader in Social Stratification*, Free Press, Glencoe, 63–75.

——(1976) *The Protestant Ethic and the Spirit of Capitalism*, Allen and Unwin, London.

——(1978) *Economy and Society: An Outline of Interpretive Sociology*, University of California Press, Berkeley.

Weinbaum, B. and Bridges, A. (1979) 'The other side of the paycheck: monopoly capital and the structure of consumption', in Eisenstein, Z. R. (ed.) *Capitalist Patriarchy and the Case for Socialist Feminism*, Monthly Review Press, New York, 190–205.

Werbner, P. (1996) 'The enigma of Christmas: symbolic violence, compliant subjects and the flow of English kinship', in Edgell, S., Hetherington, K. and Warde, A. (eds) *Consumption Matters: The Production and Experience of Consumption*, Blackwell, Oxford, 135–162.

Wernick, A. (1991a) *Promotional Culture: Advertising, Ideology and Symbolic Expression*, Sage, London.

——(1991b) 'Sign and commodity: aspects of the cultural dynamic of advanced capitalism', *Canadian Journal of Social and Political Theory*, 15(1–3), 152–169.

Whyte, W. H. (1954) 'The consumer in the new suburbia', in Clark, L. H. (ed.) *Consumer Behaviour*, New York University Press, New York, 1–14.

——(1960) *The Organization Man*, Penguin, Harmondsworth.

Wildavsky, A. (1991) 'Help, Ma, I'm being controlled by inanimate objects!', *Southern California Law Review*, 65, 241–253.

Williams, C. C. (1996) 'Consumer services and the competitive city', in Haughton, G. and Williams, C. C. (eds) *Corporate City? Partnership, Participation and Partition in Urban Development in Leeds*, Avebury, Aldershot, 119–133.

——(1997) *Consumer Services and Economic Development*, Routledge, London.

Williams, K. and Johnstone, C. (2000) 'The politics of the selective gaze: closed-circuit television and the policing of public space', *Crime, Law and Social Change*, 34, 182–210.

Williams, K., Johnstone, C. and Goodwin, M. (2000) 'CCTV surveillance in urban Britain: beyond the rhetoric of crime prevention', in Gold, J. and Revill, G. (eds) *Landscapes of Defence*, Pearson, London, 168–187.

Williams, P. (1992) 'Housing', in Cloke, P. (ed.) *Policy and Change in Thatcher's Britain*, Pergamon, Oxford, 159–198.

Williams, R. (1973) *The Country and the City*, Chatto and Windus, London.

——(1976) *Keywords: A Vocabulary of Culture and Society*, Fontana, London.

——(1980) *Problems in Materialism and Culture*, Verso, London.

Williams, R. H. (1982) *Dream Worlds: Mass Consumption in Late Nineteenth-century France*, University of California Press, Berkeley.

Williamson, J. (1987) *Consuming Passions: The Dynamics of Popular Culture*, Marion Boyars, London.

——(1992) 'I-less and gaga in the West Edmonton Mall: towards a pedestrian feminist reading', in Currie, D. H. and Raoul, V. (eds) *The Anatomy of Gender: Women's Struggles for the Body*, Carelton University Press, Ottawa, 79–115.

Wilson, E. (1991) *The Sphinx in the City: Urban Life, the Control of Disorder, and Women*, Virago, London.

——(1995) 'The invisible *flâneur*', in Watson, S. and Gibson, K. (eds) *Postmodern Cities and Spaces*, Blackwell, Oxford, 59–79.

——(2000) *The Contradictions of Culture: Cities, Culture, Women*, Sage, London.

Wilson, W. J. (1997) *When Work Disappears: The World of the New Urban Poor*, Vintage, New York.

Winstanley, M. (1983) *The Shopkeeper's World, 1830–1914*, Manchester University Press, Manchester.

Wirth, L. (1938) 'Urbanism as a way of life', *American Journal of Sociology*, 44(1), 1–24.

Wolch, J. R. (1990) *The Shadow State: Government and Voluntary Sector in Transition*, The Foundation Centre, New York.

Wolff, J. (1990) *Feminine Sentences: Essays on Women and Culture*, Polity, Cambridge.

Wood, A. (1998) 'Questions of scale in the entrepreneurial city', in Hall, T. and Hubbard, P. (eds) *The Entrepreneurial City: Geographies of Politics, Regime and Representation*, John Wiley, Chichester, 275–284.

Wrong, D. (1992) 'Disaggregating the idea of capitalism', in Featherstone, M. (ed.) *Cultural Theory and Cultural Change*, Sage, London, 147–157.

Wynne, D. and O'Connor, J., with Phillips, D. (1998) 'Consumption and the postmodern city', *Urban Studies*, 35(5–6), 841–864.

Wyrwa, U. (1998) 'Consumption and consumer society: a contribution to the history of ideas', in Strasser, S., McGovern, C. and Judt, M. (eds) *Getting and Appending: European and American Consumer Societies in the Twentieth Century*, Cambridge University Press, Cambridge, 431–447.

Xenos, N. (1989) *Scarcity and Modernity*, Routledge, London.

Yang, C. Y. (1964) 'Variations in the cyclical behaviour of advertising', *Journal of Marketing*, 28, 25–30.

Young, J. (1999) *The Exclusive Society: Social Exclusion, Crime and Difference in Late Modernity*, Sage, London.

Zukin, S. (1988) *Loft Living: Culture and Capital in Urban Change*, Hutchinson/Radius, London.

——(1990) 'Socio-spatial prototypes of a new organization of consumption: the role of real cultural capital', *Sociology*, 24(1), 37–56.

——(1995) *The Cultures of Cities*, Blackwell, Oxford.

——(1998) 'Urban lifestyles: diversity and standardization in spaces of consumption', *Urban Studies*, 35(5–6), 825–839.

Index

ESSENTIAL READING

The City Cultures Reader
Edited by Malcom Miles, Tim Hall & Iain Borden

2nd edition

Hb: 0415-302447
Pb: 0415-302455   Routledge

The City Reader
Edited by Richard T. LeGates & Frederic Stout

3rd edition

Hb: 0415-27172x
Pb: 0415-271738   Routledge

Urban Theory and the Urban Experience
Encountering the City
Simon Parker

Hb: 0415-245915
Pb: 0415-245923   Routledge

The Consumption Reader
Edited by David Clarke, Kate Housiaux & Marcus Doel

Hb: 0415-213762
Pb: 0415-213770   Routledge

Urban Nightscapes
Youth Cultures, Pleasure Spaces and Corporate Power

Paul Chatterton & Robert Hollands

Hb: 0415-283450
Pb: 0415-283469   Routledge

Information and ordering details

For price availability and ordering visit our website www.tandf.co.uk

Subject Web Address   **www.geographyarena.com**

Alternatively our books are available from all good bookshops.